ACS SYMPOSIUM SERIES **919**

Urban Aerosols and Their Impacts

Lessons Learned from the World Trade Center Tragedy

Jeffrey S. Gaffney, Editor
Argonne National Laboratory

Nancy A. Marley, Editor
Argonne National Laboratory

Sponsored by the
ACS Divisions of Environmental Chemistry, Inc. and Geochemistry, Inc.

American Chemical Society, Washington, DC

Library of Congress Cataloging-in-Publication Data

Urban aerosols and their impacts : lessons learned from the World Trade Center Tragedy / Jeffrey S. Gaffney, editor ; Nancy A. Marley, editor ; sponsored by the ACS Divisions of Environmental Chemistry, Inc. and Geochemistry, Inc.

p. cm. — (ACS symposium series ; 919)

"Developed from a symposium sponsored by the Divisions of Environmental Chemistry, Inc. and Geochemistry, Inc. at the 226th National Meeting of the American Chemical Society, New York, September 7–11, 2003"—Pref.

Includes bibliographical references and index.

ISBN-13: 978–0–8412–3916–6 (alk. paper)

1. Dust—Environmental aspects—Congresses. 2. Aerosols—Environmental aspects—Congresses 3. Air—Pollution—Congresses. 4. Construction and demolition debris—Environmental aspects—New York (State)—New York—Congresses. 5. World Trade Center Site (New York, N.Y.)—Congresses.

I. American Chemical Society. Divisions of Environmental Chemistry, Inc. and Geochemistry, Inc.. II. American Chemical Society. Meeting (226th : 2003 : New York, N.Y.). III. Series.

TD884.5.U93 2005
628.5′3—dc22

2005048306

The paper used in this publication meets the minimum requirements of American National Standard for Information Sciences—Permanence of Paper for Printed Library Materials, ANSI Z39.48–1984.

Distributed by Oxford University Press

ISBN 10: 0-8412-3916-9

PRINTED IN THE UNITED STATES OF AMERICA

Foreword

The ACS Symposium Series was first published in 1974 to provide a mechanism for publishing symposia quickly in book form. The purpose of the series is to publish timely, comprehensive books developed from ACS sponsored symposia based on current scientific research. Occasionally, books are developed from symposia sponsored by other organizations when the topic is of keen interest to the chemistry audience.

Before agreeing to publish a book, the proposed table of contents is reviewed for appropriate and comprehensive coverage and for interest to the audience. Some papers may be excluded to better focus the book; others may be added to provide comprehensiveness. When appropriate, overview or introductory chapters are added. Drafts of chapters are peer-reviewed prior to final acceptance or rejection, and manuscripts are prepared in camera-ready format.

As a rule, only original research papers and original review papers are included in the volumes. Verbatim reproductions of previously published papers are not accepted.

ACS Books Department

Contents

World Trade Center Fine Particle and Volatile Organic Emissions

World Trade Center Exposure Assessments

Aerosol Transport Issues

Indexes

Preface

Urban aerosols (fine airborne particulate matter) have received considerable attention because of their links with increased urban hospitalizations from chronic pulmonary and cardiac diseases, as well as increased daily mortality (*1, 2*). The terrorist attack on the World Trade Center (WTC) on September 11, 2001, and the subsequent tragic collapse of the twin towers resulted in unprecedented release and dispersal of potentially toxic gaseous and particulate materials in a cloud of dust and smoke that extended over lower Manhattan and other parts of New York City. A number of studies by both university and government researchers were undertaken immediately to evaluate the potential human health effects of exposure to the materials from the initial event and also from continued releases from smoldering fires and site cleanup that continued through mid December 2001. Evaluation of firefighters, cleanup workers, and community residents showed exposure-related increases in cough and bronchial hyperreactivity, while a follow-up of 182 pregnant women who were either inside or near the WTC on September 11 showed a twofold increase in small-for-gestational-age infants (*3*).

A symposium was held in New York City in September 2003 at the 226th National Meeting of the American Chemical Society (ACS), to present the findings of studies related to the WTC tragedy and to bring attention to the need for a better understanding of the chemical and physical properties of urban aerosols and their impacts. Papers were presented that described studies at the WTC site, including detailed characterizations of the particulate matter generated by the collapse, the potential health impacts of exposure, and the relationships between this single catastrophic event and everyday exposures to particulate matter in urban centers. This symposium brought together researchers from government, universities, and research institutions with the common goal of determining the type and extent of the exposures and the possible future impacts to the exposed population This book is a result of that symposium. Much of the work presented at the ACS meeting is included in the following chapters. In the interest of completeness, other work presented at the symposium but not included in detail in this volume is summarized in this Preface, along with other findings reported since September 2003.

Several studies were conducted by researchers at the New York University School of Medicine at a site located at the New York University Downtown Hospital (NYUDH), about four blocks east of the WTC disaster site known as "Ground Zero." Samples were collected from September 14 to December 27, 2001. During this period, intense fires were burning at Ground Zero, and early cleanup efforts were taking place. The total mass concentrations of particulates < 15 μm ranged from 4.4 $\mu g/m^3$ to 22.7 $\mu g/m^3$ (*4*). None of the daily mass concentrations measured at the site exceeded the 24-h limit of 65 $\mu g/m^3$ set by the U.S. Environmental Protection Agency (EPA) for particulate matter < 2.5 μm (PM-2.5). Bimodal mass concentration distributions showed one peak at a diameter of about 1.8 μm and a larger mode at 0.56 μm. Bimodal size distributions were also observed for several elements. Sulfur, which is typically found in atmospheric aerosols < 1.0 μm, was present in size fractions both above and below 1.0 μm. Calcium was predominant in the larger particulates (> 1.0 μm) and correlated well with sulfur in this size range, suggesting that gypsum board containing calcium sulfate was a source. Vanadium was found primarily in the submicron size fraction, indicating oil combustion as the main source.

Particles collected at the NYUDH site exhibited a variety of globular forms, with most appearing to be agglomerates (*5*). No ultrafine acid particulates were detected. Hourly average aerosol number concentrations were variable, with occasionally very high, brief peaks reaching values over 7×10^4 cm^{-3}. The average total aerosol mass concentration of 17 $\mu g/m^3$ at this site in mid October decreased to roughly half that value in November and to as little as 5 $\mu g/m^3$ in mid December. Since the particle size distributions were mostly bimodal, the decrease in total mass was due mainly to a decrease in the large-particle fraction. The mass concentrations of very fine particles ranged from 4.3 $\mu g/m^3$ to 0.7 $\mu g/m^3$, and the mass concentrations of ultrafine particulates (< 0.1 μm) ranged from 1.46 $\mu g/m^3$ to not detectable after November 5, 2001. In general, the fine aerosol mass and number concentrations did not differ significantly from measurements made at the same site the previous year.

Positive matrix factorization was applied to trace element compositions of PM-2.5 samples collected at the same NYUDH site (*6*). Five PM-2.5 source components were identified as (a) products of WTC fires, (b) dust related to the WTC collapse, (c) dust related to WTC demolition, (d) products of oil combustion, and (e) soil. The WTC fire impacts were greatest in September and diminished in October. The demolition-related dust increased in mid October as cleanup operations began and decreased in November. Oil combustion was a large component of the overall PM-2.5 at the NYUDH site. Total PM-2.5 concentrations averaged 35 $\mu g/m^3$ in late September and 23 $\mu g/m^3$ in October. Concentrations of PM-2.5 returned to more typical New York City levels of 18–

14 μg/m^3 in December. In general, WTC sources contributed 50% of PM-2.5 during September, 27% in October, 14% in November, and about 6% in December.

Measurements were repeated at the NYUDH site one year later, on September 25 to October 24, 2002 (*7*). The particle number concentrations at this time were 10–40 cm^{-3}, similar to those measured in 2001. The average aerosol mass concentrations were 12 μg/m^3, about half of those in September and October 2001. These results show that the data obtained in 2002 were comparable to values in November and December of 2001 at the same site and to data for other urban environments. The high concentrations of large particles and elemental carbon seen at the WTC site in 2001 persisted only in September and October and were due to the WTC collapse, fires, and cleanup work at the site.

The EPA collected gaseous samples near the WTC site, beginning on September 22, 2001, to determine the effect of Ground Zero emissions on air quality (*8*). In general, total volatile organic compounds (VOCs) were highest north of Ground Zero, with a median concentration of 551 ppb. This was more than 100 ppb above levels at other sites in lower Manhattan. The highest single measurement (over 4 ppm) was made at 290 Broadway. A variable but downward trend in VOC concentrations continued until mid November, following the trend observed for PM-2.5. Ethyl benzene was the most abundant compound measured in the September and October samples, suggesting that these samples contained excess diesel fuel from the 42,000 gal stored in the basement of the WTC, as well as diesel emissions from the large vehicles used to remove debris from the site. Biomass-burning-related compounds (furan, 2-methyl furan, 2,5-dimethyl furan, benzonitrile, benzofuran, methyl chloride, and methyl bromide) decreased to undetectable levels by December 2001. Results of these studies are posted on the EPA's Web site (*9*).

In addition to the VOC measurements, concentrations of more than 60 nonpolar and nonvolatile organic compounds were measured in lower Manhattan from September 26 to October 24, 2001 (*10, 11*). Levels of 3- to 5-ring polycyclic aromatic hydrocarbons (PAHs) were six times the average values previously measured in Los Angeles (*12, 13*). The compound 1,3-diphenyl propane, not previously reported in ambient samples, was observed in the WTC samples. This species correlates well with the high particulate mass data obtained during the same time period and could serve as a source marker for the WTC plume.

An integrated modeling system used to assess exposure to urban aerosols, MENTOR (Modeling Environment for Total Risk Studies), was adapted for New York City and the surrounding area to support assessments of exposures to and health impacts of contaminants released from the WTC collapse and fires (*14*). Preliminary estimates of the WTC plume location and dilution were used to develop 8-h average characterizations of the plume's spread over the area in the four weeks following September 11, 2001. A baseline characterization of PM-2.5 exposures and doses for the population residing within 2.5 km of the WTC site is on the Rutgers University Web site (*15*).

Air samples were obtained on October 30 and 31, 2001, by the New York State Department of Health (NYSDOH), with assistance from the New York City Department of Health and Mental Hygiene, to characterize chemical and particulate releases from the smoldering debris and cleanup activities at the WTC. Samples were collected at four locations north of Ground Zero, as well as at one location to the southeast and another to the southwest. Samples were analyzed for a wide range of compounds including acid aerosols, aldehydes, inorganic acids, total suspended particulate matter and VOCs. Concentrations of several chemicals emitted directly from the debris pile exceeded both typical background levels and urban air comparison levels. These compounds included benzene, ethyl benzene, acetone, methyl ethyl ketone, inorganic acids (e.g., hydrochloric acid, nitric acid), and ammonia. This observation supports the hypothesis that combustion products contributed to the eye, nose, and throat irritation reported by workers and residents in the WTC neighborhood. The results of this study are summarized on the NYSDOH Web site (*16, 17*).

Samples of settled dust and airborne particulates were collected by the New York City Department of Health and Mental Hygiene and the U.S. Agency for Toxic Substances and Disease Registry (ATSDR) on November 4 through December 11, 2001, to assess the compositions of (a) dusts settled on outdoor and indoor surfaces and (b) airborne dusts in residential areas around the WTC. The focus was on building materials that have irritant properties (e.g., synthetic vitreous fibers [SVF] and gypsum) and those associated with long-term health concerns (e.g., crystalline silica and asbestos). Low levels of asbestos were found in some settled surface dusts in lower Manhattan, primarily below Chambers Street. No asbestos was detected in comparison samples of indoor dust taken north of 59th Street. The lower Manhattan residential areas tended to have a greater percentage of SVF and of mineral components of concrete and wallboard in settled dust than comparison areas. Exposure to significant amounts of these materials can cause skin rashes and irritation of eyes and the upper respiratory tract, symptoms experienced by citizens and first responders. Results of this study are available on the ATSDR Web site (*18*).

Blood and urine specimens from 321 firefighters responding to the WTC fires were analyzed for 110 potentially fire-related chemicals to estimate internal dose exposures (*19*). This was the most extensive biomonitoring study ever conducted during the first few weeks of occupational exposure. Metabolites of some known products of combustion (e.g., PAHs) were present in greater amounts in exposed firefighters than in a firefighter control group. Also evident were unanticipated increases in urinary antimony and in serum heptachlorodibenzodioxin and heptachlorodibenzofuran. Comparison of exposed and control groups indicated that, although the levels were statistically elevated, they were generally lower than reference values or workplace threshold levels.

Preliminary investigations by the EPA of WTC fallen dust samples within 0.5 mi of Ground Zero (*20*) identified calcium sulfate (gypsum) and calcium carbonate (calcite) as the major components of the very coarse dust fractions ($< 53\ \mu m$). These components were also present in the fine particulate matter (PM-2.5). Levels of carbon were relatively low, indicating that combustion-derived particles did not form a significant fraction of the fallen dust near Ground Zero immediately after the collapse of the buildings. Because gypsum and calcite are known to cause irritation of the mucus membranes of the eyes and respiratory tract, inhalation of high doses of the dusts could potentially cause toxic respiratory effects.

Considering that the attack on the WTC was an unprecedented event on U.S. soil, the data and results presented in this book by the various contributors reflects in many cases their complete dedication to the science and to the country. Much of the work reported here was completed on a volunteer basis, without funding or support as the work had to begin immediately after the attack, even though funding mechanisms were not in place. The researchers were, in many cases, attempting to gather key field data in the face of a disaster unlike any previous emergency in a major U.S. urban center. These contributors are to be commended. We hope that this compendium of their results and insights into the exposures originating at Ground Zero will help the health community in its monitoring of the long-term health effects of the WTC attack on September 11, 2001.

JEFFREY S. GAFFNEY
NANCY A. MARLEY
Environmental Research Division
Argonne National Laboratory
Building 203, 9700 South Cass Avenue
Argonne, IL 60439-4843

December 12, 2004

References

1. Roberts, S. *Environ. Health Perspect.* **2004**, *112*, 309–313.
2. Pope, C.A. *Aerosol Sci. Technol.* **2000**, *32*, 4–14.

3. Landrigan, P.J.; Lioy, P.J.; Thurston, G.; Berkowitz, G.; Chen, L.C.; Chillrud, S.N.; Gavett, S.H.; Georgopolous, P.G.; Geyh, A.S.; Levin, S.; Perera, F.; Rappaport, S.M.; Small, C. *Environ. Health Perspect.* **2004**, *112*, 731–739.
4. Hazi, Y.; Chillrud, S.N.; Maciejczyk, P. *Preprints of Extended Abstracts*, Urban Aerosols and their Impact: Lessons Learned from The World Trade Center Tragedy, New York, NY, Sep 7–11, 2003; American Chemical Society: Washington, DC, 2003; paper 131.
5. Cohen, B.S.; Heikkinen, M.S.A.; Hazi, Y. *Preprints of Extended Abstracts*, Urban Aerosols and their Impact: Lessons Learned from The World Trade Center Tragedy, New York, NY, Sep 7–11, 2003; American Chemical Society: Washington, DC, 2003; paper 114.
6. Thurston, G.; Maciejczyk, P.; Lall, R.; Hwang, J.-S.; Hsu, S.-I.; Chen, L.C. *Preprints of Extended Abstracts*, Urban Aerosols and their Impact: Lessons Learned from The World Trade Center Tragedy, New York, NY, Sep 7–11, 2003; American Chemical Society: Washington, DC, 2003; paper 134.
7. Heikkinen, M.S.A.; Hsu, S.-I.; Lall, R.; Peters, P.; Cohen, B.S.; Chen, L.C.; Thurston, G. *Preprints of Extended Abstracts*, Urban Aerosols and their Impact: Lessons Learned from The World Trade Center Tragedy, New York, NY, Sep 7–11, 2003; American Chemical Society: Washington, DC, 2003; paper 132.
8. Seila, R.L.; Swartz, E.; Lonneman, W.A.; Vallero, D.A. *Preprints of Extended Abstracts*, Urban Aerosols and their Impact: Lessons Learned from The World Trade Center Tragedy, New York, NY, Sep 7–11, 2003; American Chemical Society: Washington, DC, 2003; paper 113.
9. *EPA Response to September 11*; U.S. Environmental Protection Agency. [http://www.epa.gov/wtc/voc] (accessed Dec 2004).
10. Swartz, E.; Stockburger, L.; Vallero, D.A. *Preprints of Extended Abstracts*, Urban Aerosols and their Impact: Lessons Learned from The World Trade Center Tragedy, New York, NY, Sep 7–11, 2003; American Chemical Society: Washington, DC, 2003; paper 111.
11. Fraser, M.P.; Cass, G.R.; Simoneit, B.R.; Rasmussan, R. *Environ. Sci. Technol.* **1997**, *31*, 2356–2367.
12. Swartz, E.; Stockburger, L.; Vallero, D.A. *Environ. Sci. Technol.* **2003**, *37*, 3537–3546.
13. Fraser, M.P.; Cass, G.R.; Simoneit, B.R.; Rasmussan, R. *Environ. Sci. Technol.* **1998**, *32*, 1760–1770.
14. Georgopoulos, P.G. *Preprints of Extended Abstracts*, Urban Aerosols and their Impact: Lessons Learned from The World Trade Center Tragedy, New York, NY, Sep 7–11, 2003; American Chemical Society: Washington, DC, 2003; paper 351.

15. "Plume and Microenvironmental Modeling for Contaminants Associated with the World Trade Center Fire and Collapse." Computational Chemodynamics Laboratory, Rutgers University: New Brunswick, New Jersey. [http://www.ccl.Rutgers.edu/wtc] (accessed Dec 2004).
16. "Report on Air Sampling near the World Trade Center Site" New York State Department of Health. [http//www.health.state.ny.us/nysdoh/alert/voc/airsamp.htm] (accessed Dec 2004).
17. "VOC Monitoring near the World Trade Center Site" New York State Department of Health. [http//www.health.state.ny.us/nysdoh/alert/voc/voc.htm] (accessed Dec 2004).
18. "Final Technical Report of the Public Health Investigation to Assess Potential Exposures to Airborne and Settled Surface Dust in Residential Areas of Lower Manhattan" Agency for Toxic Substances and Disease Registry. [http://www.atsdr.cdc.gov/asbestos/final-report-lowermanhattan-02-execsummary.html] (accessed Dec 2004).
19. Edelman, P.; Osterloh, J.; Pirkle, J.; Caudill, S.P.; Grainger, J.; Jones, R.; Blount, B.; Calafat, A.; Turner, W.; Feldman, ED.; Baron, B.; Lushniak, B.D.; Kelly, K.; Prezant, D. *Environ. Health. Perspect.* **2003,** *111*, 1906–1911.
20. McGee, J.K.; Chen, L.C.; Cohen, M.D.; Chee, G.R.; Prophete, C.M.; Haykal-Coates, N.; Wasson, S.J.; Conner, T.L.; Costa, D.L.; Gavett, S.H. *Environ. Health Perspect.* **2003,** *111*, 972–980.

Overview

Chapter 1

Introduction to Urban Aerosols and Their Impacts

Nancy A. Marley and Jeffrey S. Gaffney

Environmental Research Division, Argonne National Laboratory, 9700 South Cass Avenue, Argonne, IL 60439

Atmospheric aerosols, or particulate matter, may be solid or liquid, with effective diameters from ~0.002 to ~ 100 μm. They can be emitted in particulate form directly into the atmosphere (*primary aerosols*) or formed in the atmosphere by chemical reactions of gases (*secondary aerosols*). Aerosol sources include mechanical processes such as grinding or wind erosion, gas-to-particle conversion of gas-phase primary pollutants, and combustion processes which occur primarily in urban centers. The impact of aerosols on the environment, human health, and climate depends on their number concentration, mass, size, chemical composition, phase (i.e. liquid or solid), morphology, and surface properties. Outlined in this Chapter is a brief overview of the physical and chemical properties of atmospheric aerosols as they relate to their to potential impacts on the environment and human health.

Atmospheric aerosols are solid or liquid atmospheric particles with diameters of approximately 0.002–100 μm (*1*). Aerosols can be emitted in particulate form directly into the atmosphere (*primary aerosols*) or formed in the atmosphere by gas-to-particle conversion involving chemical reactions of gases that result in condensable species (*secondary aerosols*). Although primary aerosol sources can produce particles of all sizes, secondary aerosol sources produce particles in the nanometer size range (< 0.1 μm). These very small particles have very short lifetimes, because they rapidly become attached to larger particles and therefore can exist in the atmosphere at significant concentrations only if they are continuously produced. Larger particles (> 2 μm) also have relatively short atmospheric lifetimes, because they are removed rapidly by sedimentation. Particulate matter larger than 20 μm can remain airborne for only a limited time and is therefore usually restricted to the immediate vicinity of the source (*2*).

Atmospheric aerosols, therefore, can exist in a range of sizes spanning four orders of magnitude. Traditionally, the size ranges have been subdivided into three modes shown in Figure 1 (*1*). The smallest size range, composed of aerosols < 0.1 μm, is called the *Aitken range* (or nuclei range). The existence of these aerosols was discovered in 1875 by Coulier (*3*), and were thoroughly studied for the first time by Aitken at the turn of the century (*4*). These ultrafine aerosols are produced by ambient-temperature gas-to-particle conversion and also by combustion processes generating hot gases that subsequently undergo condensation (*1*). Aerosols in the Aitken range have particle diffusion coefficients > 0.001 cm^2/sec (Figure 1), and therefore they can diffuse rapidly to the surfaces of other particles (*5*). In addition, these ultrafine aerosols can act as nuclei for the condensation of low-vapor-pressure gases causing them to grow rapidly into the next size range.

Atmospheric aerosols in the *coarse range*, larger than approximately 2 μm, are usually produced by mechanical processes such as grinding or wind erosion. Thus, because of the nature of their sources, they are composed predominately of minerals and inorganics such as sand and sea salt. Also included in this size range are larger bioaerosols such as spores, pollen, and bacteria. With settling velocities > 0.01 cm/sec (see Figure 1), coarse aerosols are generally removed from the atmosphere fairly rapidly by sedimentation. However, the atmospheric transport of coarse aerosols can occur over relatively long distances by convective processes where fallout is balanced by reentrainment. Mineral dusts from western China have been detected in western North America (*6*) and Canada (*7*), and African mineral dusts have been detected in south central Florida (*8*).

Aerosols in the intermediate size range from approximately 0.1–2 μm are known as the *accumulation range*. They are so named because the particle removal mechanisms of sedimentation and diffusion to surfaces are least effective in this range causing particles of this size to accumulate in the atmosphere. Aerosols in this fine size range typically arise from coagulation of

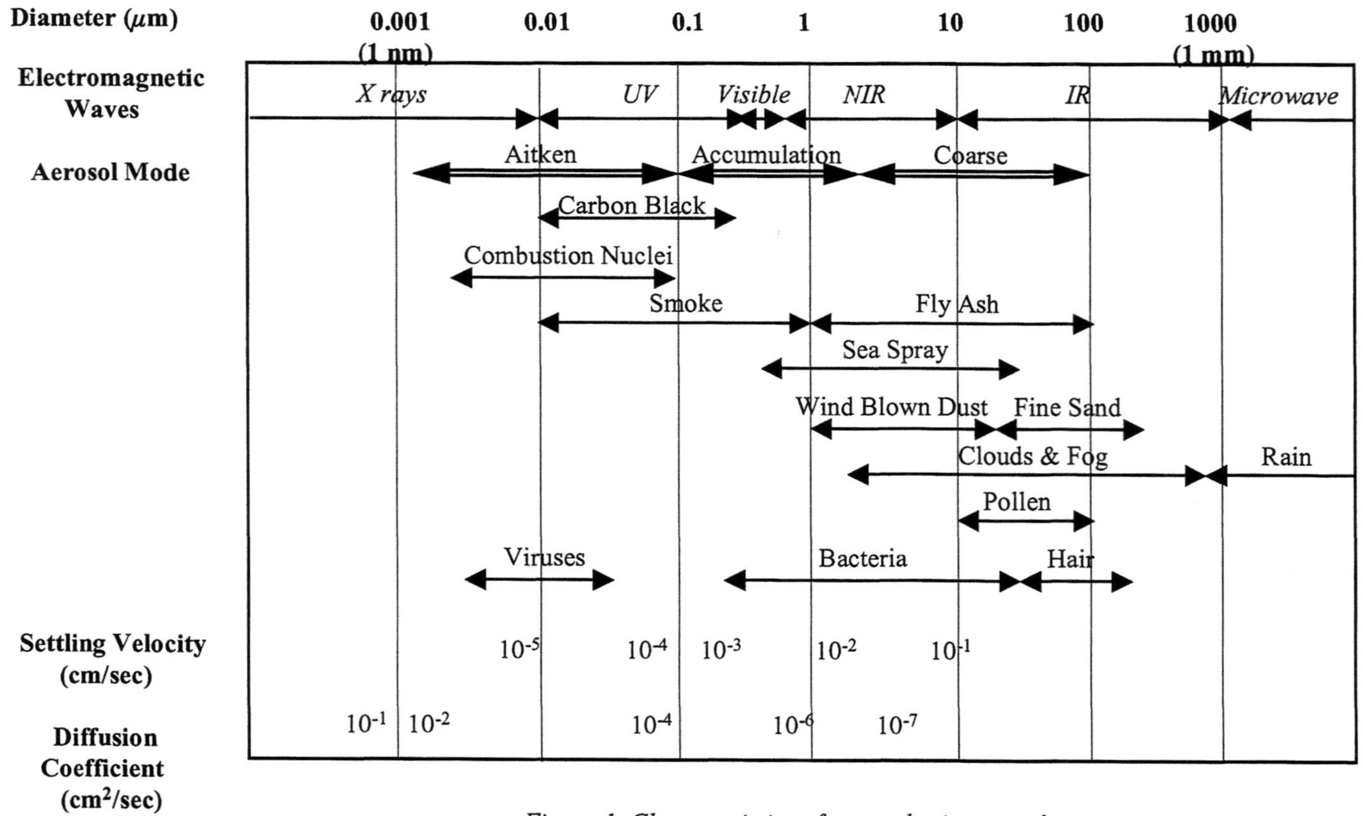

Figure 1. Characteristics of atmospheric aerosols.

smaller particles in the Aitken range or from condensation of low-volatility gases (water vapor, organics, etc.) onto existing particles.

Because of their sources, particles in the accumulation range typically contain organic compounds and soluble inorganics such as ammonium (NH_4^+), nitrate (NO_3^-), and sulfate (SO_4^{2-}). Being too small to settle out of the atmosphere, they are removed relatively slowly, primarily by incorporation into clouds and subsequent rainout. The rate of cloud droplet formation from these aerosols depends on their chemical compositions and their hygroscopicity, with the more soluble species being removed faster. Alternately, fine aerosols can be removed from the atmosphere by dry deposition after being carried to surfaces by eddy diffusion (*9*).

The properties of atmospheric aerosols that are important in determining their impacts to the environment and human health include their number concentration, mass, size, chemical composition, phase (liquid or solid), morphology, and surface properties. Currently, the U.S. Environmental Protection Agency (EPA) has two standards for particulate matter (aerosols) that are based on mass loadings. These are particulate matter < 10 µm in size (PM-10, PM10, or PM_{10}) and particulate matter < 2.5 µm (PM-2.5, PM2.5, or $PM_{2.5}$). There is also a current mandate to develop a standard for coarse particulate matter that would include the mass of particulate matter that lies between 2.5 and 10 µm [PM-C, PM-(10-2.5), or $PM_{10\text{-}2.5}$]. These EPA standards do not account for exposures to larger particulates, and they are not especially sensitive to ultrafine particle loadings. They also do not account for differences in the chemical natures of the aerosols.

Physical Properties

Since atmospheric particles are rarely spherical, their size is typically described in terms of an *effective aerodynamic diameter*, defined as the diameter of a sphere of unit density (1 g/cm^3) that has the same settling velocity as the atmospheric particle. This parameter is particularly useful, as it reflects both the removal rates of the particulates from the atmosphere as well as their deposition rates in the various parts of the human respiratory system.

The size ranges of atmospheric aerosols are typically determined by using impactors. The operation of impactors is based on the principle that, when an air stream is bent sharply, the particles in the air tend to continue in a straight line because of their inertia. In an impactor, a collection plate placed in an air flow causes the gas to stream around the plate. However, particles in the air may strike the plate and stick to it, depending on the size of the particle and the velocity of the air stream. The particle collection efficiency of an impactor (η) depends on the particle diameter (D), the flow velocity of the air (V), and the particle density (ρ) as in Equation (1).

$$\eta = (D^2 V \rho)/(18 \mu D_b) \qquad (1)$$

Here μ is the viscosity of air and D_b is related to the curvature of the air stream, which depends on the physical dimensions of the impactor (*10*).

Cascade impactors constructed of multiple stages are commonly used to size-fractionate atmospheric aerosols. Figure 2 shows the design of a Lundgren-type cascade impactor. Each stage consists of a cylinder covered with a removable collection surface. Before each stage are slits or nozzles of various sizes. The diameters of the slits become progressively smaller as the air travels through the cascade impactor, thus causing the air flow to move increasingly faster while collecting smaller particles on each successive plate, according to equation (1). A final exit filter is normally used to collect all small particles not captured by previous impactor stages.

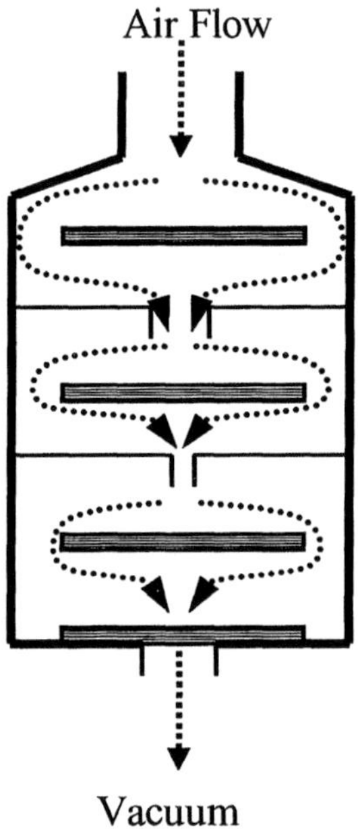

Figure 2. Operation of a three-stage Lundgren-type cascade impactor with three impactor plates and a final filter.

The size of the aerosols collected by an impactor stage is typically expressed as a *50% cutoff point* or fractionation size for that stage, defined as the diameter of monodisperse spheres of unit density, 50% of which are collected by the impactor stage being represented (*1*). The assumption is then made that each stage of the impactor will collect particles in a size range midway between its cutoff point and that of the next higher (or lower) impactor stage. The aerosol size ranges and 50% cutoff points for an eight-stage cascade impactor (Graseby-Anderson) with a flow rate of 1 cfm are shown in Table I.

Table I. Range of aerosol diameters and 50% cutoff points for each stage of an eight-stage Lundgren-type cascade impactor.

Stage	*Range (μm)*	*50% Cutoff (μm)*	*Δ D (μm)*	*Δ log D (μm)*
1	8.5–13	9	3.5	0.184
2	4.2–9	5.8	4.8	0.331
3	3.5–7.4	4.7	3.9	0.325
4	2.4–5	3.3	2.6	0.319
5	1.5–3	2.1	1.5	0.301
6	0.75–1.6	1.1	0.95	0.329
7	0.45–1	0.65	0.55	0.347
8	0.3–0.7	0.43	0.4	0.368
Final	0.01–0.3	0.1	0.3	1.477

The physical properties of atmospheric aerosols, such as number (N), mass (m), surface (S), and volume (V), are typically represented as a distribution over the various aerosol size ranges at the 50% cutoff point (D) of each impactor stage. However, because atmospheric aerosols can exist in a range of sizes spanning four orders of magnitude, it is most convenient to use logarithmic intervals ($\log D$) to display these distributions. Furthermore, because the widths of the size intervals collected by a cascade impactor are not equal, each range is normalized by the width of the interval (ΔD or $\Delta \log D$) as given in Table I. Therefore, a typical mass distribution of atmospheric aerosols collected by a cascade impactor is expressed similar to that in Figure 3, as a plot of $m/\Delta \log D$ versus $\log D$.

As noted earlier, the EPA's air quality standards for particulates (aerosols) are expressed in terms of total mass of particulate with a diameter < 10 μm per unit volume of air (PM-10) and the total mass of particulates with a diameter < 2.5 μm per unit volume of air (PM-2.5). The aerosol mass distribution in Figure 3 shows a trimodal mass distribution with most of the aerosol mass in the largest size range, typical of that seen in an urban center. The PM-10 measurement for the atmospheric aerosols in Figure 3 would be 98 μg/m^3, while the PM-2.5 measurement would be 42 μg/m^3.

Although most of the aerosol mass is found in the larger sizes, the greatest number of aerosols occur in the smallest size ranges. Plate 1 gives the number, volume, and surface distributions for an idealized urban aerosol (*1*). The number distribution shows a large peak at ~ 0.02 μm and a smaller shoulder at ~ 0.1 μm. The volume distribution, which can be related to the mass distribution by the density of the aerosol, shows two large peaks at 0.1–1.0 μm and 1–2 μm. The surface distribution shows a major peak at ~ 0.1 μm and smaller peaks at 0.01–0.1 μm and 1–10 μm. While the air quality standards are based on total aerosol mass, other properties such as aerosol size and chemical composition may be more important in determining aerosol health impacts and climate effects, and surface area may be more important in determining the effects of aerosols on heterogeneous atmospheric chemistry.

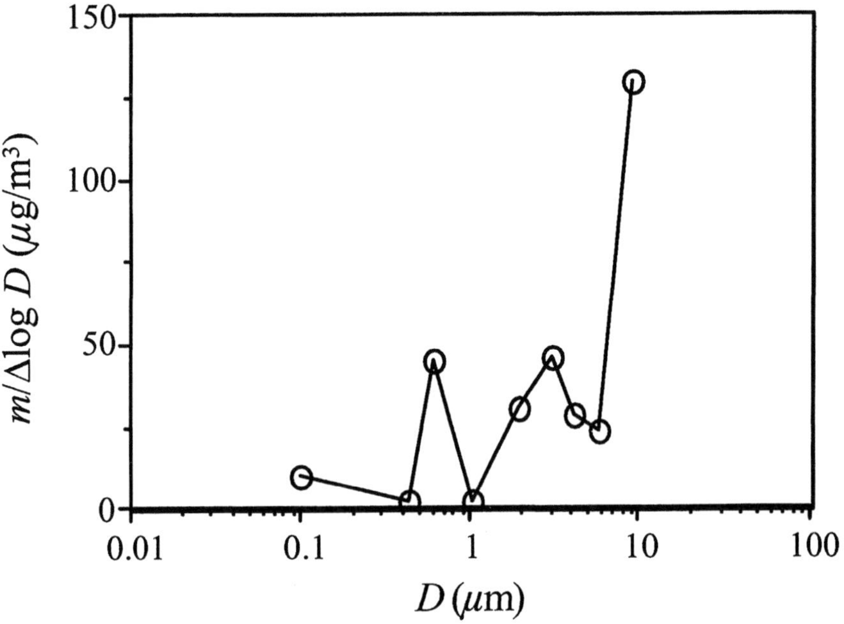

Figure 3. Aerosol mass distribution determined by Lundgren-type cascade impactor (Graseby-Anderson, Model 20-800) in Mexico City, February 21–23, 1997 (11).

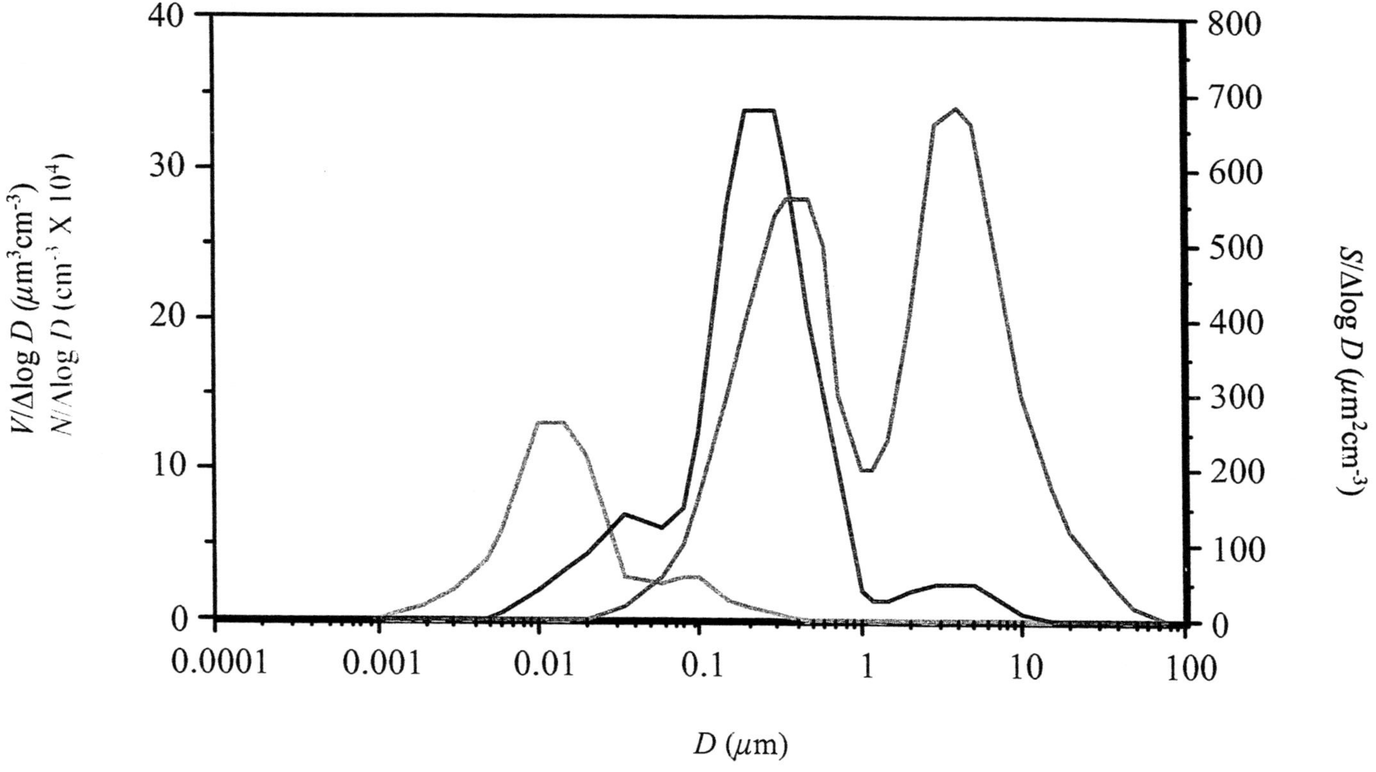

Plate1. Number (   ), volume (——), and surface (——) distributions for a model urban aerosol(1).

Chemical Properties

The chemical components that make up atmospheric particulate matter are not distributed equally among all size ranges. Instead, they tend to be found in specific size ranges, depending on the aerosol sources. The large particles (coarse range) are generated by mechanical processes such as grinding and wind erosion. These particles, therefore, are generally composed of soil elements (Si, Al, Fe, Mg, Ca, Na, K, etc.) from the continental materials and sea salt elements (Na, Cl, Mg, Ca, K, etc.) from the oceans. X-ray fluorescence has been used routinely to determine the elemental compositions of the coarse aerosols. Most of the major elements found in the coarse particles are nonvolatile and chemically inert. However, chlorine has been shown to be involved in heterogeneous surface reactions with atmospheric acidic species (*12*).

Fine particulate matter (< 2.5 μm) is generally formed by gas-to-particle conversion and combustion processes. The major components of fine particulates are therefore SO_4^{2-}, NO_3^-, NH_4^+, elemental and organic carbon, along with trace elements from combustion processes. The SO_4^{2-}, NO_3^-, and NH_4^+ aerosols are formed by the reactions of gaseous sulfur dioxide, nitrogen oxides, and ammonia in the atmosphere, while black or "elemental" carbon is produced directly by combustion (*13*).

Organic carbon aerosol compounds, found in both urban and rural areas, are composed of complex mixtures of organics. They are formed from chemical reactions in the atmosphere (e.g., multifunctional oxygenates) or from incomplete combustion processes (e.g., polycyclic aromatic hydrocarbons). The gas-phase oxidation of atmospheric organics can produce a wide variety of multifunctional organic products, some having sufficiently low vapor pressures to exist primarily in the condensed (particulate) form. In addition, semivolatile organic compounds in the atmosphere can partition between the gas and particle phases. The distribution between the two phases is important for understanding both atmospheric chemistry and human toxicology.

Both major and minor chemical constituents of atmospheric aerosols can be used as atmospheric tracers and indicators of the aerosol sources. The ratio of Al to Fe is different for various soils and mineral dusts and has been used to determine the impacts of African (*8*) and Asian (*7*) dusts in North America. Trace elements such as such as Ni and V have been used as markers for fuel oil combustion, while Se has been used as an indicator for coal-derived aerosols (*14*). In areas where leaded gasoline is still being used, Pb can be a marker for vehicle particulate emissions. However, Ba and Zn have been suggested as substitute tracers in areas where ledded fuel is no longer in use (*15*).

Black carbon measurements are commonly used to determine the amount of diesel-fuel-derived particulate matter in atmospheric aerosols. However, recent results have suggested that there may be other significant sources of black carbon aerosols (e.g., biomass smoke) in some areas (*16*). Degradation products of biopolymers, such as levoglucosan from cellulose or diterpenoids from resins, have been used as an indication of aerosols released from biomass burning (*17*). In some cases, a detailed determination of substituents can lead to the identification of the plant class being burned (*18*).

In addition to the use of marker compounds for aerosol source evaluation, the chemical constituents of aerosols can be used as tracers of their transport and as indicators of their lifetimes in the atmosphere. Beryllium-7, which becomes attached to aerosols in the upper troposphere and lower stratosphere, has been used as a tracer of the transport of aerosols from aloft (*19*). In contrast, ^{210}Pb becomes attached to atmospheric aerosols at ground level, and measurements of the ratio of ^{210}Pb to its decay products has been used to estimate the atmospheric residence times of fine aerosols (*11, 20*).

Health Effects

The amount of particulate matter that can be inhaled and subsequently absorbed in the human respiratory tract is highly dependent on particulate size as well as solubility. The smaller the particle, the deeper it can penetrate into the respiratory tract. Table II lists the factors that influence where particles in air are deposited as they are inhaled and the principal mechanisms of deposition for each size range (*21*). The mechanisms involved in particle deposition to surfaces within the human respiratory system are the same as those that govern the removal of particulate matter from the atmosphere; these are impaction, sedimentation, and diffusion.

In general, particles having aerodynamic diameters of 5–30 μm are deposited in the nasopharyngeal region (nose, nasal cavity, and throat). Impaction is the predominant deposition mechanism in this region because of the high air speeds and the many turns in this part of the respiratory tract. As with impaction sampling devices, the abrupt changes in air flow direction cause larger particles to hit the air passage walls and deposit there. Minimal absorption takes place in the nasopharyngeal region due to the cell thickness. Cilia in the upper regions will normally expel the deposited particles, with a clearance half-time of minutes (*22*). An exception to this clearance mechanism is motile bacteria and some viruses.

Smaller particles with aerodynamic diameters of 1–5 μm are deposited in the tracheobronchial region (trachea and upper bronchial tubes). Sedimentation

Table II. Factors that influence particulate deposition in the human respiratory system (*21*).

Particle Size (μm)	*Area Deposited*	*Method*	*Air Direction Change*	*Relative Air Velocity*
5–30	Nasopharyngeal	Impaction	Very abrupt	4
1–5	Tracheobronchial	Sedimentation	Less abrupt	3–2
< 1	Pulmonary	Diffusion	Calm	0–1

is the most common method of deposition in this area because the air flow has slowed enough for the particle settling velocities to become important (Figure 1). While most soluble gases can enter the blood stream in this region, deposited particles are moved back up the trachea where they are expelled.

The very small particles with aerodynamic diameters ≦ 1 μm can reach the pulmonary region, consisting of the very small airways (bronchioles) and the alveolar sacs of the lung. This region has a very large surface area, and the air flows are essentially calm. In addition, the alveoli consist of only a single layer of cells with very thin membranes that separate the inhaled air from the blood stream. Absorption in this area is quite efficient, compared to other regions. Particles reaching this area are deposited by diffusion to the surfaces of the alveoli, and the water-soluble components can pass through the cell membranes by simple passive diffusion (*21*).

Ultrafine particles and nanoparticles < 0.1 μm have been proposed by some as the biologically important ambient particles (*23*). These particles are deposited in the deep lung by diffusion and can pass through the cells lining the lung into the interstitial space. The very high surface areas of the ultrafine particles can result in a faster release of toxic compounds than would occur from larger particles of the same composition.

Clearance in the pulmonary region is accomplished by macrophages on the surfaces of the alveolar walls, which can engulf and digest particles with typical clearance half-times of 24 h. Some insoluble particles are scavenged by these macrophages and cleared into the lymphatic system. Other insoluble particles (e.g., coal dust, asbestos fibers) may remain in the alveoli indefinitely. The nature of the toxicity of deeply inhaled particles depends on whether they are absorbed or remain in the alveoli. Lipid-soluble material, such as chlorinated polycyclic aromatic compounds, if absorbed can be distributed rapidly throughout the body to various organs or fat depots. Non-absorbed material can cause severe toxic reactions in the respiratory system, taking the form of chronic bronchitis, emphysema, fibronic lung disease, and lung cancer (*24*).

Particulate air quality standards in the United States were first expressed in terms of *total suspended particulate matter (TSP)*. Since, as noted above, most of the aerosol mass is in very large particles, this measurement was heavily weighted by very large wind-blown dust and soil particles and was not a good indication of the amounts of aerosols in the respirable size ranges. The standard was then changed to the total mass of suspended particulate matter < 10 μm in size, or PM-10. Recently, the standard was modified to include particulate matter < 2.5 μm in size or PM-2.5, due to a large number of studies that have shown a relationship between increased mortality rates and the concentration of fine particulate matter (*25*). Increases of 1.5% in total daily mortality, 3.3% in deaths due to chronic obstructive pulmonary disease, and 2.1% in deaths due to ischemic heart disease have been reported for a PM-2.5 increase of 10 $\mu g/m^3$ (*26*).

Although the biological mechanism for these correlations is not as yet clear, recent studies have reported a linear correlation of daily mortality with increases in combustion-related aerosols with no evidence of a threshold. Although a 1.5% increase in deaths was reported for a 10-$\mu g/m^3$ increase of total PM-2.5, a 3% increase in deaths was seen for a 10-$\mu g/m^3$ increase in particulate emissions from traffic (*27*). Increased black smoke from 17 $\mu g/m^3$ to 67 $\mu g/m^3$ was associated with a 5% increase in daily deaths (*28*).

Black carbon (soot) is the major type of fine particulate generated by combustion processes. Black carbon has been reported to reach the deep lung and, once there, to be quite resistant to absorption (*29*). Both organic and inorganic compounds associated with black carbon have been found to have mutagenic, carcinogenic, and irritant properties. Polycyclic aromatic hydrocarbons, products of incomplete combustion, are associated with carbon soot and are known to be mutagenic and carcinogenic (*1*). A study of size-fractionated urban aerosols showed that the mutagenic activity increased with decreasing particle size, consistent with the effects of organic compounds condensed on submicron combustion particles (*30*). However, quantitation of individual compounds can account for only a fraction of the observed mutagenicity of atmospheric soot particulates (*31*).

Visibility and Climate Effects

The production of atmospheric haze that reduces visibility is the most readily apparent impact of urban air pollution. The ability to see objects clearly through the atmosphere depends on the concentrations of atmospheric particles and gasses that can absorb and scatter light in the visible wavelength region.

This scattering and absorption of visible light causes a decrease in the contrast or a change in the percieved color of distant objects. The absorption of light by gas molecules and particles is sometimes the cause of color changes in the atmosphere, but light scattering by particles is the dominant factor in visibility reduction. Light scattering by atmospheric aerosols can reduce the perceived contrast of distant objects by deflecting the light from the object out of the line of sight. In addition, during the daytime sunlight is simultaneously scattered into the line of sight, making dark objects appear lighter.

Light scattering by particles consists of three types, depending on the size of the particles relative to the wavelength (λ) of light (*1*). For particle diameters much less than λ ($D \leqq 0.03$ μm for the UV/visible region), light scattering is similar to that by gaseous molecules (Rayleigh scattering). The scattered intensity is symmetrical in all directions relative to the incident light and varies as λ^{-4}. For particles much larger than λ ($D > 10$ μm), light is scattered by reflection from the surface, diffraction around the edges, or refraction through the interior of the particle (geometric scattering). Most of the light scattered in this manner is not diverted significantly from its original path.

Particles in the intermediate size range ($D \cong \lambda$) scatter light most efficiently and in a more complicated manner, known as Mie scattering. The intensity of light (I) scattered at distance r and scattering angle θ from a particle is given by

$$I(\theta, \mathrm{r}) = \frac{I_o\, \lambda^2\, (i_1 + i_2)}{8\pi^2 r^2} \qquad (2)$$

where I_o is the incident light intensity, and i_1 and i_2 are the Mie intensity parameters for the perpendicular and parallel polarized components of the scattered light (*10*). The Mie scattering parameters are a complex function of the refractive index (m) and the size parameter ($a = \pi D/\lambda$) of the particle and the scattering angle (θ).

Most light scattering by atmospheric aerosols is due to particles in the size range 0.1–1 μm. Figure 4 gives the light scattering coefficient (σ_{sp}) as a function of particle diameter for particles with a refractive index of 1.5 and incident light at 550 nm (*1, 32*). As in Figure 4, aerosols in the accumulation range are clearly expected to dominate the scattering of visible light in the atmosphere.

Light scattering by atmospheric aerosols also affects Earth's radiative balance and climate by scattering incoming solar radiation back into space and therefore acting to cool the surface (*33*). The general cooling effect from solar light scattering by anthropogenic sulfate aerosols in the Northern Hemisphere has been estimated to be comparable in magnitude to the atmospheric warming produced by increases in carbon dioxide (*34*).

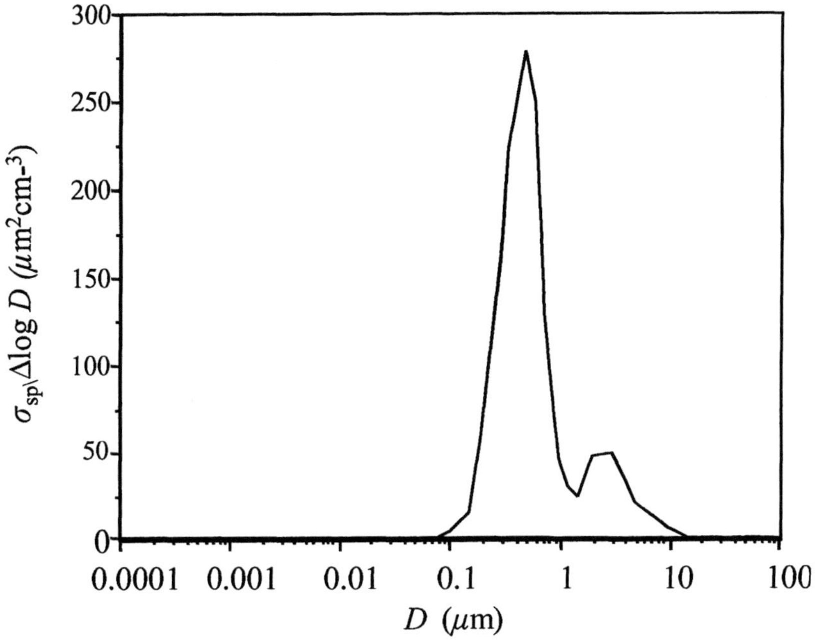

Figure 4. Optical scattering ($\lambda = 550$ nm) by atmospheric particles (σ_{sp}) of refractive index 1.5 and the volume size distribution shown in Plate 1 (*1, 32*).

droplets are effective scatterers of incoming solar radiation, limiting the flux Atmospheric aerosols can also cool Earth's surface through the formation of clouds. Since fine inorganic aerosols (e.g., SO_4^{2-}, NO_3^-) are hygroscopic, they can act as cloud condensation nuclei (CCN) and increase cloud formation. Cloud reaching the surface and leading to a net cooling effect. The extent of this effect depends on the size and concentration of cloud water droplets, which are controlled by the size, concentration, and chemical composition of the aerosols.

Depending on their chemical composition, aerosols can also absorb incoming solar radiation. This absorption of light will lead to warming of the atmosphere as the absorbed energy is converted into heat. In addition, someaerosols can absorb longwave thermal infrared radiation, acting as greenhouse species in the same manner as the greenhouse gases (*35, 36*). The dominant light-absorbing component of atmospheric aerosols is black carbon produced in combustion processes. Black carbon can absorb as much as 20–25% of incoming solar radiation and is thought to be most important in urban polluted areas where absorbing aerosol concentrations are highest (*37*).

The total effect of atmospheric aerosols on radiation can be expressed in terms of the particle extinction coefficient: $\sigma_e = \sigma_{ap} + \sigma_{sp}$. The absorption ($\sigma_{ap}$) and scattering ($\sigma_{sp}$) coefficients of the particle are derived from the real and imaginary parts of the complex refractive index, defined as $m(\lambda) = n(\lambda) - ik(\lambda)$, where n is the refractive index of the particle and k is the aerosol absorption index. The absorption index is related to the absorption coefficient as: $k = \sigma_{ap}\lambda/4\pi$. The magnitude of light scattering and absorption by atmospheric aerosols is therefore determined by the complex refractive index of the particle, which is dependent on the chemical composition of the aerosol.

The values of the index of refraction and the absorption index for some species found in atmospheric aerosols are listed in Table III. All of the aerosols listed will scatter light effectively in the visible region, as indicated by the values of n in Table III. However, the light-absorbing aerosols include diesel soot (as well as other sources of black carbon) and the iron-containing minerals. Other absorbers of visible light not listed in Table III include organics such as carbonyl compounds (*1*). The contributions of these organic carbon compounds to atmospheric light extinction are not well understood, but it appears that they can be significant in remote areas. Studies of light absorption in the U.S. national parks estimate the contribution of organic carbon and black carbon to be approximately equal (45% and 41%, respectively) (*38*).

The inorganic atmospheric aerosols SO_4^{2-} and NO_3^- also absorb in the thermal infrared region. Part of this increased absorption in the infrared is due to the absorption of the associated liquid water in this region. However, black carbon remains the major contributor to the absorbtion of radiation across the

Table III. Complex refractive index of some aerosol species in the visible (~ 550 nm) and infrared (~ 3000 cm^{-1}) wavelength ranges (*1, 39–44*).

Species	*Visible* n	*Visible* k	*Infrared* n	*Infrared* k
Water	1.33	0	1.45	0.04
NaCl	1.34	0		
$(NH_4)_2SO_4$	1.53	0		
H_2SO_4	1.44	0	1.43	0.07
NH_4NO_3	1.56	0	1.36	0.08
$CaCO_3$	1.59	0		
SiO_2	1.48	0		
Fe	1.51	1.63		
Fe_3O_4	2.58	0.58		
Diesel soot	1.68	0.56	1.6	0.7

entire spectrum, and the warming effect from these combustion-related aerosols is expected to be most important in urban areas where their concentrations are highest. It is also interesting to note that the black carbon transported from these urban centers could potentially impact the albedo in remote areas like the Arctic or Antarctic by adding a strongly absorbing substance to snow surfaces (*45*).

Urban Centers and Megacities

Urban centers have become major sources of aerosols and their precursor gases. Megacities, urban areas with more than 10 million inhabitants, are becoming more numerous as our planet's population continues to soar. In 1950, the only megacity was New York City, the site of the World Trade Center tragedy on September 11, 2001. In 1995, New York City was fifth in size among 14 megacities, with Mexico City and Tokyo contending for first place. The number of megacities projected for 2015 is 21, with Bombay, Seoul, and Sao Paolo rapidly on the rise. The number of "mini-megacities", urban areas with 5-10 million inhabitants, is projected to be 37 in 2015 (up from 7 in 1995). The rise in these urban centers is tied to a redistribution of population from principally rural environments (2/3 in 1995) to principally urban (1/2 by 2025). These megacities are major sources of aerosols that are impacting urban, regional, and global environments. Activities associated with the large population densities in megacities produce significant amounts of primary aerosols. Among the major sources are; diesel engines producing black carbonaceous soots and road dusts, construction, and industrial activities producing larger mechanically generated aerosols. The combustion-related release of gasses such as oxides of nitrogen and sulfur is also associated with these large centers of population and leads to the downwind production of significant amounts of secondary NO_3^- and SO_4^{2-} aerosols. This constant production of significant amounts of particulate matter and aerosol precursors affects urban air quality, health, and visibility in many urban centers. An example of visibility reduction in Mexico City is shown in Plate 2.

The cloud of dust produced in the initial collapse of the World Trade Center's twin towers in September of 2001 was significantly larger than what is typically observed in an urban area, as outlined in Chapter 2. Because the dust cloud and smoke from subsequent fires were followed closely by the worldwide news media, many people became aware of the exposures that were occurring in this megacity due to this catastrophic event. The fact is that significant aerosol releases and exposures are occurring continually in megacities throughout the world, albeit less dramatically than during the WTC aerosol releases. These constant releases of material will increase as the population and the number of megacities continue to increase in the future.

Plate 2. Aerial photograph of Mexico City taken in April 2003, showing the early afternoon "haze" over this megacity.
(See page 2 of color inserts.)

Plate 3. Photograph from a commercial aircraft, taken in the early afternoon in April 2003, showing the Mexico City megacity plume of aerosols leaving the urban area and its regional-scale impact. *(See page 3 of color inserts.)*

Indeed, it is also important to recognize that the significant amounts of aerosols apparent in the "smog" cloud over the megacities are transported out of these areas to the regional scales leading to impacts on weather and climate as well as on regional ecosystems. This is quite apparent from the urban "megacity" plumes seen downwind of these areas (see Plate 3). The aerosols will eventually be removed from the atmosphere by the processes outlined previously including gravitational settling, diffusional loss to other aerosol surfaces, hygroscopic uptake of water, and so forth, to be deposited in ecosystems on regional and global scales, depending on the atmospheric lifetimes of the aerosols. The use of natural radionuclide tracers has estimated this lifetime to be on order of 10–60 days for the accumulation mode aerosols, depending on their chemical compositions (*11*). Thus, it is important to understand urban centers and megacities as sources of aerosols and precursor gases as they affect human health on urban scales and also climate, weather, and ecosystems on regional scales. This has clearly been an important finding of past work addressing megacity-associated visibility reduction in rural areas and also the effects of acid precipitation as the soluble acidic nitrate and sulfate aerosols are removed by washout over regional areas.

Acknowledgements

The authors' work is supported by the U.S. Department of Energy (USDOE), Office of Science, Office of Biological and Environmental Research, Atmospheric Science Program, under contract W-31-109-Eng-38. We wish to particularly thank Mr. Peter Lunn and Mr. Rick Petty (USDOE) for their continuing encouragement and support.

References

1. Finlayson-Pitts, B. J.; Pitts, J. N. Jr. *Atmospheric Chemistry of the Upper and Lower Atmosphere*; Academic Press: San Diego, 2000.
2. Junge, C. E. *Air Chemistry and Radioactivity*; Academic Press: New York, 1963.
3. Verzar, F. *Experientia* **1959,** *1519*, 362–363.
4. Aitken, J. *Collection of Scientific Papers*; Cambridge University Press: London, 1923.
5. Seinfeld, J. H. *Atmospheric Chemistry and Physics of Air Pollution*; John Wiley and Sons: New York, 1986.
6. VanCuren, R. A. *J. Geophys. Res.* **2003,** *108*, 4623–4633.

7. McKendry, I. G.; Hacker, J. P.; Stull, R.; Sakiyama, S.; Mignacca, D.; Reid, K. *J. Geophys. Res.* **2001,** *106*, 18,361–18,369.
8. Prospero, J. M.; Olmez, I.; Ames, M. *Water, Air, Soil Pollut.* **2001,** *125*, 291–317.
9. Twomey, S. *Atmospheric Aerosols*; Elsevier Scientific: New York, 1977.
10. Finlayson-Pitts, B. J.; Pitts, J. N. Jr. *Atmospheric Chemistry: Fundamentals and Experimental Techniques*; John Wiley and Sons: New York, 1986.
11. Marley, N. A.; Gaffney, J. S.; Drayton, P. J.; Cunningham, M. M.; Orlandini, K. A.; Paode, R. *Aerosol Sci. Technol.* **2000,** *32*, 569–583.
12. Finlayson-Pitts, B. J.; Hemminger, J. C. *Phys. Chem. A.* **2000,** *104*, 11,463–11,477.
13. Gaffney, J. S.; Marley, N. A. *Scientific World* **2003,** *3*, 199–234.
14. Song, X.-H.; Polissar, A. V.; Hopke, P. K. *Atmos. Environ.* **2001,** *35*, 5277–5286.
15. Monaci, F.; Moni, F.; Lanciotti, E.; Grechi, D.; Bargagli, R. *Environ. Pollut.* **2000,** *107*, 321–327.
16. Schauer, J. J. *J. Expos. Anal. Environ. Epidemiol.* **2003,** *13*, 443–453.
17. Simoneit, B. R. T. *Environ. Sci. Pollut. Res.* **1999,** *6*, 159–169.
18. Simoneit, B. R. T.; Rogge, W. F.; Mazurek, M. A.; Standley, L. J.; Hildemann, L. M.; Cass, G. R. *Environ. Sci. Technol.* **1993,** *27*, 2533–2541.
19. Zanis, P.; Schuepbach, E.; Gaggeler, H. W.; Huebener, S.; Tobler, L. *Tellus* **1999,** *51B*, 789–805.
20. Gaffney, J. S.; Marley, N. A.; Cunningham, M. M. **2004,** *Atmos. Environ. 38*, 3191-3200.
21. Klaassen, C. D. *Casarett and Doull's Toxicology: The Basic Science of Poisons*; 3rd. ed.; MacMillan: New York, 1986.
22. Davey, B.; Halliday, T. *Human Biology and Health: An Evolutionary Approach*; Open University Press: London, 1994.
23. Oberdörster, G.; Gelein, R. M.; Ferin, J.; Weiss, B. *Inhalation Toxicol.* **1995,** *7*, 111–124.
24. Phalen, R. F. *Inhalation Studies: Foundations and Techniques*; CRC Press: Boca Raton, 1984.
25. Pope, C. A. *Aerosol Sci. Technol.* **2000,** *32*, 4–14.
26. Schwartz, J.; Dockery, D. W.; Neas, L. M. *J. Air Waste Manage. Assoc.* **1996,** *46*, 927–939.
27. Schwartz, J.; Laden, F.; Zanobetti, A. *Environ. Health Perspect.* **2002,** *110*, 1025–1029.
28. Scwartz, J.; Ballester, F.; Saez, M.; Pèrez-Hoyos, S.; Bellido, J.; Cambra, K.; Arribas, F.; Cañada, A.; Pèrez-Boillos, M. J.; Sunyer, J. *Environ. Health Perspect.* **2001,** *109*, 1001–1005.
29. Slatkin, D. N.; Friedman, L.; Irsa, A. P.; Gaffney, J. S. *Human Pathology* **1978,** *9*, 259–267.

30. Pango, P.; Zaiacomo, T. D.; Scarcella, E.; Bruni, S.; Calamosca, M. *Environ. Sci. Technol.* **1996,** *30,* 3512–3516.
31. Lightly, J. S.; Veranth, J. M.; Sarofim, A. F. *J. Air Waste Manage. Assoc.* **2000,** *50,* 1565–1618.
32. Waggoner, A. P.; Charlson, R. J. In *Fine Particles: Aerosol Generation, Measurement, Sampling, and Analysis*; Leu, B. Y. H., Ed.; Academic Press: New York, 1976.
33. Charlson, R. J.; Schwartz, S. E.; Hales, J. M.; Cess, R. D.; Coakley, J. A., Jr.; Hansen, J. E.; Hoffman, D. J. *Science* **1992,** *255,* 423–430.
34. Kiel, J. T.; Brieglleb, B. P. *Science* **1993,** *260,* 311–314.
35. Marley, N. A.; Gaffney, J. S.; Cunningham, M. M. *Environ. Sci Technol.* **1993,** *27,* 2864–2869.
36. Gaffney, J. S.; Marley, N. A. *Atmos. Environ.* **1998,** *32,* 2873–2874.
37. Herrmann, P.; Hanel, G. *Atmos. Environ.* **1997,** *24,* 4053–4062.
38. Malm, W. C.; Molenar, J. V.; Eldred, R. A.; Sisler, J. F. *J. Geophys. Res.* **1996,** *101,* 19,251–19,265.
39. Marley, N. A.; Gaffney, J. S.; Baird, J. C.; Blazer, C. A.; Drayton, P. J.; Frederick, J. E. *Aerosol Sci. Technol.* **2001,** *34,* 539–549.
40. Marley, N. A.; Gaffney, J. S.; Cunningham, M. M. *Appl. Optics* **1994,** *33,* 8041–8054.
41. Horvath, H. In *Atmospheric Acidity Sources, Consequences, and Abatement*; Radojevic, M.; Harrison, R. M., Eds.; Kluwer: New York, 1992.
42. Norman, M. L.; Qian, J.; Miller, R. E.; Worsnop, D. R. *J. Geophys. Res.* **1999,** *104,* 30,571–30,584.
43. Wagner, R.; Mangold, A.; Möhler, O.; Saathoff, M.; Schurath, U. *Atmos. Chem. Phys.* **2003,** *3,* 1147–1164.
44. Twitty, J. T.; Weinman, J. A. *J. Appl. Meteorol.* **1971,** *10,* 725–731.
45. Hansen, J.; Nazerenko, L. *Proc. Nat. Acad. Sci.*, **2004,** *101,* 423–428.

Chapter 2

An Overview of the Environmental Conditions and Human Exposures That Occurred Post September 11, 2001

Paul J. Lioy[1,2], Panos G. Georgopoulos[1,2], and Clifford P. Weisel[1,2]

[1]Department of Environmental and Occupational Medicine and [2]Environmental and Occupational Health Sciences Institute, Robert Wood Johnson Medical School (RWJMS) at UMDNJ and Rutgers, The State University of New Jersey, Piscataway, NJ 08854

The events of 9-11-01 left a series of environmental and occupational health issues to be evaluated for short term and long term health effects. A time line for the post 9-11-01 events is provided based upon the types and intensity of the emissions of dust and smoke particles and gases that were released on that fateful day and/or through the last days of the fires on 12-20-01. Summaries are presented on the nature of the plume, the major components of the plume, and the types of exposures that may have occurred outdoors and indoors that will also be essential components of future risk assessments. This overview includes information from a number of published and unpublished studies and analyses. It must be remembered, however, that the situation was a response to a terrorist attack and the subsequent disaster. Thus, the acquisition of samples and the types of analyses conducted by various organizations on those samples must be evaluated within the context of that chaotic enviornment.

Introduction

The attack on the World Trade Center (WTC) was an unprecedented act of war on American soil. The terrorists, by virtue of colliding two commercial jets into the two massive buildings ignited fires that eventually led to the collapse of both buildings producing a complex environmental disaster (*1, 2*). Because these towers collapsed at vertical speeds up to 120 miles per hour, a large fraction of millions of tons of material that composed each of the WTC towers were reduced to pulverized dust and smoke, and the debris were dispersed radially from the WTC site throughout southern Manhattan (Figure 1) (*2*). The collapse of the towers also produced an intense plume that moved southeast above Brooklyn during the rest of the day. The only fortunate part about this horrendous act of war was that the terrorists released no biological, chemical, or radiological agents during the attack. The materials that were once part of the WTC disintegrated into a large mass of particles that were primarily >10 μm in diameter and had a large variety of shapes, and fibers also >10 μm diameter or length, which contained construction debris, cement, chrysotile, cotton fibers (lint), tarry products, charred wood, soot and glass fibers (*3*). The particles had a very alkaline pH and produced irritation of the upper respiratory system (*3, 4*). In some cases the inhalation of dust and smoke yielded a variety of symptoms and health outcomes (e.g. the exacerbation of asthmatic conditions among workers and some residents) (*5–14*). Most of the immediate effects were limited to individuals who spent many hours at Ground Zero during the first few days post 9-11 or re-entered a building before it was professionally cleaned for dust contamination. The general population of southern Manhattan and individuals who had commuted to New York City on that day did not show the same frequency of acute effects, as the rescue workers and firefighters (*5*).

Nature of the Post 9-11 Problems

It is clear that individuals who were in southern Manhattan during the event were acutely exposed to at least lifetime doses of inhalable and respirable particles (*3,5*). Photographs show individuals completely covered by the dust (Figure 2). These particles were primarily >10 μm in diameter (Figure 3), a size range not normally or traditionally measured by the U.S. Environmental Protection Agency (EPA). Further, the particles were composed of glass fibers, cement dust (high pH), other construction materials, and silica, (*3,4*) materials not typically present in ambient air. A summary of the major individual and classes of materials found in the initial dust and smoke is shown in Table I (*3*).

Figure 1. Collapse of the WTC showing the dust cloud moving away from Ground Zero(15).

Figure 2. The suspended inhalable dust after the collapse of the WTC and exposed individuals (16).

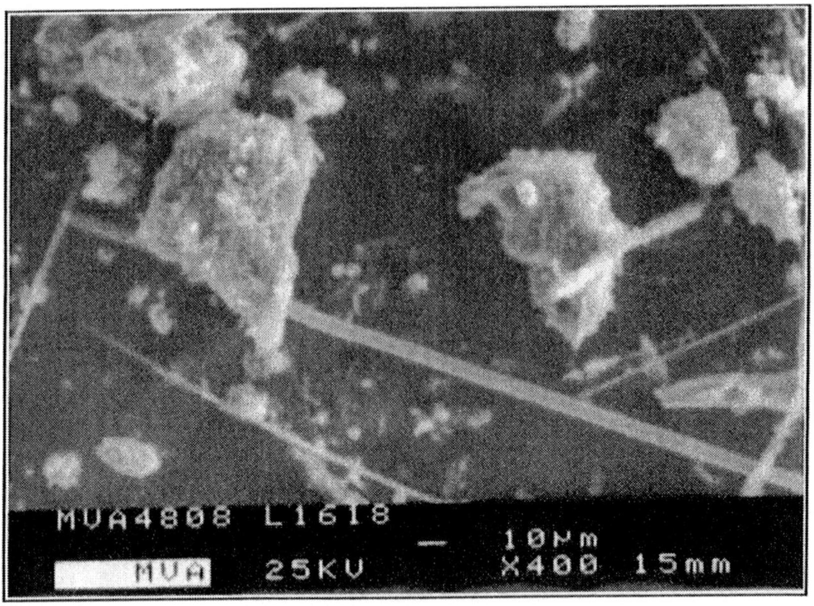

Figure 3. Micrograph of the WTC Dust (Analyses by J. Millette, MVA, Ga.; for P. J. Lioy).

However, it was, and still is, impossible to compare the exposures to these suspended and resuspendable materials with current environmental health based standards since impossible to compare the exposures to these suspended and resuspendable materials with current environmental health based standards since there are no standards for atmospheric mass loadings of particles >10 µm in diameter and/or the types of materials that composed this very complex mixture. The basic reason is that all current National Ambient Air Quality Standards (NAAQS) for particles are focused on long term health effects caused by exposure to atmospheric particles <10 µm and < 2.5 µm in diameter that primarily reach the thoracic and gas exchange regions of the lung, respectively (see Chapter 1) (*17*). Most of the immediate health effects recorded during the aftermath of the WTC collapse were in the extra thoracic and nasopharyngeal regions of the respiratory system, which requires information on the inhalation, deposition and health risks caused by respirable particles and fibers above 10 µm

Table I. Summary of the Major Components of the Initial Total Outdoor Dust and Smoke Released from the Collapse of the WTC (*3*).

Component	*Units*	*Range*
pH	units	9.3 – 11.5
Non fiber	%	37 – 50
Glass Fiber	%	40 – 41
Cellulose	%	9 – 20
Chrysotile asbestos	%	0.8 – 3.0
Aerodynamic Diameter:		
<2.5 μm	%	0.9 – 1.3
2.5 – 10 μm	%	0.3 – 1.4
10 – 53 μm	%	35 – 47
>53 μm	%	52 – 64
Selected Metals:		
Titanium	mg/g	1.5 – 1.8
Zinc	mg/g	1.7 – 3.0
Magnesium	mg/g	0.6 – 0.8
Lead	mg/g	0.14 – 0.5
Selected Organics:		
Polycyclic aromatic hydrocarbons	%	>0.1
Phthalates	ppm	100's
Incomplete combustion products	ppm	100's
Jet fuel	ppm	10's
PCBs, PBDEs and pesticides	ppb	<100

Note: Non fibers are composed of cement and carbon. PCBs are polychlorinated biphenyls. PBDEs are polybrominated diphenyl ethers.

in diameter (*18*). This sort of information would be useful in protecting the public from the acute exposures derived from the inhalation of construction and building demolition debris, and uncontrolled piles of roadway and construction debris (*19*).

The long-term health consequences of the inhalation of high concentrations of these specific compounds for a very short period of time (hours to days) remain to be identified over the next 10-20 years. The compounds of concern, however, include asbestos, polycyclic aromatic hydrocarbons (PAHs) and other organic compounds (*2,3,5,20*). The primary populations of concern for these effects would still be the highly exposed rescue and recovery workers (e.g.

firefighters) (*5,7*); especially those individuals not wearing any respiratory protection (*7,19*).

The events of 9-11-01 led to a number of exposed populations, previously identified by Lioy *et al.* (*3*). Many individuals were exposed outdoors to the resuspended dust and smoke during the days following the collapse, but others were periodically exposed to the emissions of gases and fine particles emitted by the fires (smoldering and active) that burned until December 20, 2001 (*22-24*). During the weeks and months after 9-11, fires burned at various locations with variable intensity based upon the nature of the recovery operations being conducted at Ground Zero. Further, the nature of the exposure and the types of people that could have inhaled this material changed once the local situation moved from a rescue to a recovery operation. (e.g. more aggressive removal of construction debris and spikes in the fires).

Time Course of Post 9-11 Emissions and Exposure Conditions

Considering the ongoing environmental health concerns about the aftermath of 9-11, we have been able to determine that there were at least four periods or *categories* of *outdoor* exposure conditions after the collapse of the WTC. These exposure categories have been identified previously by Lioy *et al.*, and Landrigan *et al.* (*3, 5*) and are summarized in a theoretical conceptual outdoor exposure profile that is illustrated in Figure 4. During the first few hours after the event (category 1), everyone in the immediate vicinity of the WTC was exposed to and inhaled excessive doses of total particulate matter (inhalable thoracic, and fine particles) and gases released by the disintegration and collapse of the towers, and the intense fires. We have no quantitative information on the actual atmospheric concentrations of gases and particles released because nothing was being measured in the vicinity of WTC at the time. This situation is to be expected since it would have been improbable to have measurements of the compounds of concern being made at the time of, and at locations surrounding the WTC towers. However, as seen in the videos taken during and after the collapse by many people, visibility was substantially reduced by the suspended and resuspended particles along the streets of lower Manhattan, (Figure 2). The initial levels of gases would have probably been in the high ppb to low ppm range, since post event grab samples collected by the EPA at the site found volatile organic carbon (VOC) levels that ranged from non-detectible to the 1-5 ppm range for the period that began about a week post 9-11 (*25*). The particle concentrations would have been in the range of 100,000's $\mu g/m^3$. During the next 24-48 hrs (category 2), the smoke released by the initial fires remained significant, but would have generated particles with a smaller size distribution and lower concentration. Concurrently, the settled dust and smoke was

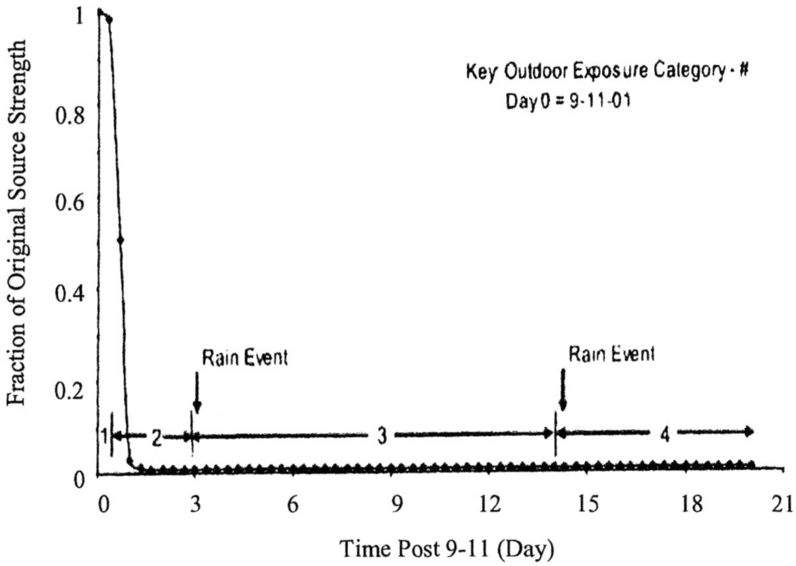

Figure 4. Conceptual sigmoidal decay model vs time of smoke and dust released during and after 9-11.

repeatedly resuspended in the air during the initial days after 9-11 throughout southern Manhattan (3). This was due in part to the movement of the emergency vehicles and the intense efforts to search for and rescue survivors buried under the debris at Ground Zero. Initially it was assumed that over 12,000 people might have been trapped, requiring a massive rescue effort. Thus, the second period of time for potential exposure (category 2) lasted from the latter part of day 1 through day 3. Initially the winds were west and northwest, but on September 12, 2001 the wind and the WTC plume rotated a full three hundred and sixty degrees from Brooklyn to New Jersey, and then to northern Manhattan and back to Brooklyn (see Figure 5). For category 3, which included day 3 through approximately day 15, the rescue efforts continued. However, the initial re-suspendable settled dust and smoke concentrations diminished because approximately one inch of rain fell in southern Manhattan on Friday, the 14th of September (*5*, 23, *24*). Unfortunately, the rain did not wash all the outdoor dust and smoke away, and it did not extinguish the fires. After the rain, the settled dust and smoke were still present at various locations in southern Manhattan. In fact, there were major efforts by the EPA and others to remove the settled dust

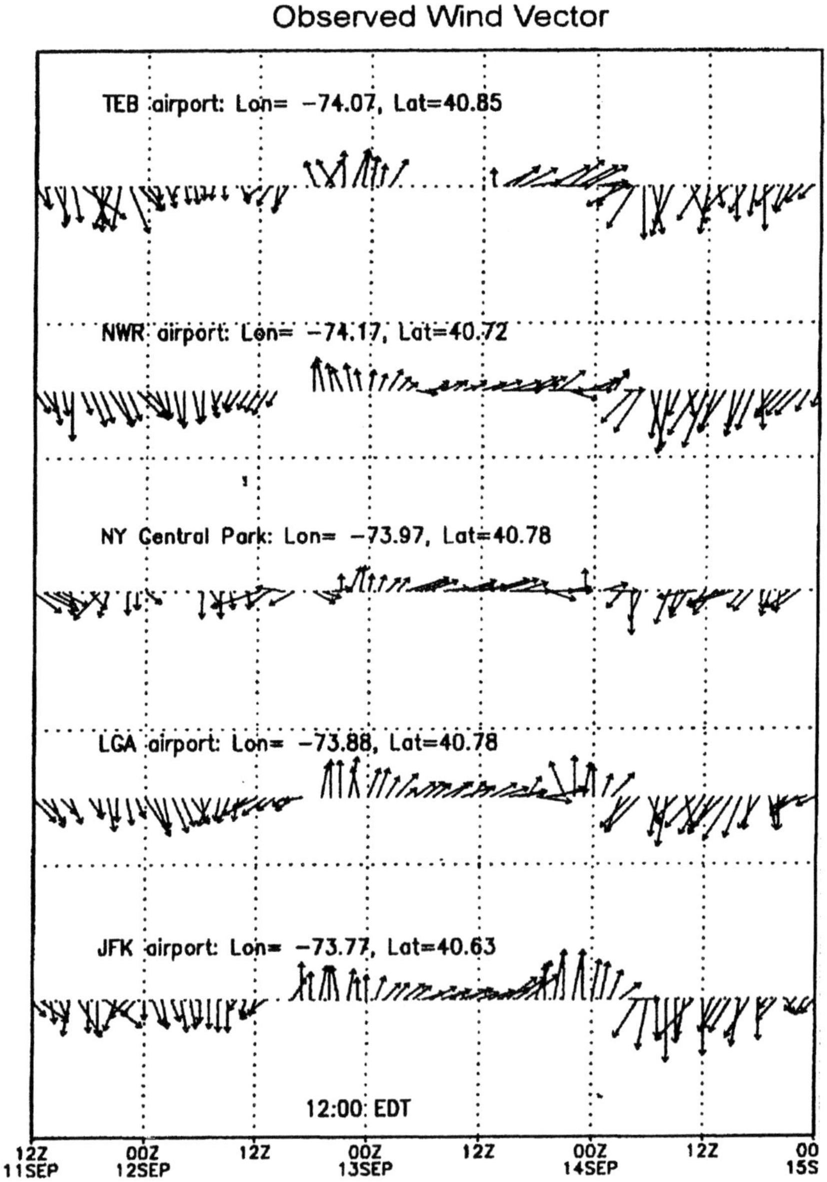

Figure 5. Wind vectorsiImpacting NYC and the metropolitan area for 1-12-01.

from the outdoor and indoor areas in and around Wall Street (NY Stock Exchange) prior to the 17th of September, when the stock market was scheduled to reopen. The category 3 post-WTC collapse period continued until the second rain event that occurred on September 25, 2001. At about this time the rescue efforts ended and recovery efforts began at Ground Zero. The end of the rescue effort then changed the character and types of activities conducted in and around Ground Zero. The second rain event washed much, but not all, of the remaining outside settled dust and smoke away. However, smoldering fires and debris removal activities continued at Ground Zero. The next exposure period (category 4) started about September 26th. It was characterized primarily by smoke and fumes emitted from the fires that burned at various locations across the 16-acre site until December 20, 2001 (*5,23,24,26*). These smoldering fires burned at various temperatures and with varying degrees or intensity of emissions. One reason for the variation of emissions during the category 4 time period was the effort to recover bodies and to begin removal of debris from the sites on diesel powered trucks. As the material was removed, areas of combustion were uncovered resulting in intermittent releases of combusted material into the air above Ground Zero. Throughout this time period the wind direction associated with the NYC Metropolitan area varied, and the smoldering fire plumes did not remain over a single area. In addition, the particles released by the fires were now much smaller in size than released previously, similar in size to particles generated during uncontrolled combustion but different in composition from typical combustion sources. The composition of the gases released would be affected in a similar manner.

Due to the buoyancy of fire plumes and the mild dry weather that dominated the metropolitan area from September 12 through November 01, 2001 the contaminants emitted at Ground Zero rose away from areas in the city and above Brooklyn, thus reducing the potential for "contact" with the materials in the plume and of associated exposures among most downtown Manhattan workers and residents (*2,22*). In October 2001 there were some days (e.g. October 18) where the evening subsidence of the atmosphere led to a ground level inversion, which increased the concentrations of fine particles and gases for short periods of time (*23*).

Needless to say, the nature and the level of emissions were not consistent with typical air pollution sources. There was no constant source of material being burned and released at the WTC site during this time period with the location and intensity of the emissions changing from day to day and hour to hour. Research completed has suggested that the concentration of fine particles in the smoke from the smoldering fires was reduced to background NYC urban levels by the end of October. The fires, however, still burned until December 20th, which marked the end of the time period of concern (*23,24,26*). Samples were still collected in and around Ground Zero well into 2002. This covered the time period until the recovery efforts ended in the spring and emissions were primarily combustion from trucks and resuspension of particles associated with the removal of debris.

Finally there was a category 5 of exposure, which is not identified on the graph shown in Figure 4. This category includes personal exposures to indoor air. Such exposures would be comprised of the indoor settled and resuspendable dust and in some locations would have lasted until homes or buildings with significant levels of the initial dust and smoke were cleaned up professionally. During the collapse of WTC, the compression caused by the collapse caused the dust to be pushed out radially from the 16 acre site into many areas of southern Manhattan below Canal Street and deposited a layer of dust in many buildings. The deposited dust could be found in a layer that varied from less than a millimeter thick to, in some cases, over 3 inches thick (*27,28*). The result was contamination of the floors and walls of apartments, offices, and open interior spaces, including the mechanical air handling systems and personal items and furnishings in each location. The degree of contamination of these indoor spaces, both commercial and residential, were variable with places that were closest to Ground Zero receiving the highest impact. Work published by Yiin *et al.* (27) and Offenberg *et al.* (28), has shown that the dust and smoke deposited indoors was very similar in composition to the material that settled outdoors. The main difference between these materials was that the indoor particles generally had less material in the very large particle range, above 53 um in diameter, (Table II). The level of inhalation exposures indoors would be directly related to the amount of time one stayed in contact with the dust and smoke

Table II. Comparison of Selected Components of WTC Settled Dust in Indoor and Outdoor Samples.

Component	*Units*	*Indoor*	*Outdoor*
Aerodynamic Diameter:			
<2.5 μm	%	0.40 – 0.80	0.88 – 1.33
2.5 – 10 μm	%	0.20 – 2.30	0.30 – 0.40
10 – 53 μm	%	20.1 – 78.5	34.6 – 46.6
>53 μm	%	19.1 – 79.1	52.2 – 63.6
Selected Metals:			
Pb	mg/g	0.05 – 0.25	0.14 – 0.50
Ti	mg/g	0.8 – 1.6	1.5 – 1.8
As	μg/g	2 – 3	1 – 2
U	μg/g	ND	3.9 – 4.0

NOTE: ND is not detectable.

while working or living in a poorly cleaned location or not wearing respiratory protection during the clean up. For some individuals these were the major opportunities for exposure to emissions from the WTC.

The Challenges to Exposure/Risk Assessment

There is no simple way to assess the risk to any subgroup of the population that was exposed to the gases and particles released by the WTC fires and building collapse. For instance, few people were exposed to the same materials for a prolonged period of time or repetitive periods of time and these exposures were derived from contact with a very complex mixture of debris. The group coming closest to a continuous exposure to gases and particles were the fire fighters, police, and emergency rescue workers that were at Ground Zero during the first few days (*5,7,24*) and possibly the indoor cleanup workers (*3*). Those individuals who did not wear respirators all the time while working at and around the site, or did not wear them properly, were exposed to significant levels of resuspended dust and smoke particles. These particles and gases were generated by the intense burning fires (>1000 °F) and particles generated by the removal of materials that were associated with the efforts to rescue buried individuals within the debris. The individuals exposed to these materials and the locations and durations of the exposures are not documented. The issues concerning exposure have to be examined not only with respect to whether or not individuals were at Ground Zero, close to Ground Zero, or within areas impacted by the plume on 9-11 and beyond. They also include the personal activities (e.g. returning to un-cleaned buildings or walking in the areas around Ground Zero) and whether or not personal protection was used by workers indoors or outdoors etc. (*21*).

Satellite photography and net cam photography have shown that the initial plume was very buoyant (>1000 feet). This was helpful in reducing exposures among many people since the concentrations of gases and particles released in the plume would rise above the roofline and the concentrations would be reduced at ground level in NYC. This is not to say that individuals who were living in high-rise apartments at some locations downwind of the collapsed WTC would not receive exposures; in fact they would, but the exposure patterns would be intermittent.

In terms of risk assessment, it is clear from the data acquired from the dust and smoke measurements from ambient air monitoring and from modeling reconstruction of the plume that the people most exposed to outdoor emissions as a result of the attack on WTC were those working within or near Ground Zero within the first 24 to 48 hours (*3, 5, 7, 22, 29*). Risk assessments for such individuals would be associated with acute exposures for primarily acute or sub-acute effects. These same people may also be at risk for long term effects caused by large doses. The risk assessments, however, must estimate the likelihood that a short time dose, which led to a cough, would translate to non-reversible effects and long-term health consequences, especially in susceptible individuals. Long term health risks from intermittently breathing smoke released in the days and weeks after the collapse need to be quantified based upon the extensive databases on air quality and exposure acquired by many organizations. The highest uncertainties for a risk assessment are due to difficulties in estimating exposures that can lead to health consequences caused by the initial collapse and the indoor settled and re-suspended dust (*5, 24, 29, 30*). There is no general consensus about when individuals could and did go back into the homes. Further, some people went back into the outdoor areas around Ground Zero just after the attack, or returned to their homes without respirators to recover personal items. Subsequently, some homes were cleaned professionally by the homeownes and others were cleaned using much more rudimentary procedures (*24,30*). Thus, the degree and the intensity of exposures need to be documented in a manner that establishes potential profiles for indoor and outdoor conditions. However, the lack of data makes the reconstruction of individual exposure profiles very difficult.

Individuals who may have experienced acute health effects may or may not be susceptible to any potential chronic health effects. They require risk assessments using exposure scenarios that are relevant to the time, duration, and location of those individuals indoors as well as outdoors. Also, there is a need to incorporate data on the background concentrations for some of these compounds to provide a point of reference in homes that did not receive large amounts of dust, or to establish a benchmark for how clean residences would normally be in NYC. The EPA has recently completed such an investigation (*31*).

Conducting risk assessments for the WTC will require a great degree of care with respect to the definition or the characterization of the exposures for survivors within southern Manhattan and those downwind of the smoldering plume and/or the initial dust and smoke plume. This point was discussed in the initial risk assessment completed by the U.S. EPA on the aftermath of the collapse (24). Any future risk assessment will have to, as a consequence, be separated into risks that may have been derived for both acute and chronic health effects. Longer term risk associated with the exposure to polycyclic aromatic hydrocarbons (PAHs), other products of incomplete combustion, or asbestos, will need to be qualitatively and quantitatively estimated based upon the information collected by many organizations on the composition of the contamination and on potential exposures. However, this cannot be a traditional lifetime based exposure and risk assessment. The reason, as shown in Figure 4, is that the initial exposures were much higher than considered in traditional risk assessments. The source of the emitted material didn't remain constant in strength of emissions over time and, as discussed above, the character of the material emitted by the source also changed over time (*5, 23*). Thus the characterization of exposures has to take into account the variable intensity of the source in order to define the concentrations and levels of exposure. Such efforts must take into account the localized character of the complex plumes that resulted from the emissions. This is not a small feat since for the first few days after 9-11-01 a majority of the emission mass was associated with inspirable particles in the size ranges above 2.5 and 10 μm in diameter. These particles, which also included alkaline and fibrous material, could in fact cause irritation in the upper airways and was probably the cause of the WTC cough (*3,7*). In contrast, during a majority of the days post 9-11, from approximately September 25 until December 20, 2001, the emissions were mostly fumes and smoke from the smoldering fires (*22-24, 26*).

Future Analyses

Toxicological studies have been completed by the EPA for the 2.5 micron fraction of the dust and smoke (*13*). Detailed acute and lifetime toxicological studies on the total mass of the dust and smoke is being done through a collaboration between the National Institute of Environmental and Health Sciences, Environmental and Occupational Health Sciences Institute (NIEHS/EOHSI), and collaborations from New York University (NYU) and the University of Rochester, which are unpublished at the time of this paper. It is clear, however, that this information will eventually provide a better picture of the potential chronic long-term health effects associated with the inhalation of the WTC dust initially released after the collapse. In addition, the traditional risk assessment on the samples taken in the ambient air post 9-11 in downtown Manhattan will be very useful in establishing exposure profiles for individuals who may have lived in southern Manhattan and had been periodically exposed to the plumes caused by the continuing fires. However, all these situations will require the use of the plume location and distribution estimates being generated by Drs. Georgopoulos and Huber at EOHSI and Research Triangle Park (RTP), respectively; to establish the frequency and the potential intensity of exposure and development of exposure indices for individuals that were exposed to the plume (*22*). These analyses will obviously build upon the initial risk and exposure assessments completed by the EPA and others post 9-11 (*24*). They will be of value in the development of epidemiologic investigations that a) look at the health care of individuals exposed immediately after the event to quantitate both health outcomes and the potential exposures; and b) try to establish exposure indices for individuals who in fact become registered by either NYC or the other organizations to determine whether or not they truly were in harms way after the event. One such example is the analysis being conducted to establish exposure response relationships among 187 women who were pregnant at the time of the collapse and were in the vicinity of ground zero during one or more periods from the time of the collapse through October 8, 2001 (*32*). Each woman provided a diary for the times and locations that she was within five zones (estimated by EOHSI using descriptions from the US EPA) potentially impacted by the WTC plume. These zones were associated with radial distances from ground zero. Such location information is fairly weak with respect to establishing exposure, unless it is coupled with results that indicate the location of the plume with respect to the location of the woman at any particular period of time. Thus, to improve on the characterization of the exposure in the epidemiological analyses, an exposure index was calculated over time from 9-11 through October 8th, based upon the plume movement and the reasonable pattern of change in the relative emission intensity estimated by Georgopoulos *et al.* and matched for outdoor and indoor locations over the same time intervals from the diary of each woman (*32*). In the future, these types of profiles may become valuable in attempting to reconstruct exposures for specific individuals that become part of the WTC registry, being established by the Agency for Toxic Substances and Disease Registry (ATSDR) and the City of New York to assist in providing a basis for defining exposure and response relationships.

Acknowledgements

The authors wish to thank all colleagues who collaborated with us in dealing with the aftermath of the Terrorist attack. They devoted much time and effort in support of the national emergency. The many individuals are listed among the WTC references associated with this manuscript, #2, 3, 5, 20, 21, 22, 27, 28, 32 also G. Stenchikov, W. Li, and N. Lahoti for calculations leading to Figures 4 and 5. We also wish to remember all the souls lost, and thank the many volunteers who assisted in dealing with the aftermath. Support for this research was provided through and EPA University Partnership (CR827033-), and an NIEHS Center Grant Supplement (ES05022-1551).

References

1. Caludio, L. *Environ. Health Perspect.* **2001,** *109*, A528-536.
2. Vette, A.; Gavett, S.; Perry, S.; Heist, D.; Huber, A.; Lorber, M.; Lioy, P.; Georgopoulos, P.; Rao, S.T.; Petersen, W.; Hicks, B.; Irwin, J.; Foley, G. *Environ. Management.* **2004,** *Feb.* 14-22.

3. Lioy, P. J.; Weisel, C. P.; Millette, J. R.; Eisenreich, S.; Vallero, D.; Offenberg, J;. Buckley, B.; Turpin, B.; Zhong, M.; Cohen M. D.; Prophete, C.; Yang, I.; Stiles, R.; Chee, G.; Johnson, W.; Porcja, R;. Alimokhtari, S.; Hale, R. C.; Weschler, C.; Chen., L. C. *Environ. Health Perspect.* **2002,** *110*, 703-714.
4. Oktay, S. D.; Brabander, D.J.; Smith, J.P.; Kada, J.; Bullen, T.; Olsen, C.R. *Eos Trans.* **2003,** *AGU 84*, 24-25.
5. Landrigan, P; Lioy, P.J.; Thurston, G.; Berkowitz, G.; Chen, L.C.; Chillrud, S.N.; Garett, S.H.; Georgopoulos, G.; Geyh, A.S.; Levin,S.; Perera, F.; Rappaort, S.M.; Small, C. *Environ. Health Perspect.* **2004,** doi:10.1289/ehp.6702.
6. Banauch, G. I.; Alleyne, D.; Sanche, R.; Olender, K.; Cohen, H. W.; Weiden, M.; Kelly, K. J.; Prezant, D. J. *Am J. Respir. Crit. Care Med.* **2003,** *168*, 54-62.
7. Prezant, D. J.; Weiden, M.; Banauch, G. I.; McGuinness, G.; Rom W.N.; Aldrich, T.K.; Kelly, K. J. *N. Engl. J. Med.* **2002,** *347*, 806-815.
8. Rom, W. N.; Weiden, M.; Garcia, R.; Yie, T.A.; Vathesatogkit, P.; Tse, D.B. *Am. J. Respir. Crit. Care Med.* **2002,** *166*, 797-800.
9. Beckett, W. S. *Am. J. Respir. Crit. Care Med.* **2002,** *166*, 785-786.
10. Berkowitz, G. S.; Wolff, M. S.; Janevic, T. M.; Holzman, I. R.; Yehuda, R.; Landrigan, P. J. *JAMA.* **2003,** *290*, 595-596.
11. Galea, S.; Ahern, J.; Resnick, H.; Kilpatrick, D.; Bucuvalas, M.; Gold, J.; Vlahov, D.. *N. Engl. J. Med.* **2002,** *346*, 982-987.
12. Galea, S.; Resnick, H.; Ahern, J.; Gold, J.; Bucuvalas, M.; Kilpatrick, D.; Stuber, J.; Vlahov, D. *J. Urban Health* **2002,** *79*, 340-353.
13. Gavett, S. H.; Haykal-Coates, N.; Highfill, J. W.; Ledbetter, A. D.; Chen, L. C.; Cohen, M. D.; Harkema, J. R.; Wagner, J. G.; Costa, D. L.. *Environ. Health. Perspect.* **2003,** *111*, 981-991.
14. Levin, S.; Herbert, R.; Skloot, G.; Szeinuk, J.; Teirstein, A.; Fischler, D.; Milek, D.; Piligian, G.; Wilk-Rivard, E.; Moline, J. *Am. J. Ind. Med.* **2002,** *42*, 545-547.
15. *"EPA Response to September 11"* [http//www.epa.gov/wtc/pictures/images; world_trade_center_dust_cloud_.jpg] US. Environmental Protection Agency.
16. "World Trade Center Photo Repository" [http//www.phatmax.net/wtc/?s=trade_groundAP410x267.jpg] mac@phatmax.net, October 4, 2003.
17. *US EPA, Air Quality Criteria for Particulate Matter.* National Center of Environmental Assessment; Office of Research and Development, Research Triangle Park, NC. 1996, EPA600/AP93-004aF-oF.
18. *ACGIH, TLV's and BEI's – 2003*, Cincinnati, OH, 2003, 73-76.

19. Beck, C.M.; Geyh, A.; Srinivasa, A.; Breysse, P.N.; Eggleston, P.A.; Buckley, T.J. *J Air & Waste Management Assoc.* **2003,** *53*, 1256-1264.
20. Offenberg, J.H.; Eisenreich, S.J.; Chen, L.C.; Cohen, M.D.; Chee, G.; Prophete, C.; Weisel, C.; Lioy, P. J. *Environ. Sci. Technol.* **2003,** *37*, 502.
21. Lioy, P. J.; Gochfeld, M. *Am. J. Ind. Med.* **2002,** *42*, 560-565.
22. Huber, A.; Georgopoulos, P.; Gilliam, R.; Stenchikov, G.; Wang, S.; Kelly, B.; Feingersh, H. *Environ. Management.* **2004,** 35-40.
23. Thurston, G.; Maciejczyk, P.; Lall, R.; Hwang, J.; Chen, L.C. *Epidemiology.* **2004,** In Press.
24. *U.S. EPA, Exposure and Human Health Evaluation of Airborne Pollution from the World Trade Center Disaster*, National Center for Environmental Assessment. Office of Research and Development. 2002, Tech. Report No. EPA/600/P-2/002A.
25. *"EPA Response to September 11"* [http://www.epa.gov/wtc/voc] US. Environmental Protection Agency.
26. Cahill, T.A.; Cliff, S.S.; Perry, K.D.; Jimeniz-Cruz, M.; Bench, G.; Ueda, D.; Shackrlford, J.F.; Dunlap, M.; Meier, M.; Kelly, P.B.; Riddle, S.; Selco, J.; Leifer, R. *Aerosol Sci. Technol.* **2004,** *38*, 165-183.
27. Yiin, L.-M.; Millette, J. R.; Vette, A.; Ilacqua, V.; Quan, C.; Gorczynski, J.; Kendell, C.; Chen, L.C.; Weisel, C.; Buckley, B.; Yang, I.; Lioy, P.J. *J. of Waste Management Assoc.* **2004,** *54*, 515-528.
28. Offenberg, J.; Eisenreich, S.J.; Gigliotti, Chen, L.C.; Xiong, Q-L.; Zhong, M.; Gorczynski, J.; Yiin, L-M; Illacqua, V.; Lioy, P.J. Persistent Organic *J. of Exp. Anal. and Environ. Epidem.* **2004,** *14*, 164-172.
29. Lorber, M.; Gibb, H.; Grant, L.; Pinto, J.; Lioy, P. *Environ. Management.* **2004,** 27-29.
30. *O.I.G. – EPA, Evaluation Report – EPA Response to the World Trade Center Collapse: Challenges, Successes and Areas for Improvement,* Report 2003-P-00012, Washington, DC August, 2003.
31. *US EPA Region II. World Trade Center Background Study Report, Interim Final,* New York, NY, April, 2003.
32. Wolf, M.S.; Teitelbaum, S.L.; Lioy, P.J.; Santella, R.; Georgopoulos, P.; Li., W.; Berjiwutzm, G. *ISEE Abstract.*, NYC Conference, August 2004.

World Trade Center Dust Characterization

Chapter 3

Spectroscopic and X-ray Diffraction Analyses of Asbestos in the World Trade Center Dust: Asbestos Content of the Settled Dust

Gregg A. Swayze, Roger N. Clark, Stephen J. Sutley, Todd M. Hoefen, Geoffrey S. Plumlee, Gregory P. Meeker, Isabelle K. Brownfield, K. Eric Livo, and Laurie C. Morath

Denver Microbeam Laboratory, U.S. Geological Survey, MS 964, Box 25046 Denver Federal Center, Denver, CO 80225

On September 17 and 18, 2001, samples of settled dust and airfall debris were collected from 34 sites within a 1-km radius of the WTC collapse site, including a sample from an indoor location unaffected by rainfall, and samples of insulation from two steel beams at Ground Zero. Laboratory spectral and x-ray diffraction analyses of the field samples detected trace levels of serpentine minerals, including chrysotile asbestos, in about two-thirds of the dust samples at concentrations at or below ~1 wt%. One sample of a beam coating material contained up to 20 wt% chrysotile asbestos. Analyses indicate that trace levels of chrysotile were distributed with the dust radially to distances greater than 0.75 km from Ground Zero. The chrysotile content of the dust is variable and may indicate that chrysotile asbestos was not distributed uniformly during the three collapse events.

Introduction

Spectroscopy is a useful tool for evaluating environmental contaminants at scales ranging from hand specimens to those available from airborne and orbital remote sensing platforms (*1*). Spectroscopic analysis frequently differs from x-ray diffraction (XRD) analysis in its sensitivity to materials, and often provides complimentary information. Reflectance spectroscopy over the visible to near-infrared (VIS-NIR) wavelength range (0.4 - 2.5 μm) is particularly sensitive to hydroxyl (OH) and water-bearing materials (*2*), organic C-H containing compounds, such as oil, paper, and wood, carbonate-bearing rocks such as marble and limestone (*3*), water-bearing sulfates such as gypsum found in wallboard and concrete (*4*), and iron-containing compounds such as hematite – a pigment used to color bricks red (*2*). Spectral reflectance is defined here to be the ratio of reflected to incident light relative to a diffuse standard, with reflectance values varying between 0 and 100%. Laboratory spectroscopy differs from airborne imaging spectroscopy in three ways: 1) individual samples are analyzed under controlled conditions using artificial illumination; 2) there is little interfering atmosphere to block portions of the electromagnetic spectrum; and 3) spectral resolution and signal-to-noise ratio can be much higher. Consequently, laboratory spectral measurements can have much higher sensitivity to the individual materials in a complex sample compared to remote spectral measurements.

On September 16, 2001, five days following the collapse of the World Trade Center (WTC) Towers, hyperspectral data were collected over Ground Zero with the Airborne Visible/Infrared Imaging Spectrometer (AVIRIS) (see Chapter 4). These data were collected to rapidly assess the potential asbestos hazards of the dust that settled over lower Manhattan (*5, 6*). Within two days of the overflight, field samples of settled dust and airfall debris were collected from lower Manhattan, including samples from two indoor locations unaffected by rainfall, and samples of insulation from two steel beams at Ground Zero, to verify imaging spectrometer mineral maps. A U.S. Geological Survey (USGS) study then underway at Libby, Montana (*7*), indicated that laboratory spectral measurements could detect an absorption feature diagnostic of potentially asbestiform minerals that is at a wavelength often obscured by atmospheric absorptions in airborne remote sensing data. Laboratory reflectance measurements and XRD analyses of the WTC dust, beam-insulation, and concrete samples were made. Samples were also analyzed for a variety of mineralogical and chemical parameters using scanning electron microscopy and chemical analyses (see Chapters 5 and 12) (*8*).

Method

Sample Collection

A two-person USGS team (authors T.M. Hoefen and G.A. Swayze) collected samples of settled dust and coarser airfall debris from 35 localities within a 1 km radius centered on the WTC Ground Zero site on September 17 and 18, 2001 (Figure 1). Samples collected outdoors were exposed to wind and precipitation during a rain storm on the night of September 14 prior to collection. One sample (WTC01-20) was collected indoors near the gymnasium in the World Financial Center across from the WTC site on West Street. A sample of dust (WTC01-36) blown by the collapse into an open window of an apartment, located 30 floors up and 0.4 km southwest of the center of the WTC site, was also acquired a few days later. Two samples of insulation coatings (WTC01-8 and WTC01-9) were collected from steel beams that had been removed from the debris pile of the WTC. Samples of concrete (WTC01-37A and WTC01-37B) were collected from the WTC debris at the same location as WTC01-8 and WTC01-9, respectively.

Many of the streets bordering the collection locations had been cleaned or were in the process of being cleaned at the time of sample collection. Given these limitations, the collection of dust samples was restricted to undisturbed areas such as window ledges, car windshields, flower pots, building entrances, covered steps, or from sidewalks adjacent to walls where the dust was sheltered from the weather and cleanup process. In many cases, the samples formed compact masses suggestive of having been dampened by rain and subsequently dried during the intervening 3-4 days. Samples were gathered by scooping with a nitrile-glove covered hand, and then placed in doubled plastic ziplock freezer bags. When possible, several handfuls of material were collected from different locations within a few-square-meter area and combined into one sample. The collection locations were identified by street intersections.

Spectral Measurements

Reflectance spectra of the samples were measured in a laboratory High Efficiency Particulate Air filter (HEPA) fume hood with an Analytical Spectral Devices (ASD) Full Range Spectrometer® over the wavelength range from 0.35 -2.5 μm using a halogen lamp for illumination and Spectralon® panel for reference (*1*). The ASD spectrometer has 5 nm spectral resolution from 0.35 - 1.0 μm and 11 nm spectral resolution from 1.0 - 2.5 μm. The entire sample was first poured from the plastic sample bag onto white paper, then the sample was mixed with a spatula leaving a relatively flat pile about a centimeter thick for

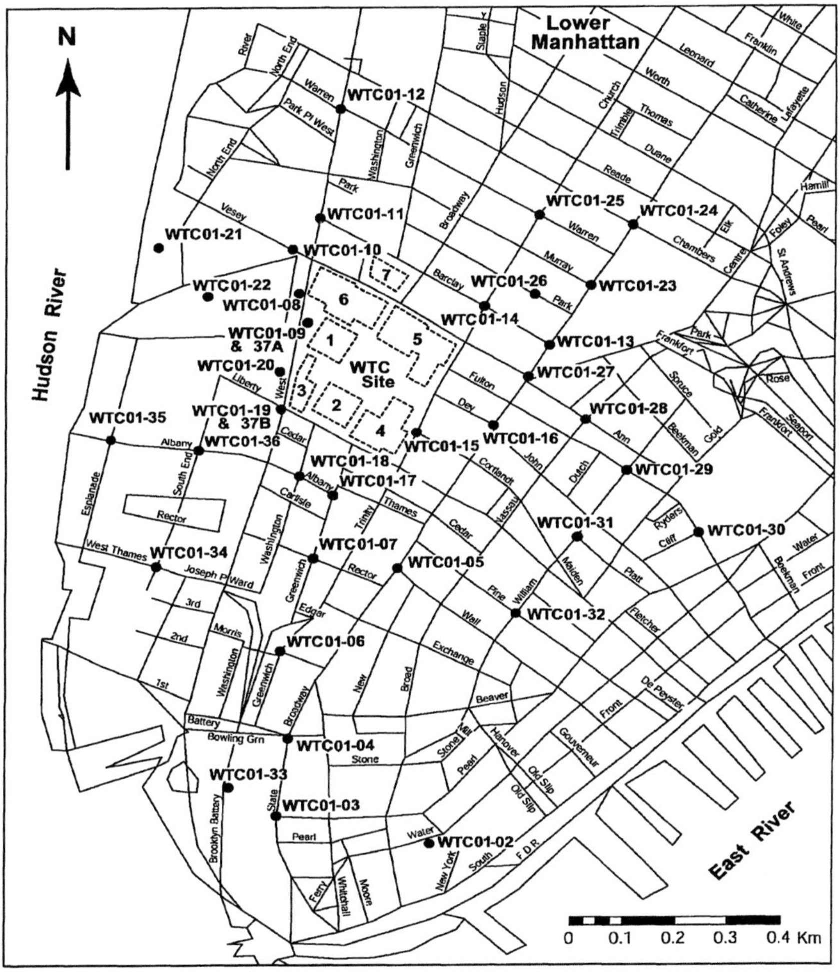

Figure 1. Location of settled dust samples collected in lower Manhattan on September 17 and 18, 2001 for this study and those reported in Chapters 4, 5 and 12. The original locations of the WTC are indicated by the dashed outlines.

spectral measurement. By mixing the sample, we hoped to avoid possible inadvertent effects of particle sorting that may have occurred during transport or pouring from the bag.

Given that VIS - NIR reflectance spectroscopy detects materials down to a few millimeters, in most cases, beneath the surface of the dust, ten spectra of the pile were measured, using a six second integration time for each spectrum, and then the pile was re-mixed before collecting an additional ten spectra, to expose previously unmeasured material at the surface. The spectrometer optical fiber was held a few centimeters above the pile and moved constantly in an elliptical manner to spatially average the surface of all but the edges of the pile. This method allowed about 40-60% of the entire sample volume to be spectrally characterized. Spectra of each dust sample were averaged and corrected to absolute reflectance (*9*). Low levels of noise observed in the averaged spectra indicate that a very high signal-to-noise ratio of 28,000:1 was achieved, based on the standard deviation of reflectance values in a flat portion of the spectral average of a relatively dark (23% reflectance) dust sample (WTC01-31) near 1.38 μm. Sample splits for analyses were obtained by the cone and quartering method (*10*).

X-Ray Diffraction Measurements

Most crystalline materials have unique x-ray diffraction patterns that can be used to differentiate between materials. X-ray powder diffractometry was used to characterize the mineralogy of the WTC samples. Three gram aliquots of the research splits were dry pulverized with a mortar and pestle to an average particle size of about 50-150 μm. About one gram of each specimen was then packed in an aluminum sample holder and analyzed with a Scintag X-1 Automated Diffractometer® fitted with a spinning sample holder using copper (Cu) K-alpha radiation (*11*). Complete XRD analyses for all samples are available on the USGS website and are summarized in Chapter 5 (5).

Results and Interpretation

Spectra of the dust samples, in general, look nearly identical with reflectance levels that vary between 20 - 45% and strong absorption edges between 0.35 - 0.8μm (Figure 2A). There are only weak spectral absorptions, apart from the absorption edge, in the electronic absorption region (0.35 -1.35 μm) of the spectra indicating that the dust contains only low abundances of materials with Fe or other transition elements. However, there are numerous spectral absorptions in the vibrational spectral region (1.35 - 2.5 μm) principally at 1.45, 1.75, 1.94, 1.97, and from 2.2 - 2.4 μm (Figure 2). The strongest spectral feature is at 1.94 μm and is due to structural and adsorbed water. A weaker feature at

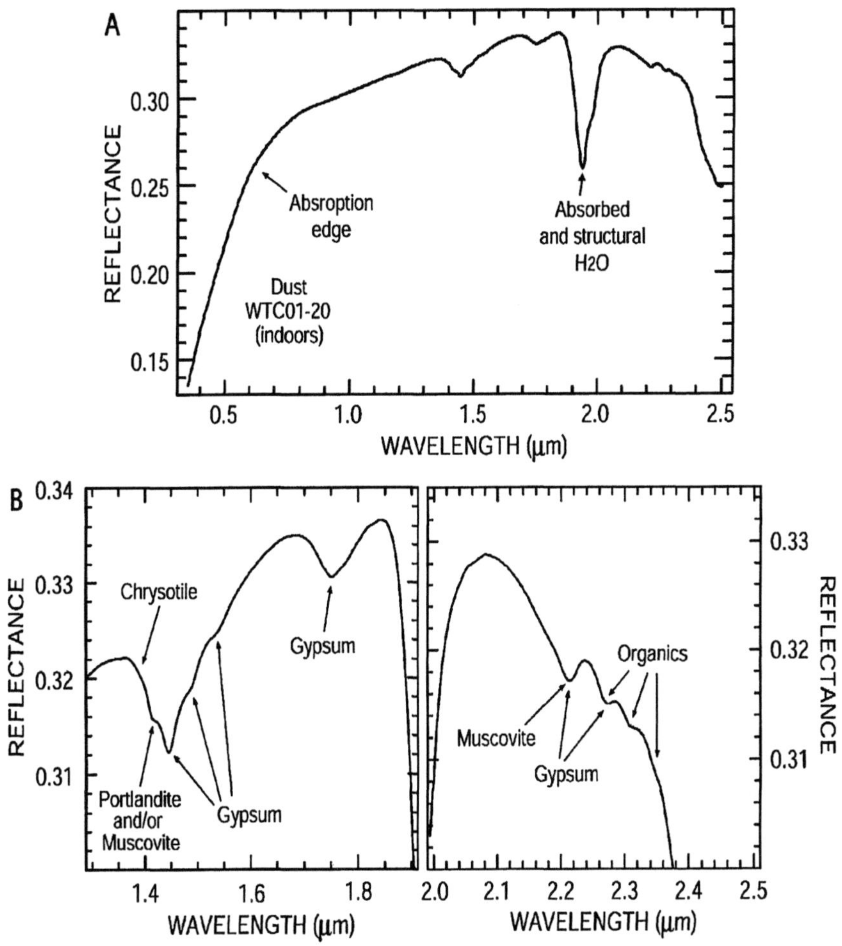

Figure 2. (A) Reflectance spectrum of a sample of WTC dust collected indoors near the "gym" immediately adjacent to Ground Zero. (B) Details of absorptions in the vibrational region.

1.45 μm is due to water and/or OH, and varies in depth from 2 - 13%. Spectral features at 1.75μm are even weaker with depths of only a few percent. Spectral features in the 2-μm region are weakest and vary from a fraction of a percent up to 1% depth.

Absorption features can be diagnostic of some materials contained in the dust. Gypsum, from pulverized wallboard and concrete (12), has the strongest spectral features, accounting for most of the bands observed in the vibrational portion of the spectra. This mineral has three diagnostic absorption features that form a stair-step like triplet between 1.42 - 1.54 μm recognizable in nearly all dust samples (Figure 2B). It also has absorption features at 1.75, 1.95, 1.975, 2.17, 2.217, and 2.268 μm, which closely match the positions and geometry of the absorption features observed in the dust. Figure 3 shows spectral estimates of gypsum based on the relative band depths of its 1.75 μm absorption in the dust muscovite/illite, both mineral components of the WTC concrete. Hydroxyl-related absorption features from other mineral components of the pulverized concrete could theoretically occur at or near this wavelength. Variable levels of muscovite/illite were found in two thirds of the dust samples, while portlandite was found in three samples (including the two indoor samples) using XRD analysis. Additional absorptions located at 2.307 and 2.343 μm are due to C-H stretches from organic materials (Figure 2B).

Electron microprobe analyses indicate fibrous glass with an elemental composition closely matched by slag wool, a type of mineral wool insulation, in all dust samples analyzed by this method (see Chapter 5). Slag wool has only weak spectral absorptions and is virtually transparent over the 0.35- 2.5 μm wavelength range. Organic materials associated with slag wool may contribute to some of the spectral features observed in the dust (*5*). A non-asbestos coating (WTC01-9) taken from a steel beam removed from the debris pile west of WTC Tower 1 has relatively strong 2.307 and 2.343 μm absorptions that match the positions of similar, but weaker, bands in all of the dust samples. Although the positions of these bands are compatible with C-H absorptions in most organic materials (e.g., plastics, paper, fabrics), washing of sample 9 with methanol was sufficient to significantly reduce the strength of these absorptions, indicating they are caused by a soluble organic material, possibly oil used as a dust suppressant in insulation containing slag wool (*5*).

Spectra of both beam coatings (samples 8 and 9) have an absorption at 1.725 μm that closely matches the position of a C-H absorption found in oil. Use of slag wool was widespread in the WTC Towers (*13*) as part of fireproof coatings on steel beams and the undersides of floors, and possibly in ceiling tiles. The apparent widespread use of slag wool and its brittle nature may explain its presence in all of the dust samples as a volumetrically significant component. Figure 4 shows spectral estimates of organic materials in the dust based on the relative band depths of the C-H combination absorption at 2.31 μm. Samples have relatively weak bands near Ground Zero and stronger bands further away. Because grain size should decrease outward, this pattern suggests that the organic material responsible for the absorption is more abundant further from the source, perhaps because it coats pulverized slag wool or other materials which were widely distributed during the collapse.

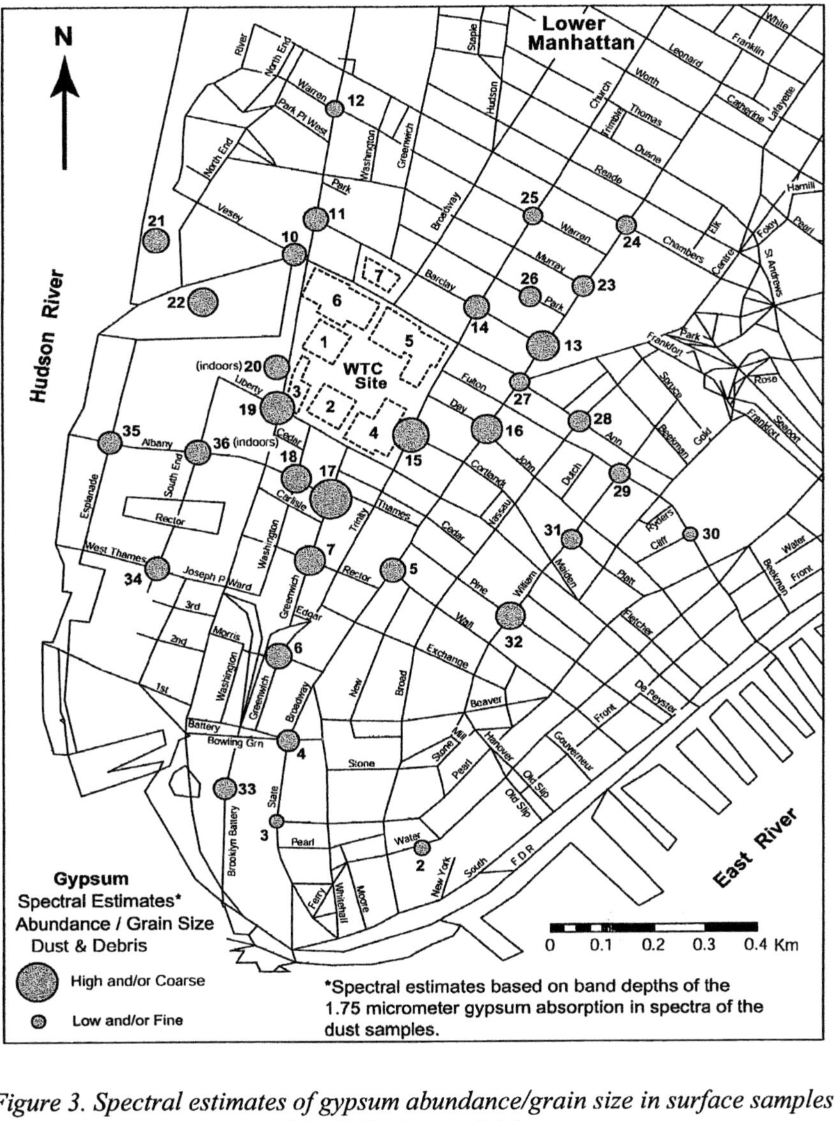

Figure 3. Spectral estimates of gypsum abundance/grain size in surface samples of the WTC dust and debris.

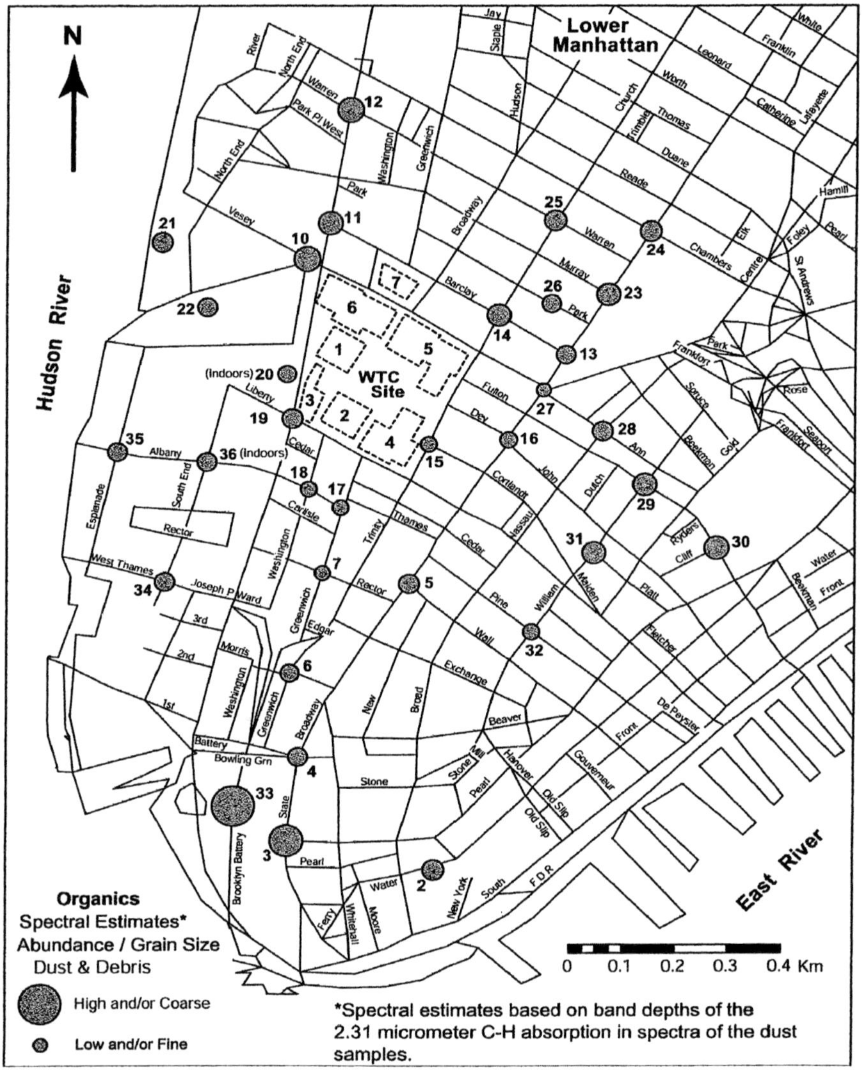

Figure 4. Spectral estimates of organic material abundance/grain size in surface samples of the WTC dust and debris.

Possible Sources of Asbestos in the WTC Dust

Asbestos refers to a fibrous crystal form that is characteristic of several different minerals (*14*). The two main types of asbestos used commercially are chrysotile, a fibrous member of the serpentine mineral group, and several fibrous amphibole minerals. Inhalation of asbestos has been associated with asbestosis, mesothelioma, and lung cancer (*15*). Some possible sources of asbestos in the collapsed buildings at the WTC include the fireproof coatings sprayed onto steel beams, on the undersides of floors, and in the elevator shafts, having been used up to the 38 floor of Tower 1 (16-18).

Sample 8 was taken from a coating on a steel beam removed from the debris pile west of WTC Tower 1. A spectrum of this sample shows absorption features at 1.385 and 2.323 μm, which match in position and overall shape to those of chrysotile asbestos (Figure 5). In sample 8, the position of the 2.32-μm absorption is consistent with that of chrysotile, but the band width is somewhat wider possibly indicating the presence of overlapping 2.307- and

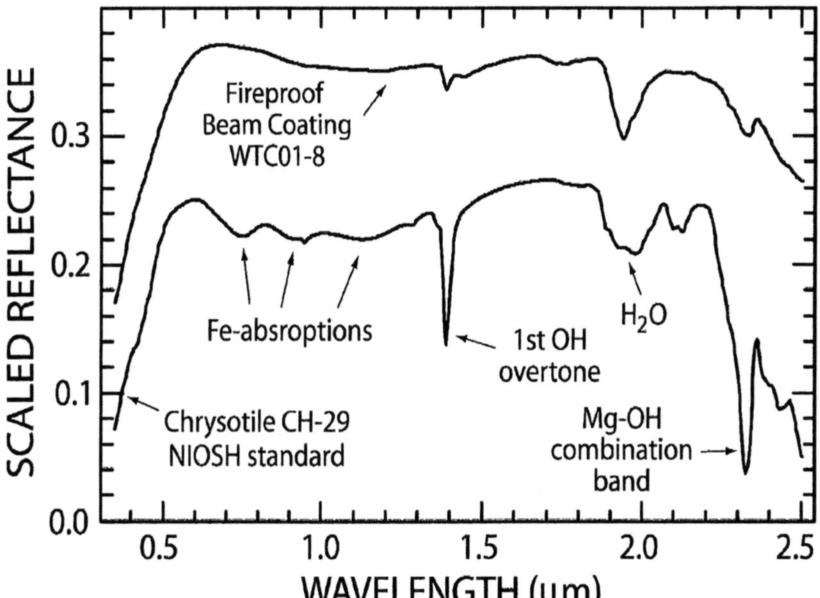

Figure 5. Reflectance spectrum of the fireproof coating collected from a steel beam at Ground Zero compared with a spectrum of pure chrysotile asbestos (offset downward 0.3 units for clarity).

2.343-μm absorptions from dust suppressant oil (*13*) or other organic compounds. X-ray Diffraction analysis and the depth of these spectral absorptions is consistent with chrysotile forming up to 20 wt% of this coating material (*5, 16*).

Other potential sources of asbestos in the dust include vinyl floor tiles (typically 10-15% chrysotile), insulation on pipes, and amosite asbestos insulation possibly from boilers (*13, 18*). Spectroscopy and XRD analyses of the two concrete samples did not detect any measurable chrysotile. Asbestos abatement and insulation repairs in WTC Towers 1 and 2 involved replacement of easily-accessible original fireproof coatings that may have contained vermiculite (*16*). Vermiculite has the potential to contain amphibole asbestos (19, *20*). However, only minute amounts (<0.02 wt%) of fibrous amphibole were found in a dust sample collected by Chatfield and Kominsky (*21*) north of WTC Building 7. Vermiculite was not detected in either sample of the beam coatings collected at Ground Zero by spectroscopy or XRD. Vermiculite was detected in several dust samples during scanning electron microscope (SEM) and visual examinations (see Chapter 5) (*8*). Trace levels of amphibole were found in samples 22 and 33 by XRD analysis, but it is unknown if these were fibrous. Observations by SEM described in Chapter 5 detected chrysotile, but not fibrous amphiboles, in all dust samples examined (approximately one half of the samples collected).

Spectral Detection of Chrysotile

To better constrain the abundance of chrysotile asbestos in the dust, mixtures of dust and chrysotile were prepared. Sample 6 was selected as a representative dust sample with a relatively low level of asbestos (~0.1 wt%) based on high spectral resolution measurements, which resolved the chrysotile spectral features in detail (Figure 6). At this spectral resolution the chrysotile feature is resolved into multiple absorptions, which can be seen weakly at 1.38 and 1.388 μm in upper spectrum of Figure 6B. Measurements of the relative depth of the 1.388-μm absorptions indicated the presence of approximately 0.1 wt% chrysotile in dust sample 6. This high resolution spectrometer (Nicolet 760 Magna Fourier Transform Infrared Spectrometer®) was not used to measure the suite of WTC dust samples because of its small sample size and long measurement time requirements. Known amounts of a well characterized, very finely ground, National Institute of Occupational Safety and Health (NIOSH) chrysotile standard (*22*) were added to splits of sample 6 to create mixtures with

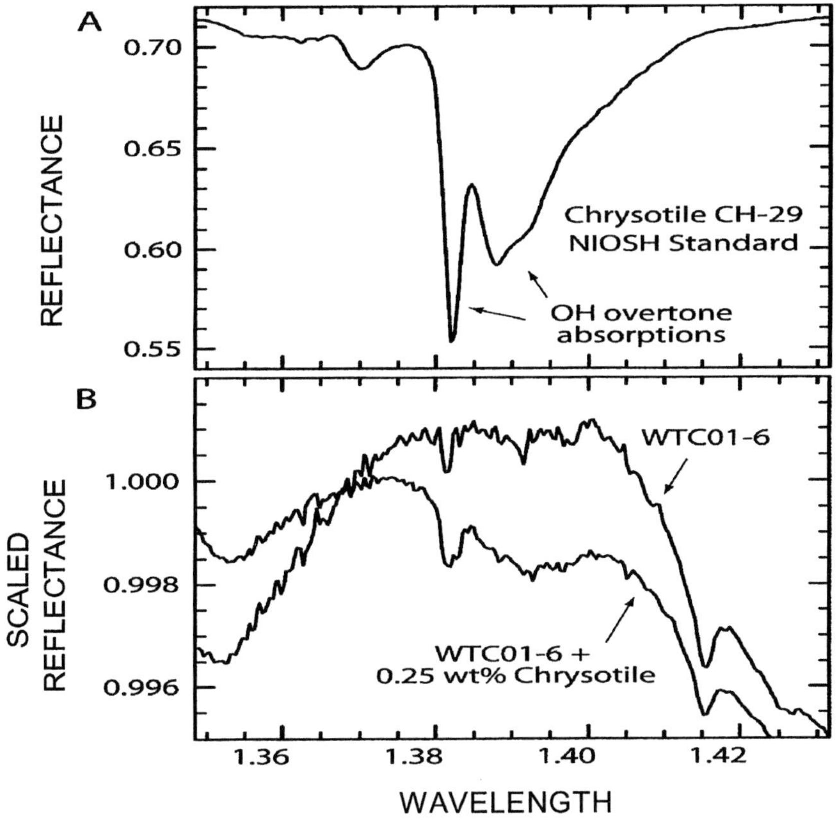

Figure 6. High resolution (4 cm^{-1}) reflectance spectra of (A) pure chrysotile and (B) sample 6 alone and with 0.25 wt% added chrysotile. Spectra were scaled to allow comparison in Figure 6B.

0.25, 0.5, 1.0, 2.5 and 5.0 wt% chrysotile. These mixtures were then measured to provide a series of spectra that could be directly compared to those of the dust samples.

At the coarser spectral resolution of the ASD spectrometer, chrysotile has two strong absorptions at 1.385 and 2.323 μm due to the first OH stretch overtone and a Mg-OH combination band, respectively (Figure 5) (*1*). Even though the 2.323-μm chrysotile absorption is the stronger of the two bands, its wavelength position is between stronger C-H absorptions at 2.307 and 2.343 μm and overlaps with a 2.317-μm absorption from pulverized concrete, all of which tend to mask the presence of chrysotile in the dust samples at levels below ~1 wt% in laboratory spectra. The weaker 1.385-μm chrysotile absorption occurs in a relatively less cluttered spectral region, providing a means of estimating chrysotile content. However, at concentrations less than 0.5 wt% the 1.385μm absorption feature of chrysotile cannot be differentiated from a similar absorption feature of lizardite (a non-asbestiform variety of serpentine) at the spectral resolution of the ASD spectrometer.

Figure 7A shows spectra of constructed dust plus chrysotile mixtures with straight-line continua removed from the 1.4-μm region. Continuum removal consists of dividing the dust spectrum by a line (or curve) fit to the top shoulders of an absorption. In this way the depth of the absorption is normalized relative to 100% reflectance and interference from broader, nearby absorptions is

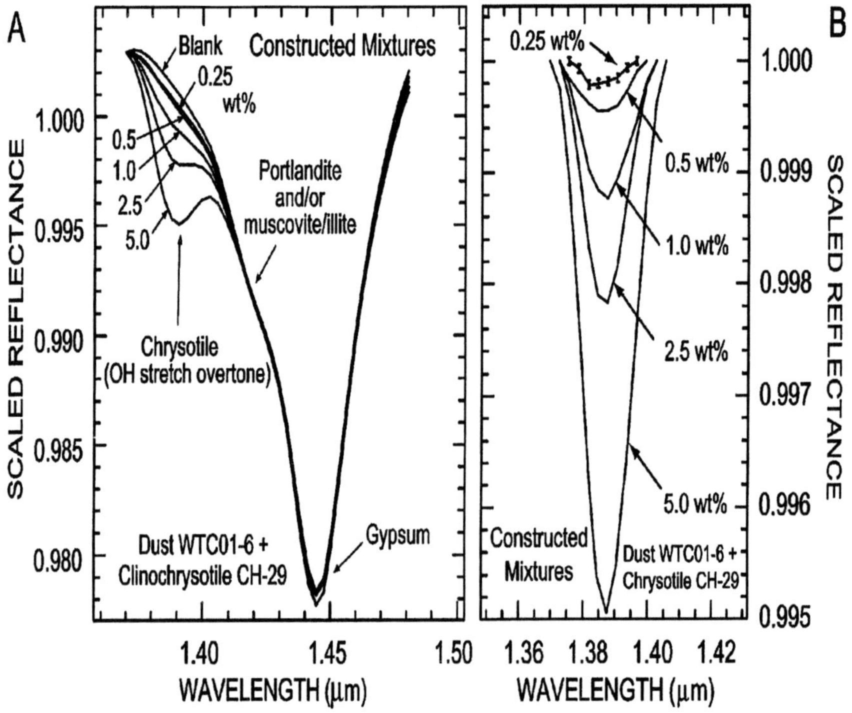

Figure 7. (A) Continuum-removed spectra of the OH stretch-overtone region of WTC dust sample 6 alone (blank) and with a NIOSH chrysotile asbestos standard added (0.25-5%). (B) Spectra are centered on the chrysotile 1.385-μm absorption and divided by a curved continuum.

minimized (*23*). This process helps make the absorptions of interest more comparable for samples of variable composition. Note the growth of the 1.385-μm chrysotile absorption from a mere inflection at 0.25 wt% to a full fledged band at 5 wt%, with increasing chrysotile content. Because the chrysotile absorptions are relatively weak, even for the 5 wt% mixture (about 0.5 % band depth), diagnostic spectral measurements must be done with a high signal-to-noise ratio.

Figure 7B shows curved-continuum-removed spectra of the mixture series centered on the spectral region of the chrysotile 1.385-μm absorption. Continuum removal of the mixture spectra produces distinct absorption bands even at the 0.25-wt% level. The error bars shown for the 0.25 wt% sample in Figure 7B are ± 1 standard deviation of the spectral noise derived by averaging 20 spectral measurements, and are representative of the measurement precision for the other spectra in this wavelength region.

Continuum removal over the wavelength region corresponding to the 1.385-μm absorption provided a means of estimating the amount of serpentine/amphibole/talc present in each dust sample based on the depth of the 1.385-μm absorption when compared to the mixture series spectra (Figure 8). Band depths used in this correlation were normalized for each mixture by dividing the 1.385-μm absorption band depth by the original reflectance value of the spectrum, in this spectral region, as described by Clark (*24*). To more accurately represent actual chrysotile contents in the correlation, the 0.25, 0.5, 1.0, 2.5, and 5.0 wt% chrysotile values of the mixture series were numerically adjusted to reflect an additional 0.1 wt% chrysotile originally present in sample 6, represented by the leftmost point on the graph in Figure 8. A standard error of ± 0.2 wt% is an estimate of the uncertainty in a single value of the weight percent chrysotile calculated using the best-fit equation shown in Figure 8 for dust sample 6.

Some amphiboles (e.g., fibrous or non-fibrous varieties of tremolite, richterite, hornblende) and talc have spectral absorptions in the 1.39μm region and their presence at levels < 0.5 wt% could potentially produce weak absorptions similar to those of chrysotile at the spectral resolution of the ASD spectrometer and, thus, add to the estimate of chrysotile. Alternatively, XRD analysis can detect amphiboles and talc without interference from chrysotile and the presence of amphibole was detected in dust samples 22 and 33 at trace levels. The potential presence of amphiboles and talc at concentrations below 0.5 wt% in the other dust samples means that our spectral estimates are upper limits for the abundance of chrysotile.

Other potential sources of error in the estimates of chrysotile in the WTC dusts depend on the grain size, albedo (brightness), and composition of the other dust particles. Because diffuse reflectance is a non-linear process, decreasing the

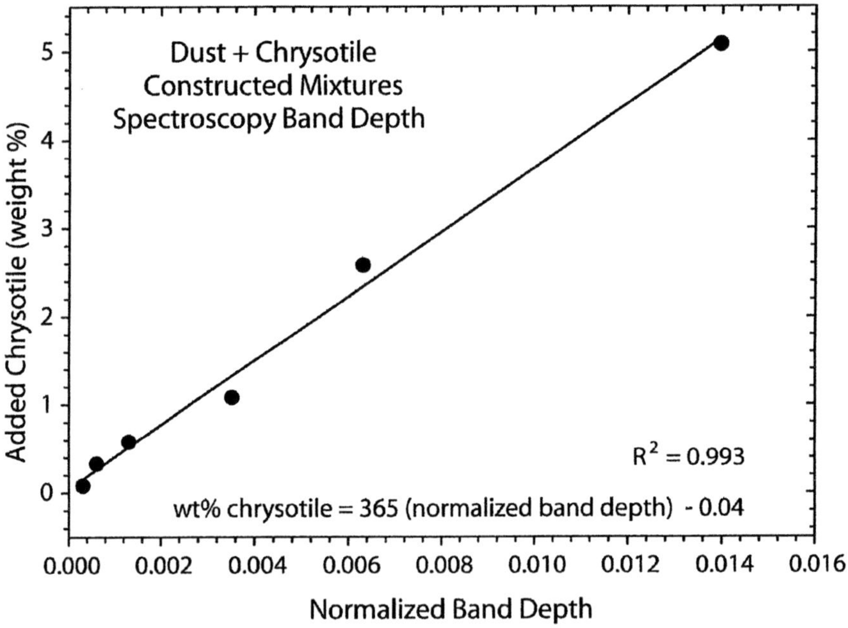

Figure 8. Correlation between normalized band depth of the 1.385-μm absorption and chrysotile content of mixtures constructed using dust sample 6 with variable amounts of the chrysotile NIOSH standard added. R^2 is the correlation coefficient for the best fit line.

grain size of the dust particles results in more scattering causing a decrease in the strength of all the absorptions (*23*), including the chrysotile absorption, thus, decreasing the spectral sensitivity to chrysotile content. Because most of the dust samples have an albedo similar to that of the sample used to create the mixtures, errors in estimating chrysotile content will probably be smaller than those due to grain-size variations. Considering these potential sources of error, we estimate that application of the spectral calibration (Figure 8) to the other dust samples, apart from sample 6, to have an abundance error of approximately ± 0.3 wt %.

Spectra of dust samples that contain spectrally significant concentrations of portlandite (from pulverized concrete) and muscovite/illite have wide 1.42-μm absorptions that can conceal chrysotile absorptions that may be present at 1.385 μm (*5*). Figure 9 shows a spectrum of a 1 wt% chrysotile + dust mixture compared with a spectrum of dust sample 2, which contains a relatively high

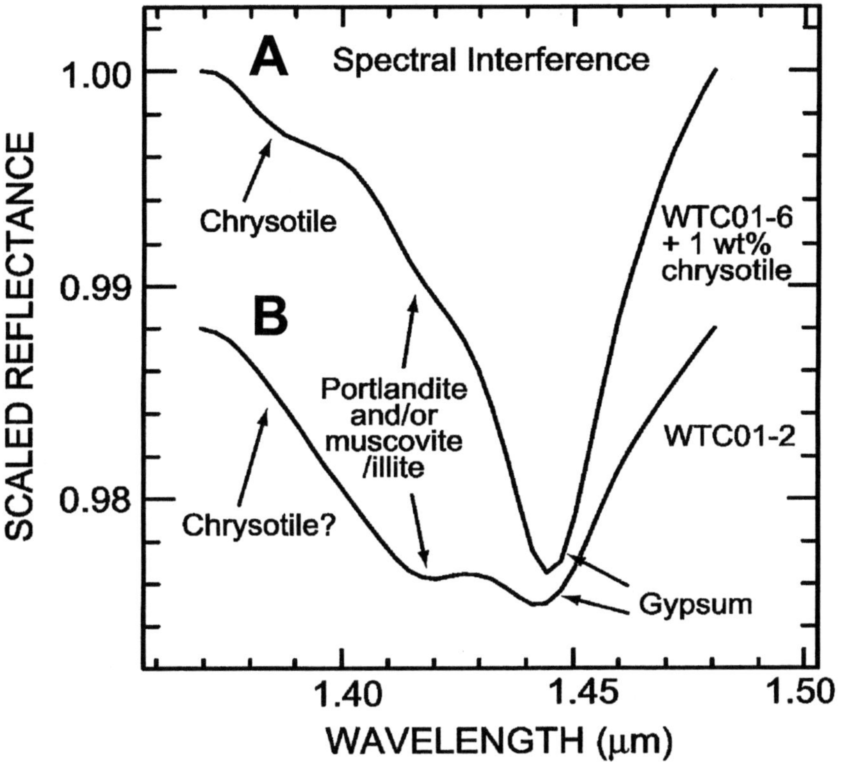

Figure 9. Spectra of dust samples that contain (A) trace and (B) higher concentrations of portlandite and/or muscovite/illite, which can interfere with spectral determination of asbestos content. Spectra are continuum removed and offset vertically for clarity.

concentration of portlandite and/or muscovite/illite and no spectrally detectable chrysotile. The band depth of the 1.385-μm chrysotile absorption in spectrum A is 24 times greater than the standard deviation of the noise in this spectrum (see Figure 7B). Spectroscopy could not be used to estimate chrysotile content of five dust samples because of interference from nearby spectral absorptions of portlandite and/or muscovite/illite. However, in many of the outdoor dust samples portlandite reacted with carbonic acid in rain water from thunderstorms on September 14 to form calcite (see Chapter 12), thus reducing portlandite's spectral interference with the nearby chrysotile absorption feature.

X-Ray Diffraction Detection of Chrysotile

The same mixture series was also characterized using XRD to provide a series of diffractograms that could be directly compared to those of the dust samples. Figure 10A shows x-ray diffractograms of mixtures of sample 6 and a NIOSH chrysotile standard. Figure 10B shows x-ray diffractograms of sample 8, from the steel beam fireproof coating, with up to 20 wt% chrysotile (*8*), of sample 28 with a trace of serpentine, and of sample 6 in which no serpentine. was detected with XRD analysis. A diffraction peak diagnostic of chrysotile occurs near 12 degrees two theta on the slope of a much larger gypsum peak at 11.7 two theta in the mixture series (Figure 10A). At concentrations below a few weight percent, XRD analysis cannot be used to differentiate chrysotile from the

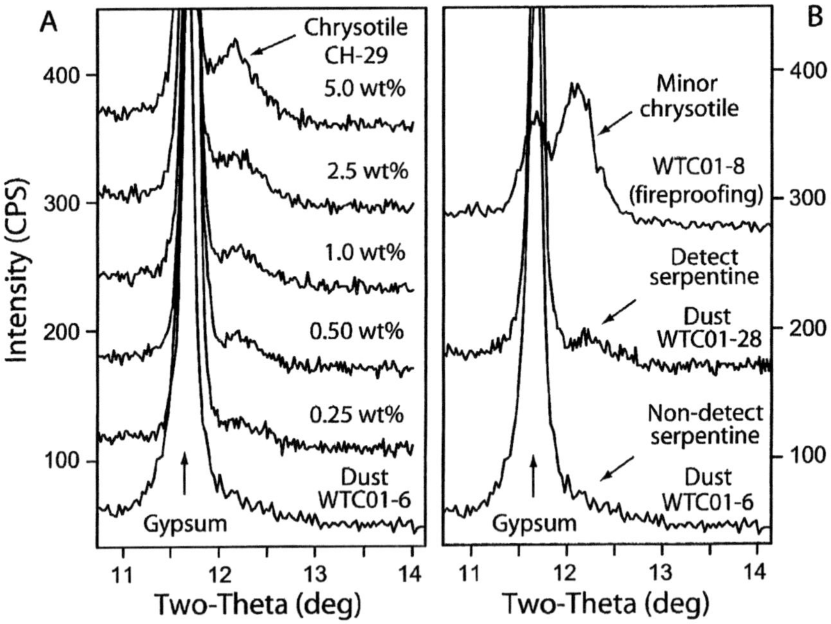

Figure 10. (A) X-ray diffractograms of dust sample 6 with the NIOSH chrysotile standard added. (B) X-ray diffractograms of a chrysotile-bearing beam coating and two dust samples. Diffractograms are offset vertically For clarity. CPS = counts per second; deg = degrees.

other serpentine minerals in the WTC dust samples. The mixture series diffractograms (Figure 10A) show possible detection levels of chrysotile down to the 0.25-wt% level. Because of the limited time available for analysis of the dust samples during the weeks following the WTC disaster, these mixtures were made using a hand-ground split from sample 6 so homogeneity, grain size, and mixing procedures did not meet the accepted criteria for quantitative XRD analysis. However, these diffractograms can be used as guides to determine possible quantities and detection limits for low concentration levels of chrysotile.

Intra-Sample Variation of Chrysotile Content

One method of estimating the variability of chrysotile within a dust sample is to make replicate measurements on different portions of the sample. Because of the limited time available for analysis of the dust samples during the weeks following the WTC disaster, only one set of XRD measurements was made on all but two of the dust samples. However, twenty spectral measurements were made on separate portions of each dust sample. Figure 11A shows variations in the normalized band depth of the 1.385 μm absorption for each spectral measurement of dust sample 28. The band depth error bar shown in Figure 11A represents ± 1standard deviation of the channel-to-channel noise integrated over the wavelength interval of the 1.385μ:m absorption. These measurements may indicate that serpentine/amphibole/talc is not distributed homogeneously within this sample. This observation is also supported by two replicate XRD measurements of dust sample 28, in which serpentine was detected in one run but not the other. X-ray diffraction measurements were made on hand-ground 3 gram splits of each dust sample, whereas spectroscopic measurements were made on approximately half of the entire sample volume (35 – 200 grams of material).

Figure 11B shows the cumulative average for estimates of serpentine/amphibole/talc contents of dust sample 28 based on band depths shown in Figure 11A and the spectral regression shown in Figure 8. The abundance error bar (± 0.3 wt%) represents the uncertainty in estimates of serpentine/amphibole/talc contents of the dust. Note that the spectrally derived cumulative average for serpentine/amphibole/talk content of sample 28 becomes relatively stable after the first five measurements. If it is assumed that chrysotile accounts for most of the 1.385-μm absorption seen in sample 28, then the spectral estimates may be more representative of the bulk chrysotile contents of the dust than are those derived from the more limited number of XRD measurements.

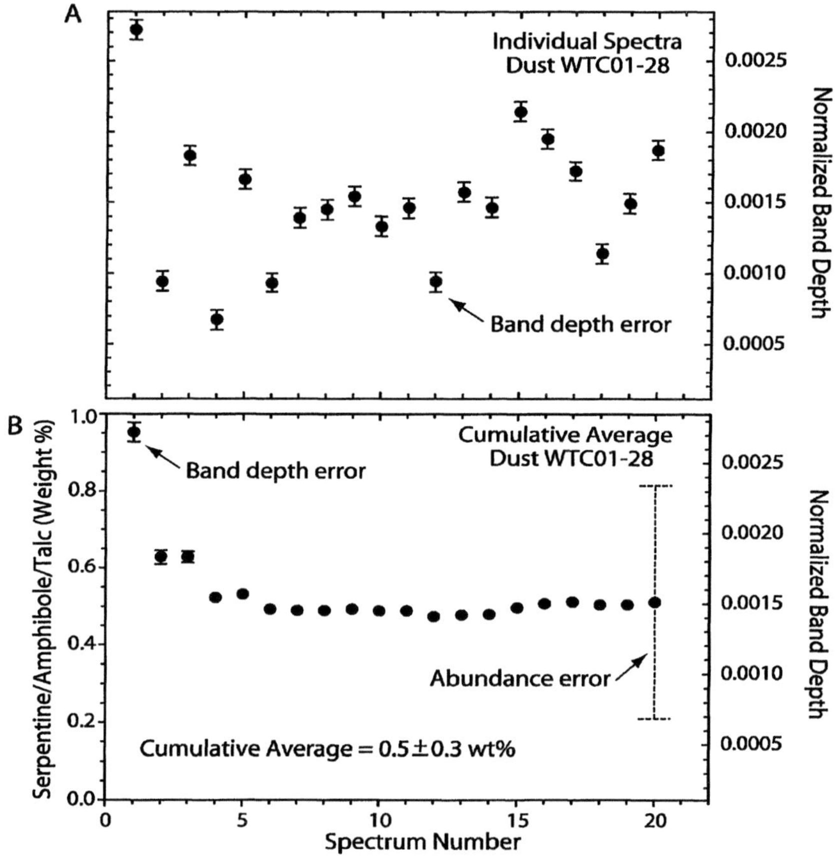

Figure 11.(A) Estimates of normalized band depth of the 1.385-μm absorption feature in dust sample 28 for each of the 20 spectra measured. (B) The cumulative average for estimates of serpentine/amphibole/talc contents of dust sample 28.

Chrysotile Asbestos Contents of WTC Dust Samples

Other studies have shown that chrysotile is a constituent of the WTC dust (*8,21, 25-27*). Nevertheless, the spectral and XRD techniques described above cannot be used to evaluate chrysotile content, at low levels, without the potential for overestimation when other serpentine minerals, amphiboles, or talc are present. Therefore estimates provided by these methods should be considered to be upper limits for the chrysotile content of the WTC dust.

Except for the five samples with significant spectral interference, spectroscopy was used to estimate the serpentine/amphibole/talc content in all of the dust samples. These results are shown in Figure 12. The largest circles in Figure 12 corresponds to ~1 wt% serpentine/amphibole/talc assuming the correlation shown in Figure 8 can be applied to dust sample 14. Scanning electron microscopy indicates that chrysotile is present, at unspecified levels, in the "below detection" samples examined (see Chapter 5). At concentrations below ~0.5 wt%, spectroscopy cannot distinguish between the 1.385-μm absorption features caused by chryostile, lizardite, antigorite, amphiboles, or talc at the spectral resolution used for these measurements. Levels of these minerals were above 0.3 wt% in 14 of the dust samples. Dust samples 13 and 14 had the highest spectrally derived concentration of serpentine/amphibole/talc at ~1 wt%. Samples with spectral interference were collected on the periphery of our collection area where the dust deposits were thin. This distribution may be evidence of compositional segregation of portlandite and/or muscovite/illite as the dust clouds settled.

X-ray diffraction analysis detected serpentine in 16 of the dust samples (Figure 13), including 3 of the 5 dust samples determined to have significant spectral interferences. The largest circle shown in Figure 13 corresponds to ~1 wt% serpentine using the mixture series shown in Figure 10A as guides. Scanning electron microscopy indicates that chrysotile is present, at unspecified levels, in the XRD non-detect samples examined (see Chapter 5).

There is only moderate agreement between the spectral and XRD estimates when they are considered on an individual sample basis. For instance, there were 6 samples, apart from those with spectral interference, where XRD detected serpentine but spectroscopy did not. This apparent discrepancy may be a consequence of the heterogeneous nature of the dust, different levels of sensitivity of each analytical method to materials in the dust, and different quantities of dust measured by each method. The problem of detecting unevenly distributed chrysotile in the WTC dust is likely to plague any comparison of analytical methods.

When the results of the two methods are combined, variable levels of serpentine, amphibole, and/or talc were detected in over two-thirds of the dust samples. Neither spectroscopy or XRD detected potential chrysotile concentrations higher than ~1 wt% in any of the 33 dust samples. Examination of these samples with other analytical techniques is needed to further substantiate the quantification levels shown on these maps. It is important to note that the dust and debris samples were collected within a few-square-meter area at sites shown on Figure 1, and that the circles shown in Figures 12 and 13 were made larger than the areas sampled for illustrative purposes only.

The combined results from the two analytical techniques used in this study suggest that potential chrysotile concentrations in the dust are higher to the west

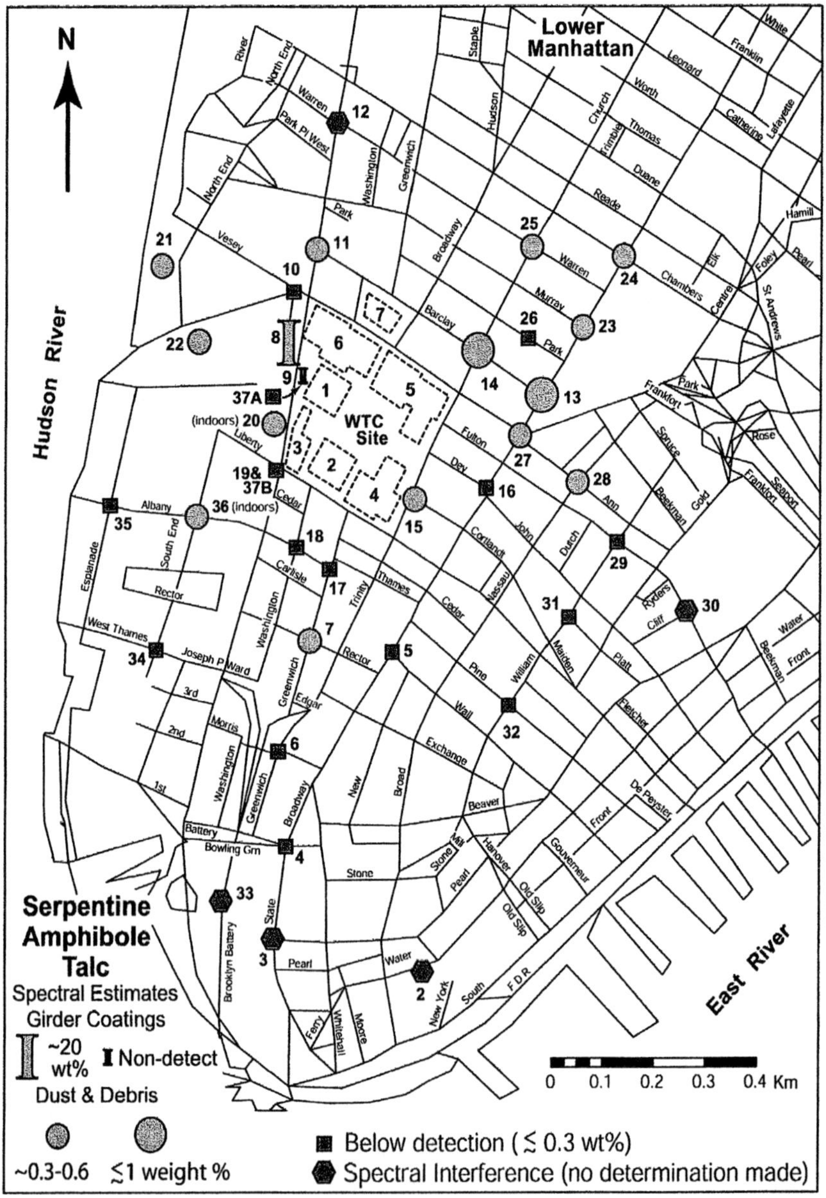

Figure 12. Spectral estimates of serpentine/amphibole/talc concentrations in settled dust samples in lower Manhattan. Information shown on this map should not be viewed as a replacement for detailed, standardized sampling and test measurements for asbestos.

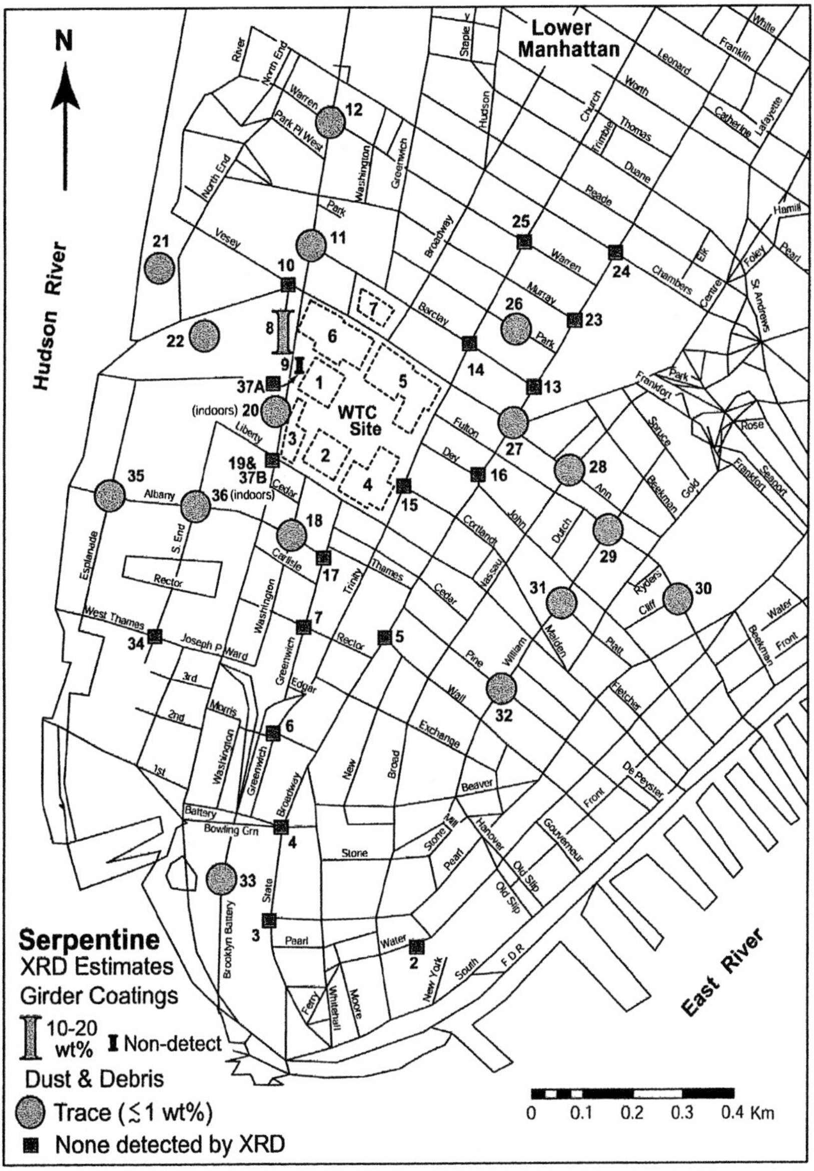

Figure 13. X-ray diffraction estimates of serpentine concentrations in settled dust samples in Lower Manhattan. Information shown on this map should not be viewed as a replacement for detailed, standardized sampling and test measurements of asbestos.

and east of Ground Zero and lower south of Ground Zero. If true, certain aspects of this pattern may be explained by noting that the WTC dust blanket was formed by three chronologically and spatially separate dust dispersal events, caused by the initial collapse of Tower 2, followed by Tower 1, and finally Building 7 (*28*). At some locations, the dust blanket may have been composed of three dust layers. These layers may not have completely overlapped each other, leaving areas where dust from the collapse of one building compositionally dominated the dust blanket. Because use of chrysotile-bearing fireproof insulation at the WTC was discontinued in 1969, prior to major construction of Tower 2 (*16, 28*), the dust produced by its collapse may have been relatively lower in chrysotile compared to dust produced from the collapse of Tower 1. It may be possible that the relatively lower chrysotile content of the dust south of Ground Zero may reflect this area's proximity to Tower 2, which did not use chrysotile-bearing fireproof insulation.

Conclusions

Laboratory spectral and x-ray diffraction analyses of WTC settled dust samples detected potential trace levels of chrysotile asbestos in about two-thirds of the dust samples (23 out of 33 samples), but at concentrations at or lower than ~1 wt%, well below the detection level (~5 wt%) of the AVIRIS imaging spectrometer (*6*). Scanning electron microscopy indicates that chrysotile is present, at unspecified levels, in all of the non-detected samples examined (see Chapter 5). Information shown in Figures 12 and 13 should not be viewed as a replacement for detailed standardized sampling and test measurements for asbestos. One sample of a beam coating material contained up to 20 wt% chrysotile asbestos. Carbonic acid in rain water may have reacted with some of the portlandite, making spectral detection of chrysotile easier with the portable spectrometer. Increasing the spectral resolution to 4 nm in the 1.38-μm wavelength region would increase the portable spectrometer's sensitivity to chrysotile content while minimizing interference from other minerals. Spectra with a very high signal-to-noise ratio (28,000:1) were needed to estimate potential chrysotile content in the dust down to the approximately 0.3 wt% detection level. The need for this and, perhaps, even finer levels of detection present an extraordinary challenge for imaging spectrometers designed for emergency response. In addition to airborne imaging spectroscopy, laboratory spectroscopy was a necessary component of the spectral analysis of the dust because its sensitivity level extends well below that of AVIRIS (*6*) and it allowed measurements in spectral regions where atmospheric gases obscure the surface. Analyses indicate that trace levels of chrysotile were distributed with the dust radially to distances greater than 0.75 km from Ground Zero. These data also suggest that the distribution of chrysotile may not have been equal in all directions due to differences in sources of the dust.

In the years since our original study of the WTC dust (*5*), higher-spectral-resolution full-range portable spectrometers with battery operated contact light probes have been developed. These spectrometers could be used to screen thousands of spots for asbestos contamination over large, potentially-hazardous areas in a matter of days, producing data that could be spot checked with traditional evaluation methods. Automated identification software (*9*) could be used to search the spectra for signs of contamination real-time, producing results immediately after each spectral measurement.

Acknowledgments

Sam Vance was instrumental in getting us permission to sample at Ground Zero. Reviews by James Crock, James Crowley, Daniel Knepper, P. Fenter improve the manuscript greatly. The USGS Minerals Program funded this research. Use of trade names is for descriptive purposes only and does not constitute endorsement by the Federal Government.

References

1. Clark, R. N.; King, T. V. V.; Klejwa, M.; Swayze, G. A.; Vergo, N. *J. Geophys. Res.* **1990,** *95*, 12,653-12,680.
2. Hunt, G. R. *Geophysics* **1977,** *42(3), 501-513.*
3. Hunt, G. R.; Salisbury, J. W. *Mod. Geol.* **1971,** *2*, 23-30.
4. Hunt, G. R.; Ashley, R. P. *Econ. Geol.* **1979,** *74*, 1613-1629.
5. Clark, R. N. ; Green R. O.; Swayze, G. A.; Meeker, G.; Sutley, S.; Hoefen, T.M.; Livo, K. E.; Plumlee, G.; Pavri, B.; Sarture, C.; Wilson, S.; Hageman, P.;Lamothe, P.; Vance, J. S.; Boardman, J.; Brownfield, I.; Gent, C.; Morath, L.C.; Taggart, J.; Theodorakos, P. M.; Adams, M. *U. S. Geological Survey, Open File Report OFR-01-0429,* 2001. [http://pubs.usgs.gov/of/2001/ofr-01-0429/].
6. Clark, R.N.; Swayze, G.; Hofen, T.; Livo, E.; Sutley, S.; Meeker, G.; Plumlee, G.; Brownfield, I.; Lamothe, P.J.; Gent, C.; Morath, L.; Taggart, J.; Theodorakos, P.; Adams, M.; Green, R.; Pavri, B.; Sarture, C.; Vance, S.; Boardman, Paper presented in ACS Symposium: *Urban Aerosols and their Impact: Lessons Learned from The World Trade Center Tragedy*, New York City, Sept. 10, 2003. *ACS Division of Environmental Chemistry Preprints of Extended Abstracts, paper 110,* **2003,** *43 (2)*, 1324.

7. Clark, R. N.; Hoefen, T. M.; Swayze, G. A.; Livo, K. E.; Meeker, G. P.; Sutley, S. J.; Wilson, S.; Brownfield, I. K.; Vance, J. S. *U.S. Geological Survey Open-File Report 03-128,* 2003. [http://pubs.usgs.gov/of/2003/ofr-03-128/ofr-03-128.html].
8. Meeker, G. P.; Sutley, S. J.; Brownfield, I. K.; Lowers, H. A.; Bern, A. M.; Swayze, G. A.; Hoefen, T. M.; Plumlee, G. S.; Clark, R. N.; and Gent, C. A., Paper presented in ACS Symposium: *Urban Aerosols and their Impact: Lessons Learned from The World Trade Center Tragedy*, New York City, Sept. 10, 2003. *ACS Division of Environmental Chemistry Preprints of Extended Abstracts, paper 133,* **2003,** *43 (2)*, 1354.
9. Clark, R. N.; Swayze, G. A.; Livo, K. E.; Kokaly, R. F.; Sutley, S. J.; Dalton, J. B.; McDougal, R. R, Gent, C. A. *J. Geoph. Res.* **2003,** *108(E12)*, 5131, doi:10.1029/2002JE001847, 44 p.
10. Lewis, D. W. *Practical Sedimentology*; Hutchinson Ross Publishing Company, 1984; 229 p.
11. Klug, H. P.; Alexander, L. E. *X-Ray Diffraction For Polycrystalline And Amorphous Materials*; John Wiley & Sons, 1974; 992 p.
12. World Trade Center indoor environmental assessment: selecting contaminants of potential concern and setting health-based benchmarks. Contaminants of Potential Concern (COPC) Committee, World Trade Center Indoor Air Task Force Working Group, May 2003, 81p.
13. Hyman Brown, WTC construction manager, oral comm., 2001.
14. Ross, M. In The Environmental Geochemistry of Mineral Deposits: Processes, Techniques, and Health Issues; Plumlee, G. S.; Logsdon, M. J., Eds.; *Soc. Econ. Geol.*, 1999; *6A*, 339-356.
15. Van Oss, C. J.; Naim, J. O.; Costanzo, P. M.; Giese, R. F. Jr.; Wu, W.; Sorling, A. F. *Clays and Clay Mins.*, **1999,** *47(6)*, 697-707.
16. Progress report on the federal building and fire safety investigation of the World Trade Center Disaster. *NIST Special Publication 1000-3*, 2003; 122 p.
17. Reitze, W. B.; Nicholson, W. J.; Holaday, D. A.; Selikoff, I. J. *J. Am. Ind. Hyg. Assoc.*, **1972,** *33*, 178-191.
18. Eric Chatfield, Asbestos Researcher, Chatfield Technical Consulting Limited, oral comm., 2001.
19. Van Gosen, B. S.; Lowers, H. A.; Bush, A. L.; Meeker, G. P.; Plumlee, G. S.; Brownfield, I. K.; Sutley, S. J. *U.S. Geol. Survey Bull. 2192*, **2002,** 8p.
20. Meeker, G. P.; Bern, A. M.; Brownfield, I. K.; Lowers, H. A.; Sutley, S. J.; Hoefen, T. M.; Vance, J. S. *Am. Min.* **2003,** *88*, 1955-1969.
21. Chatfield, W. J.; Kominsky, J. S. Summary Report: Characterization of particulate found in apartments after destruction of the World Trade Center. Report requested by: "Ground Zero" Elected Officials Task Force. 2001; 42p.

22. NOSH/R.I., NOSH Reference Material Data Reports: Chrysotile (CH-29), 1978.
23. Clark, R. N.; Roush, T. L. *J. Geophys. Res.* **1984,** *89*, 6329-6340.
24. Clark, R. N. *J. Geophys. Res*. **1983,** *88*, 10,635-10,644.
25 Lily, P. J.; Weasel, C. P.; Gillette, J. S.; Eisenreich, S.; Vallero, D; Offenberg, J. *Environ. Health Perspect.*, **2002,** *110*, 703-714.
26. Gillette, J. S.; Boltin, R.; Few, P.; Turner, W. Jr. *Microscope*, **2002,** *50(1)*, 29-35.
27. World Trade Center residential dust cleanup program, draft final report, March, 2004,[http://www.epa.gov/region02/wtc/draft_final_report.pdf].
28. World Trade Center Building Performance Study. Federal Emergency Management Agency, 2004, [http://www.fema.gov/library/wtcstudy.shtm].

Chapter 4

Environmental Mapping of the World Trade Center Area with Imaging Spectroscopy after the September 11, 2001 Attack

The Airborne Visible/Infrared Imaging Spectrometer Mapping

Roger N. Clark[1], Gregg A. Swayze[1], Todd M. Hoefen[1], Robert O. Green[2], K. Eric Livo[1], Gregory P. Meeker[1], Stephen J. Sutley[1], Geoffrey S. Plumlee[1], Betina Pavri[2], Chuck Sarture[2], Joe Boardman[3], Isabelle K. Brownfield[1], and Laurie C. Morath[1]

[1]Denver Microbeam Laboratory, U.S. Geological Survey, MS 964, Box 25046 Denver Federal Center, Denver, CO 80225
[2]Jet Propulsion Laboratory, 400 Oak Grove Drive, Pasadena, CA 91109
[3]Analytical Imaging and Geophysics LLC, 4450 Arapahoe Avenue, Suite 100, Boulder, CO 80303

The Airborne Visible/Infrared Imaging Spectrometer (AVIRIS) was flown over the World Trade Center area on September 16, 18, 22, and 23, 2001. The data were used to map the WTC debris plume and its contents, including the spectral signatures of asbestiform minerals. Samples were collected and used as ground truth for the AVARIS mapping. A number of thermal hot spots were observed with temperatures greater than 700° C. The extent and temperatures of the fires were mapped as a function of time. By September 23, most of the fires observed by AVIRIS had been eliminated or reduced in intensity. The mineral absorption features mapped by AVARIS only indicated the presence of serpentine mineralogy and not if the serpentine has asbestiform.

Introduction

Spectroscopy is a tool that detects chemical bonds in molecules (solid, liquid or gas) through absorption (or emission) features in the spectrum of the material. Imaging spectroscopy obtains a spectrum for every spatial pixel in an image format. The Airborne Visible / Infrared Imaging Spectrometer (AVIRIS), a hyperspectral remote sensing instrument, was flown by the Jet Propulsion Laboratory (JPL/NASA) over the World Trade Center (WTC) area on September 16, 18, 22, and 23, 2001. The AVIRIS sensor obtains an ultraviolet to near-infrared (0.37 to 2.50 μm) spectrum for each pixel in a spatial array. If materials are present in sufficient abundance and are spectrally active in the AVIRIS ultraviolet to near-infrared spectral region, their spatial locations can be mapped in far greater detail than traditional ground sampling methods. A single AVIRIS flight can measure spectra for approximately 25 million sample locations.

The purpose for this study was to provide data to map materials that characterize the environment around the WTC, including asbestos, which was feared by first responders to have been potentially spread by the collapse of the WTC. The U. S. Geological Survey (USGS) effort was begun in response to requests from other Federal agencies. In responding to this request, we mobilized a 2-person team (authors T.M. Hoefen and G.A. Swayze) to the WTC site to obtain calibration data in order to correct the AVIRIS data to surface reflectance and collect samples to provide ground truth to verify the AVIRIS mapping analyses (see Chapter 3). Samples were collected of dusts and air fall debris from more than 34 localities within a 1-km radius of the WTC on the evenings of September 17 and 18, 2001. This sample set provided the needed ground truth for the AVIRIS data as well as samples for more detailed laboratory analyses. Of the samples collected, all but two were outdoor samples. Two samples were collected from indoor locations that were presumably not affected by rainfall (there was a rainstorm on September 14). Two samples of material coating a steel beam in the WTC debris were also collected. The sample set and the spectroscopic and mineralogic analyses are presented in Chapter 3 while the geochemistry of the samples is presented in Chapter 12 and the microbeam analyses is presented in Chapter 5 (*1*).

The USGS ground crew also carried out on-the-ground reflectance spectroscopy measurements during daylight hours to field calibrate the AVIRIS remote sensing data. Radiance calibration and rectification of the AVIRIS data were done at JPL/NASA. Surface reflectance calibration, spectral mapping, and interpretation were done at the USGS Imaging Spectroscopy Lab in Denver. The dust/debris and beam-insulation samples were analyzed for a variety of mineralogical and chemical parameters using Reflectance Spectroscopy (RS) and

X-Ray Diffraction (XRD) (see Chapter 3), Scanning Electron Microscopy (SEM) (Chapter 5), (*1*) chemical analysis, and chemical leach test techniques (Chapter 12) in USGS laboratories in Denver, Colorado.

Imaging Spectroscopy Data Analysis

The AVIRIS instrument measures upwelling spectral radiance in the visible through near-infrared. The instrument has 224 spectral channels (bands) with wavelengths from 0.37 to 2.5 μm (micrometers) with sufficient spectral resolution to characterize diagnostic spectral features in materials (*2, 3*). AVIRIS was flown at 2 altitudes in the WTC area for this study to give pixel sizes of approximately 2 and 4 meters. Reported here are the higher resolution 2-meter/pixel data. Because AVIRIS measures reflected sunlight, it generally cannot detect materials deeper than can be seen with the human eye. For most solid materials this optical penetration is measured in millimeters.

The AVIRIS instrument was flown by NASA/JPL over the WTC area on Sept. 16, 18, 22, and 23, 2001, after the attack on the WTC. Collection of data on Sept. 18 and 22 was hampered by clouds, whereas the 16th and 23rd were clear.

The 13 gigabytes of September 16th data were sent to the USGS in Denver on September 17-18th. Atmospheric absorptions, instrument response and the solar spectrum were removed using ground calibration that employs spectra of large, uniform areas in New Jersey outside the WTC debris zone. The calibration used a theoretical model of atmospheric transmittance followed by residual corrections using areas whose reflectance properties were measured by the ground crew. The specific methodology is presented elsewhere (*4*).

While several calibration sites were initially tried, the best site was found to be the top level of a concrete parking garage in New Jersey (*5*). Because AVIRIS data were delivered to our laboratories less than 24 hours from acquisition, we used the preliminary radiative transfer model calibration to explore calibration sites. Portions of the parking garage site had undesirable organic absorptions, presumably due to oil from cars, even though there was no visual clue to its presence. Therefore, the field crew was guided in real time via cell phone to the best, spectrally neutral, areas of the site. This was accomplished by extracting spectra from the AVIRIS preliminary calibrated data (from our lab in Denver) for pixels in and around the ground crew, interpreting composition and informing the ground crew what the composition was at and around their location. They were then directed to a spectrally neutral site that allowed for a more precise calibration free of residual and anomalous absorption features. The field crew sent data via the internet to our labs where the data were reduced and information fed back to the ground crew. Several iterations resulted in an excellent calibration. Normally, such a calibration takes a month or more, but the near real-time feedback reduced calibration time to about 2 days. The reflectance calibration spectra and locations are given elsewhere (*7*)

Once calibrated, the data were analyzed for the presence of specific spectral features to identify materials expressed in the spectrum at each pixel in the image. The methods employed, from calibration to imaging spectroscopy analysis (Tetracorder analysis system), to verification of results follow the procedures and methods established previously (*6-9)*. Characterization of reflectance spectra of asbestiform minerals of concern in the WTC debris can be found in Chapter 3 (*5, 9, 10*).

Thermal Hot Spots

Results of the AVIRIS remote sensing data analysis and interpretations show the distribution and intensity of thermal hot spots in the area in and around the World Trade Center on September 16, 18, and 23, 2001. Plate 1 shows false color images of the core affected area around the WTC extending from 5 to 12 days after the collapse. Hot spots appear orange and yellow in the images. Dozens of hot spots are seen on September 16th and 18th. The image on the 18th appears dark because of clouds which blocked the sunlight but not the light emitted by the fires. Analysis of the data indicates temperatures greater than 700° C. Over 3 dozen hot spots are identified in the core zone. By September 23, only 4, or possibly 5, hot spots are apparent in the image, with temperatures cooler than those on September 16. There are other red/orange spots in Plate 1 in the area south of the World Trade Center zone. These areas are hot spots from chimneys or heating exhaust vents and are normal and do not represent other uncontrolled fires. Plate 1 also shows vegetated areas as green. Water appears blue, and the smoke from the fires appears as a light blue haze. White and lighter blue areas are rooftops, roads, and concrete as well as dust and debris from the collapsed buildings. Dust, probably more than a few millimeters thick (the optical depth), appears in shades of brown around the core WTC area on the 16th. The key in Plate 1 (lower right) shows the hot spot locations listed in Table I.

Data collected on the 16th were processed, interpreted and released to emergency response teams on the September 18, 2001. On September 18th fire fighters changed from a rescue to a recovery operation and our field team observed increased water being put on the fires. The images show significant thermal hot spots on September 16 and 18 (see Table I), but by September 23rd most of the hot spots had cooled or the fires had been put out.

On the September 16, 2001 image (Plate 1), large areas around the WTC show brownish colors, indicating the debris. On September 20, 2001 there was

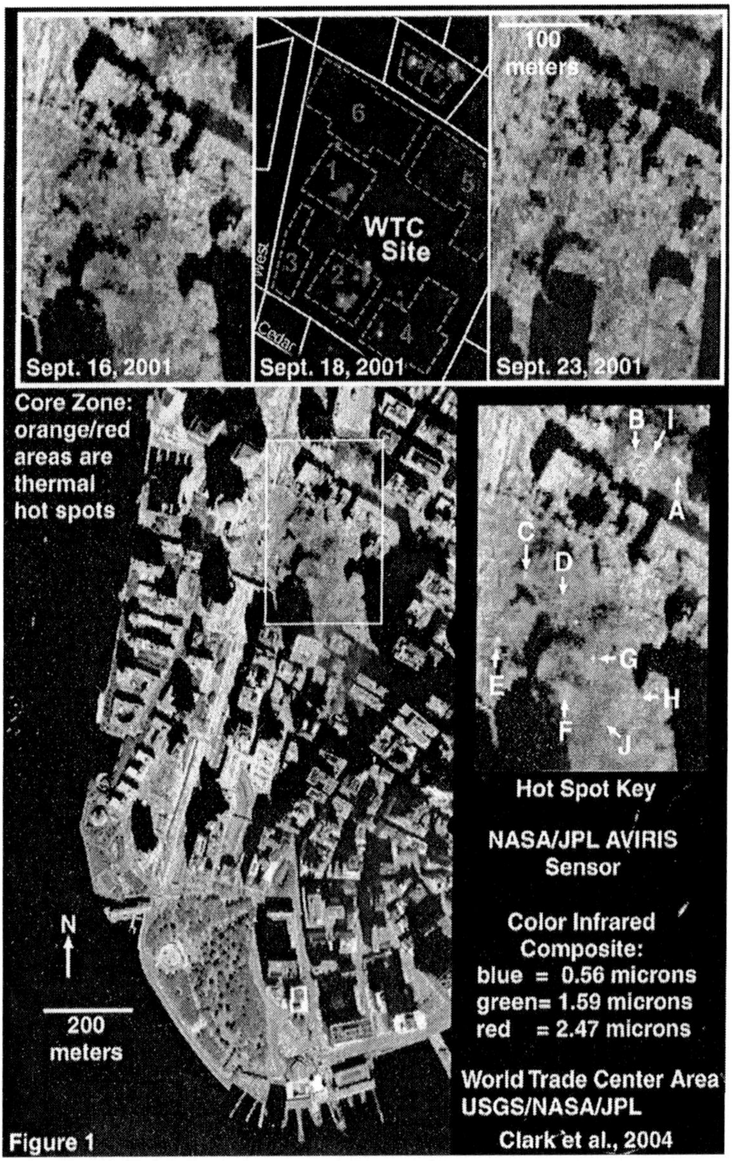

Plate 1. False color images of the core affected area around the WTC extending from 5 to 12 days after the collapse. Hot spots appear orange and yellow. The key shown at the right corresponds to the hot spot locations listed in Table 1. (See page 4 of color inserts.)

Table I. Temperature and Size of the Thermal Hot Sspots Detected by AVIRIS Remote Sensing on September 16, 18 and 23, 2001.

				Sept. 16		*Sept. 18*		*Sept. 23*	
Hot Spot	*Lat*	*Lon*	*Max Area*	*T*	*A*	*T*	*A*	*T*	*A*
A	47.18	41.43	430	730	0.6	730	1.8	<180	—
B	47.14	43.53	120	560	0.08	630	2.4	<180	—
C	42.89	48.88	130	630	0.8	330	0.6	<150	—
D	41.99	46.94	180	520	0.8	430	1.0	<180	—
E	40.58	50.15	130	440	0.4	<150	—	<150	—
F	38.74	46.70	930	430	0.4	530	1.3	<130	—
G	39.94	45.37	740	750	0.04	530	1.4	330	1.3
G	39.94	45.37	740	—	—	—	—	200	2.7
H	38.60	43.51	80	550	0.08	<170	—	<170	—
I	46.94	42.75	230	630	0.08	330	0.7	330	1.0
I	46.94	42.75	230	430	3.2	—	—	180	3.0
J	37.68	44.91	20	240	4.0	630	1.1	<180	—

NOTE: Lat is north lattitude in decimal seconds after 40° 42′. Lon is west longitude in decimal seconds after 74° 00′. Positions are estimated to be ± 6 m (18 ft.). T is temperature in °C as the highest recorded in a 4-m^2 pixel covering part of the pixel. Max Area is the total size of the hot spot in m^2. A is the area in m^2 of the listed temperature.

NOTE: Temperatures listed as "<" are upper limits and represent no detection. Hot spot J grew to 60 m^2 on September 18.

another significant rain storm that washed away more of the dusty debris. Reduction of the dust/debris distribution is apparent in the September 23 image of Plate 1, and can be attributed to the cleanup effort along with the rain.

Remote measurement of temperature is difficult because the source area of the thermal emission can be less than the field of view of the measuring instrument. In that case a thermal sensor has an ambiguous solution: a hotter temperature of a smaller area or lower temperature of a larger area can result in the same total received thermal radiation. A spectrometer, however, overcomes the ambiguity problem because the shape of the thermal spectrum can be used to derive a unique temperature, and the intensity gives the area of the emitting source (*11*). If a large enough spectral range is measured, a temperature range and the area of each hot spot can be derived. In the near-infrared spectral range of AVIRIS, reflected solar radiation also contributes to the signal. The solution to the generalized problem involving all these effects is given elsewhere (*11, 12*)

We calculated temperatures of the hot spots using two methods: calibrated radiance (*12*), and calibrated reflectance data (*11*) . In calibrated reflectance data, thermal radiation is the Planck response divided by the solar spectrum. This reflectance method has several advantages: 1) the data can be corrected for atmospheric absorption and scattering, 2) the reflected solar component can be readily assessed and compensated for, and 3) the ratio of the Black-Body response to the solar spectrum produces a very steep curve that is readily distinguishable from reflected sunlight and reflectance of surface materials (*5, 11*).

Temperature derivations from reflectance calibrated data are illustrated in Plate 2, where hot spot C is shown on two different dates. Hot spot C is found to have a 630° C temperature over 20% of a pixel, or 0.8 m^2 on September 16, but has cooled to 330° C on the September 18. It was not detectable on September 23. While the highest temperature is given for a single pixel, the area indicated as hot, greater than about 130° C, is larger. In the case for location C, the hot area covered ~130 m^2 on September 16. . The small rise near 2.5 µm in the spectrum from September 18 indicates another thermal component at a temperature lower than 330° C. Temperature data for several spots as a function of time are given in Table I. For temperatures in the 500-800° C range, temperature accuracy is estimated to be ± 30° C and the area ± 5%. For smaller spots, like spot G in Table I, the temperature accuracy is similar, but the accuracy on such small areas is approximately +5%, -0.5%. For example, decreasing the temperature to 720° C (from 750° C) on spot G increases the fractional area to about 5%.

Examination of the data in Plate 1 and Table I shows several trends. Most notable is the large decrease in temperatures from the 18th to the 23rd as a result of the increased fire fighting effort begun on the 18th. Some fires increased in area and/or temperature from the 16th to the 18th (spots A, B, F, G, J). A significant fire, J (Plate 1), appears intense on the 18th, barely shows on the 16th, and does not appear on the 23rd. The fires/hot spots observed in the AVIRIS data represent detections of fires deep in the rubble piles, as seen through holes in the rubble. Even though AVIRIS showed that near surface fires and temperatures were significantly reduced from the 16th to the 23rd, fires deeper underground were reported to burn and/or smolder for another 2 months. While the peak temperatures indicate only small areas, the images in Plate 1 and the maximum area of higher detectable temperatures (Table I) indicate that the fires affected a large area. Summing the maximum area in the region of each spot plus other hot areas in Plate 1, approximately 3,000 to 4,000 m^2 were affected. This area decreased to < 200 m^2 of detectable temperatures (hotter than 130 to 180° C) by September 23.

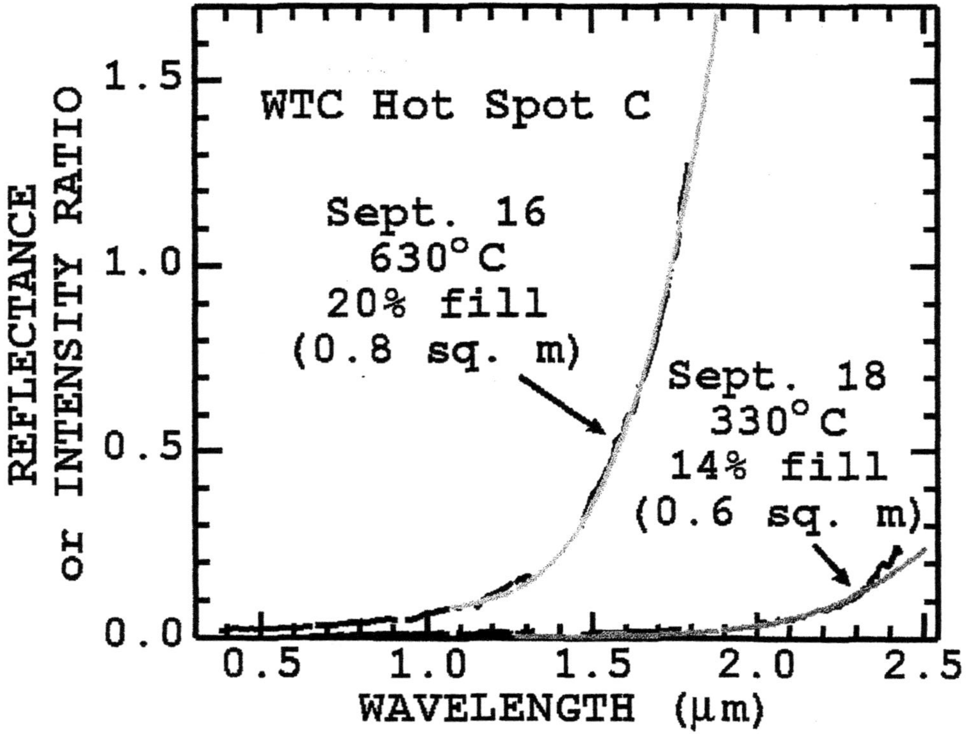

Plate 2. AVIRIS spectra (black) for thermal hot spot C on September 16 and 18 shown with the fitted thermal response (orange, red). The spectral shape is used to constrain temperature and the intensity constrains area.
(See page 5 of color inserts.)

Maps of Dust and Debris

The AVIRIS data were analyzed with our spectral feature analysis system (*8*). Spectral features from a spectral library of numerous materials (9) as well as spectral features observed in WTC field samples described in Chapter 3 were used in the mapping (*5*). The main difficulty in mapping materials from a demolished building is that the materials have the same composition as the rest of the city. However, debris plumes can be interpreted by image context, for example, crossing roads and other barriers that would normally show a change in composition. Maps of organics, minerals and building components were presented earlier (5).

The extent of the WTC dust plumes (3 plumes: from buildings 1, 2, and 7) derived from the AVIRIS data on September 16 was mapped using the spectral properties of 7 representative field samples. The spectra of the WTC field samples shown in Plate 3 show a general range of spectral characteristics. Samples WTC01-37B and WTC01-37Am are concrete that display ferrous iron absorptions (most likely due to the minerals in the aggregate). The spectrum of concrete minus the aggregate is shown in sample WTC01-37A(c). The cement shows less ferrous iron absorption. Pervasive in the spectra of the debris is gypsum, with variations in abundance of other components. Dominant spectral features include gypsum (sample GDS 524) presumably from drywall, iron as Fe^{2+} (e.g. sample WTC01-37B) presumably due to aggregate in the concrete, as well as water and hydroxyl absorptions typical in spectra of the dust and debris (*5*).

The map showing the WTC dust and debris plume as detected by AVIRIS, Plate 4, shows locations of materials with spectral shapes similar to the spectra of field samples of dust and debris collected around the lower Manhattan area. These materials/minerals include common building materials. Therefore, the maps may include materials in buildings not associated with the WTC collapse. While the minerals mapped in any one location may or may not be associated with the WTC event, a pattern is seen that appears to show the distribution of materials related to the WTC collapse. Furthermore, the debris map agrees with on scene observations 2 days after the imaging spectroscopy data were acquired. Building materials are similar all over the city (e.g. concrete occurs in many locations). However, the pulverized dust/debris does have some general spectral characteristics that can be used to indicate the extent of the WTC plume.

We mapped for these general spectral shapes over the spectral range 0.5 to 2.4 μm. The ultraviolet part of the spectrum was not included because scattering from smoke significantly affects shorter wavelengths. The maps should not be interpreted as indicating that these specific materials were mapped, only that the spectral shape is similar. However, the red and yellow colors in Plate 4 indicate areas showing more ferrous-like absorptions, and this is apparent in the WTC

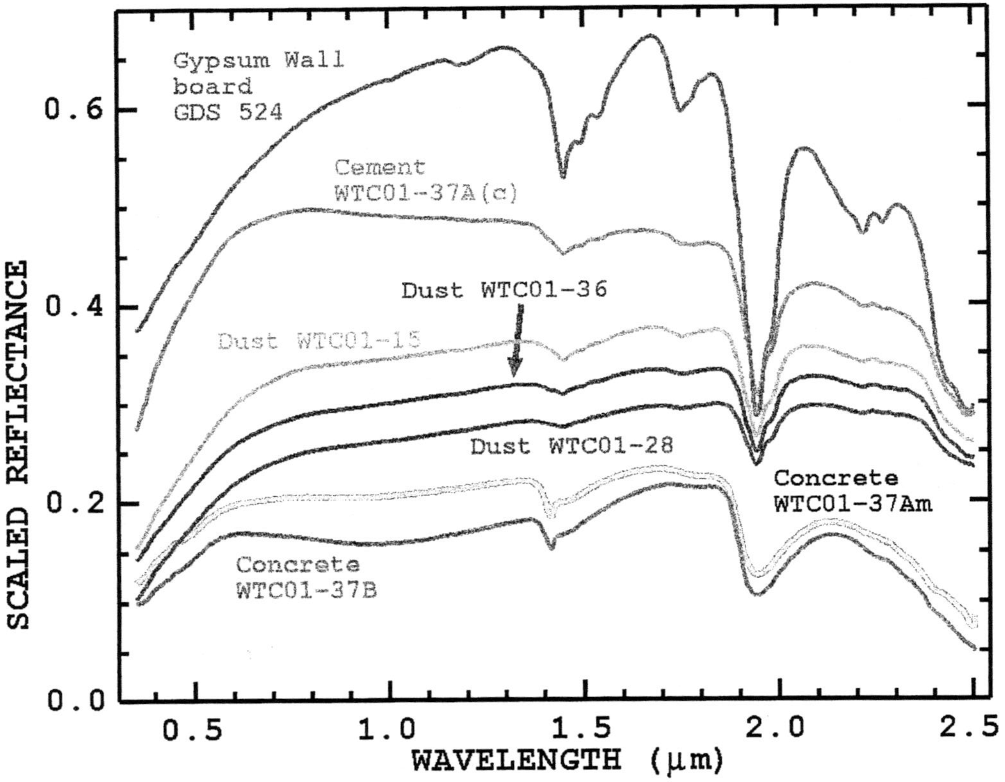

Plate 3. Spectra of the WTC field samples used to map the WTC plume.
(See page 5 of color inserts.)

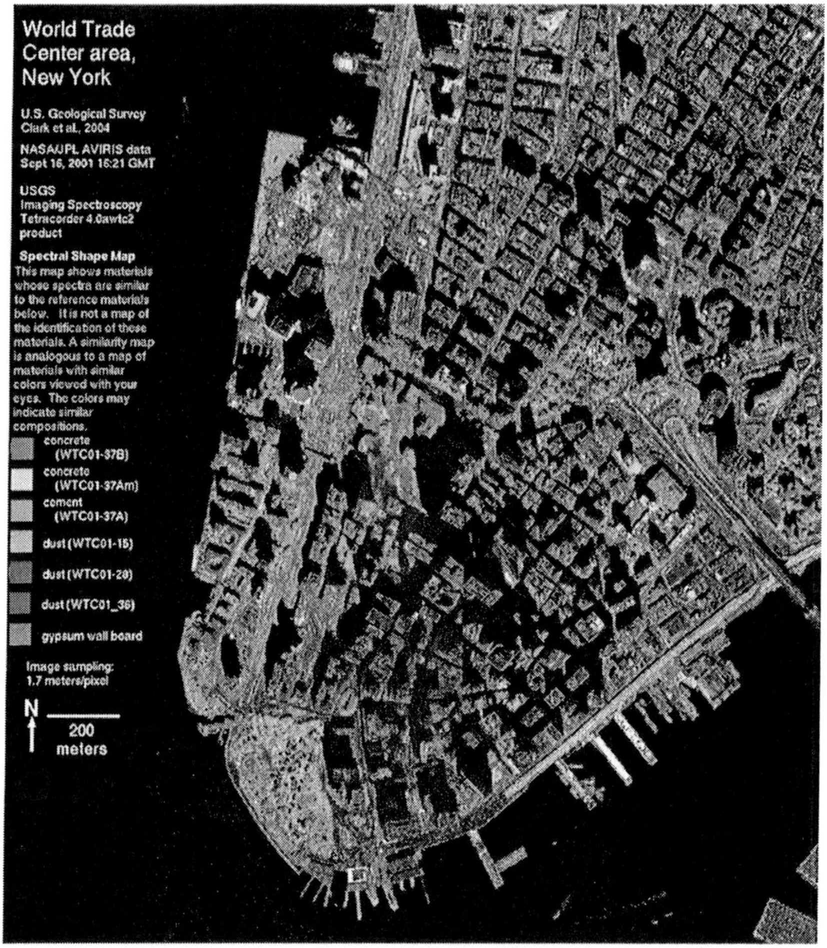

Plate 4. Map showing the WTC dust and debris plume as detected by AVIRIS indicating an asymmetry in the dust and debris distribution, with more iron bearing materials to the south by southeast.
(See page 6 of color inserts.)

core zone where many steel beams are present. Green copper roofs have a similar absorption to ferrous minerals, so they would map as red or yellow in Plate 4. The plume map in Plate 4 indicates an asymmetry in the dust/debris distribution, with more iron bearing materials to the south by southeast. It is difficult to locate the outer boundary of the dust and debris because of the problem of matching common materials throughout the city.

During the era when the WTC was built, industrial asbestos was reportedly used. Therefore, detection of asbestiform mineralogy warrants further investigation. Building materials may also contain trace serpentine and amphibole minerals that are naturally occurring but not asbestiform. To the extent that the top few millimeters contain mineralogy that is representative of the underlying material, spectral mineral maps may be used to assess potential asbestos distribution in the WTC debris. It is also possible that asbestos-bearing debris is buried beneath the surface that AVIRIS was not able to detect.

Spectroscopy identifies mineralogy by detecting absorptions, due to molecular bonds, at a characteristic wavelength and with a diagnostic band shape. The grain size of a mineral affects the intensity and, to some extent, the shape of its spectral absorptions, but spectroscopy may not be sensitive to whether or not a mineral has an asbestiform shape (long needle-like shapes with diameters less than a micrometer), especially at the relatively low spectral resolution of AVIRIS. The spectral properties of asbestiform minerals are described in Chapter 3 (*5, 9, 10*). Spectral features near 2.3 μm were used to map these minerals in the AVIRIS data. Organics have absorption features near this position so spectral shape is key to distinguishing between materials whose absorptions occur at similar positions. Also, organic materials have additional absorptions not seen in spectra of asbestiform minerals (*5, 9, 10*).

The derived AVIRIS mineral map, Plate 5, shows possible weak absorption features indicative of minerals that may occur with asbestiform morphology. The map shows relatively isolated colored pixels indicating areas that may warrant further investigation. The lower detection limits are probably in the few percent range for the AVIRIS data. Detection limits are discussed further in Chapter 3 (*10*). Note that in shadows in Plate 5, there is not enough light for spectroscopic determination of surface mineralogy, so it is unknown if these locations have potential asbestiform mineral-bearing debris.

Laboratory spectroscopy of the WTC samples described in Chapter 3 shows, in all but one case (sample WTC01-08), only very weak, if any, spectral features from chrysotile because the occurrence of chrysotile asbestos is present in trace abundances less than about 1 wt %. Such abundances are not detectable in this AVIRIS data. Thus, it is not surprising that the AVIRIS map, Plate 5, shows little asbestiform mineralogy. The lack of large contiguous clusters of colored pixels at the WTC site, indicates that spectral mineral mapping has NOT detected widespread concentrations of asbestos in the debris above a few

Plate 5. Spectral reflectance map keyed to minerals that can have asbestiform morphology showing only scattered possible occurrences at the surface around the WTC area. (See page 7 of color inserts.)

percent, mostly leaving just the black and white base image (see Plate 4 for examples of contiguous clusters of colored pixels). The AVIRIS map (Plate 5) does show pockets (or small clusters of pixels) of serpentine (closest spectral match is chrysotile). The strengths of the spectral signatures, as described in Chapter 3, indicate levels from a few percent, up to the approximately 20% level observed in sample WTC01-08 (*5*). Again, some or all of these pixels may be non-asbestiform serpentines in building materials.

Mineralogic characterization of field samples (Chapter 3) suggests that only trace levels of chrysotile asbestos were present in some of the samples from the WTC area, with one exception. The exception is a coating from a steel beam (sample WTC01-08) which showed chrysotile levels that could be as high as about 20%. It should be noted that the field sampling sites do not coincide with AVIRIS pixel locations that indicate possible chrysotile or amphibole presence. But many field sample locations whose samples contain trace asbestos occur near locations in the AVIRIS map that indicate higher levels of serpentine minerals.

The AVIRIS dust plume data, field sample data and field observations were combined to show the extend of the WTC dust plume on September 16, 2001 (Plate 6). Locations of asbestiform mineral spectral signatures from Plate 5 are shown as small orange spots in Plate 6. Note the asbestiform mineral signatures may indicate non-asbestiform amphibole and serpentine group minerals that are part of building materials. The locations of AVIRIS derived amphibole and serpentine group minerals is representative, and not exact because viewing angles of the sensor displaces positions of the tops of tall buildings relative to the ground. Therefore, buildings appear to tilt at the edge of the scan lines compared to the center. For more accurate position of any one spot, the data in Plate 5 should be used.

The correlation of mineralogically characterized field samples, described in Chapter 3 and AVIRIS mapped serpentine and amphibole mineralogy appears to show serpentine in a broad east-west trend, in both the AVIRIS and the laboratory results (Plate 6). Due to the lower limit of detection of the AVIRIS instrument for asbestos minerals, the WTC AVIRIS data set is probably in the few weight percent range, as noted above. The detection of asbestos levels below this detection limit and the implications of low levels of asbestos on human health are beyond the scope of this report.

The fact that field sampling showed high levels of chrysotile asbestos in some of the steel beam coatings (up to 20%) indicated the need for dust control during beam removal. However, video from news reports of the cleanup activity showed that many of the steel beams no longer have the insulation coating. Where did the coatings go? It is probably distributed in the dust and debris. Thus the possibility exists that there may be other pockets with high levels of chrysotile, and such pockets may be indicated in the AVIRIS data.

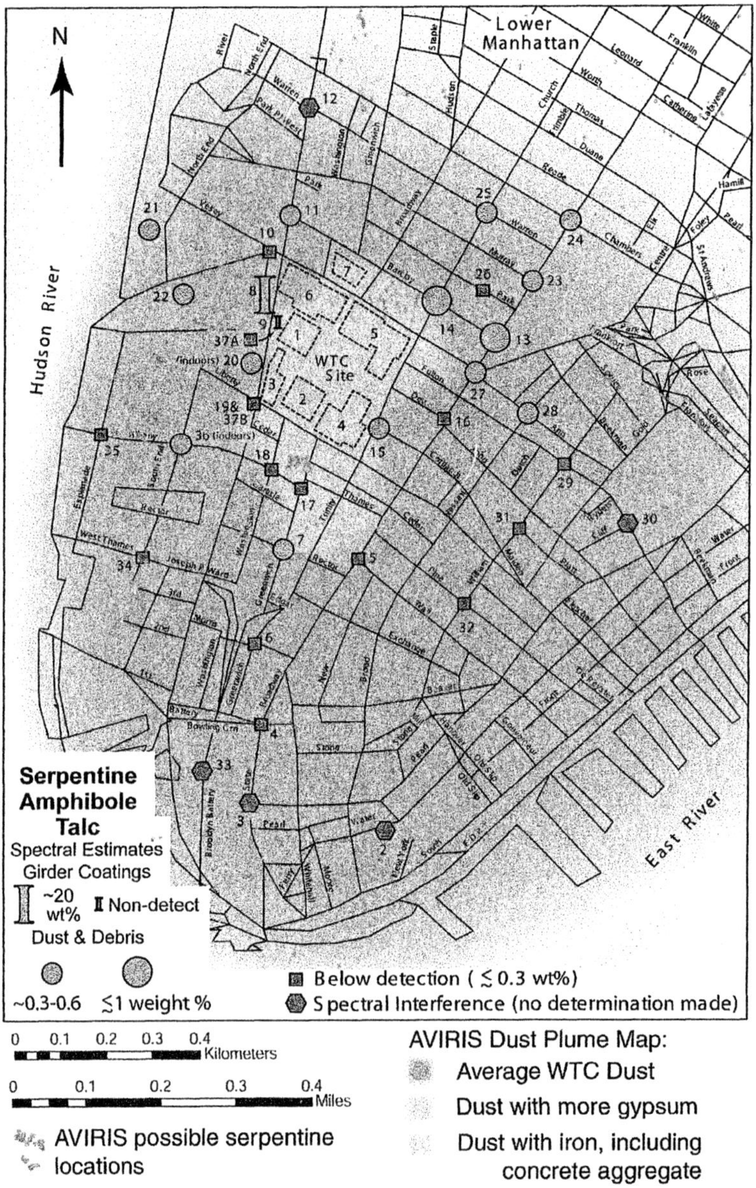

Plate 6. The combination of AVIRIS asbestiform mineral spectral signatures (orange spots), field sample data, and field observations to show the extent of the WTC dust plume on September 16, 2001. (See page 8 of color inserts.)

Conclusions

The AVIRIS data collected on September 16, 2001, revealed a number of thermal hot spots in the region where the WTC buildings collapsed. Analysis of the data indicated temperatures greater than 700° C in these hot spots. Over 3 dozen hot spots of variable size and temperature were present in the core zone of the WTC, covering 3,000 to 4,000 m^2. By September 23, most of the fires that were observable from an aircraft had been eliminated or reduced in intensity.

The AVIRIS mineral maps do not show widespread distribution of chrysotile or amphibole asbestos at the few-percent detection limit of the instrument at the ground surface. AVIRIS mapping keyed to the detection of minerals that may occur in asbestiform habits has identified isolated pixels or pixel clusters (each pixel is approximately 2m x 2m) in the area around the WTC. In these areas, potentially asbestiform minerals might be present in concentrations of a few percent to tens of percent. Some spectral absorption strengths in the AVIRIS data are similar to those observed in spectra of the chrysotile asbestos-bearing beam coating. The absorption features mapped by AVIRIS only indicate the presence of serpentine mineralogy and not if the serpentine has asbestiform. Non-asbestiform serpentine minerals occur naturally in rocks that may have been used in building materials. The AVIRIS maps may indicate areas of higher concentrations of asbestos or simply areas of non-asbestiform mineralogy and field sampling and laboratory analysis would be needed to confirm the presence of any asbestos. The AVIRIS maps show the surface minerals that are part of building materials.

The AVIRIS mineral maps show a few isolated pixels of amphibole minerals, but these pixels are isolated with no clusters like those seen in the chrysotile pixels. The few amphibole pixels that were mapped are at a statistical noise level in the WTC area and similar to the pixel noise level mapped throughout the city. The absorptions mapped by AVIRIS only indicate the presence of amphibole mineralogy, which can occur naturally (non-asbestiform) in rocks that are used in building materials, and field sampling of those pixels would be necessary to confirm the presence of asbestos. The AVIRIS maps of serpentine chrysotile and amphibole mineralogy are consistent with laboratory analyses of the field samples.

Laboratory analyses and the AVIRIS mapping results indicate that the dusts are variable in composition, both on a fine scale within individual samples and on a coarser spatial scale based on direction and distance from the WTC. Replicate mineralogical and chemical analyses of material from the same sample, as described in Chapters 3 and 12, revealed variability that presumably is due to the heterogeneous mixture of different materials comprising the dusts (*5*). The spatial variability is observed at large scales of tens of meters to centimeter and smaller scales. The AVIRIS mapping suggests that materials with higher iron content settled to the south-southeast of the building 2 collapse center. Chrysotile may occur primarily (but not exclusively) in a discontinous pattern radially in west, north, and easterly directions perhaps at distances greater than 0.75 km from ground zero.

Although only trace levels of chrysotile asbestos have been detected in the dust and air-fall samples studied to date, the presence of up to 20 % chrysotile in material coating steel beams in the WTC debris and the potential areas indicated in the AVIRIS mineral maps indicates that asbestos could be found in localized concentrations.

The AVIRIS has shown it can be an effective tool. While its compositional sensitivity is less than some laboratory analyses, AVIRIS provides near complete spatial coverage that is not possible with physical sampling of such a large site. The AVIRIS also provides data on specific locations that could be used to focus field sampling to sites of high interest that might otherwise be missed.

The original maps and images used in this paper are available on the U.S. Geological Survey Spectroscopy Lab website at their full resolution (*13*).

Acknowledgments

This research was funded by the U.S. Geological Survey Mineral Resources program. The National Aeronautics and Space Administration funded the AVIRIS data acquisition.

References

1. Meeker, G. P.; Sutley, S. J.; Brownfield, I. K.; Lowers, H. A.; Bern, A. M.; Swayze, G. A.; Hoefen, T. M.; Plumlee, G. S.; Clark, R. N.; Gent, C. A. Paper presented in ACS Symposium: *Urban Aerosols and their Impact: Lessons Learned from The World Trade Center Tragedy*, New York City, Sept. 10, 2003. *ACS Division of Environmental Chemistry Preprints of Extended Abstracts, paper 133,* **2003,** *43 (2)***,** 1354.
2. Green, R. O.; Conel, J. E.; Carrere, V.; Bruegge, C, J;. Margolis, J. S.; Rast, M.; Hoover, G. in *Proceedings of the Second Airborne Visible/Infrared Imaging Spectrometer (AVIRIS) Workshop*, JPL Publication 90-54, p. 15-22, 1990.
3. Green, R. O.; Roberts, D. A.; Conel, J. A. *Proceedings of the Sixth Annual JPL Airborne Earth Science Workshop*, JPL Publication 96-4, 135, 1996.

4. Clark, R. N.; Swayze, G. A.; Livo, K. E.; Kokaly, R. F.; King, T. V. V.; Dalton, J. B.; Vance, J. S.; Rockwell, B. W.; Hoefen, T.; McDougal, R. R. in *Proceedings of the 10th Airborne Earth Science Workshop*, JPL Publication 02-1, 2002. [http://speclab.cr.usgs.gov/PAPERS.calibrationtutorial].
5. Clark, R. N.; Green, R. O.; Swayze, G. A.; Meeker,G.; Sutley, S.; Hoefen,T. M.; Livo, K. E.; Plumlee, G.; Pavri, B.; Sarture, C.; Wilson, S.; Hageman, P.; Lamothe, P.; Vance, J. S.; Boardman, J.; Brownfield, I.; Gent, C.; Morath, L. C.; Taggart, J.; Theodorakos, P. M.; Adams, M. *U. S. Geological Survey, Open File Report OFR-01-0429*. 2001. [http://pubs.usgs.gov/of/2001/ofr-01-0429].
6. Clark, R.N.; Roush, T.L., *J. Geophys. Res.* **1984**, *89*, 6329-6340.
7. Clark, R.N. *Manual of Remote Sensing*; A.N. Rencz, ed.; John Wiley and Sons: New York, 1999; p 3-58.
8. Clark, R. N.; Swayze, G. A.; Livo, K. E.; Kokaly, R. F.; Sutley, S. J.; Dalton, J. B.; McDougal, R. R.; Gent, C. A. *J. Geophysical Research,* **2003,** *108(E12),* 5131, doi:10.1029/2002JE00184]. [http://speclab.cr.usgs.gov/PAPERS/tetracorder].
9. Clark, R. N.; Swayze, G. A.; Wise, R.; Livo, K. E.; Hoefen, T. M.; Kokaly, R. F.; Sutley, S. J. *USGS Open File Report 03-395*, 2003. [http://pubs.usgs.gov/of/2003/ofr-03-395/ofr-03-395.html].
10. Clark, R. N.; Hoefen, T. M.; Swayze, G. A.; Livo, K. E.; Meeker, G. P.; Sutley, S. J.; Wilson, S.; Brownfield, I. K.; Vance, J. S. *U.S. Geological Survey Open-File Report 03-128*, 2003. [http://pubs.usgs.gov/of/2003/ofr-03-128/ofr-03-128.html].
11. Clark, R.N., *Icarus* **1979**, *40*, 94-103.
12. Green, R. O.; Clark, R. N.; Boardman, J.; Pavri, B.; Sarture C, *Proceedings of the 10th Airborne Airborne Earth Science Workshop*, JPL Publication 02-1, [ftp://popo.jpl.nasa.gov/pub/docs/workshops/ftp://popo.jpl.nasa.gov/pub/docs/workshops/02_docs/02_docs/].
13. "USGS Spectroscopy Lab - World Trade Center USGS environmental assessment" [http://speclab.cr.usgs.gov/wtc] U.S. Geological Survey.

Chapter 5

Materials Characterization of Dusts Generated by the Collapse of the World Trade Center

Gregory P. Meeker, Stephen J. Sutley, Isabelle K. Brownfield, Heather A. Lowers, Amy M. Bern, Gregg A. Swayze, Todd M. Hoefen, Geoffrey S. Plumlee, Roger N. Clark, and Carol A. Gent

Denver Microbeam Laboratory, U.S. Geological Survey, MS 964, Box 25046 Denver Federal Center, Denver, CO 80225

The major inorganic components of the dusts generated from the collapse of the World Trade Center buildings on September 11, 2001 were concrete materials, gypsum, and man-made vitreous fibers. These components were likely derived from lightweight Portland cement concrete floors, gypsum wallboard, and spray-on fireproofing and ceiling tiles, respectively. All of the 36 samples collected by the USGS team had these materials as the three major inorganic components of the dust. Components found at minor and trace levels include chrysotile asbestos, lead, crystalline silica, and particles of iron and zinc oxides. Other heavy metals, such as lead, bismuth, copper, molybdenum, chromium, and nickel, were present at much lower levels occurring in a variety of chemical forms. Several of these materials have health implications based on their chemical composition, morphology, and bioaccessibility.

Introduction

The tragedy of September 11, 2001 produced a dust cloud that was visible from space, and covered much of lower Manhattan in millimeters to centimeters of extremely fine powdered material. This material was inhaled and ingested by thousands of people on the day of the attack and for several days afterward. In addition, thousands of apartments and offices were contaminated by the dust through open windows, airshafts, and openings around doors and windows. The short-term medical effects of this exposure were manifested in what became known as the World Trade Center (WTC) cough, documented as respiratory and other health problems among many of those who were exposed (*1, 2*). The long-term medical effects, if any, of this event may not be known for many years. Several studies have examined various components of the dusts generated by the collapse of the WTC (*3, 4, 5, 6, 7*). The relationship of any medical problems arising from exposure to this material can best be evaluated with knowledge not only of the bulk chemistry, but also of the chemical and mineral phases present in the dust. It is those phases that determine how the dust will react within the body and how bioaccessible the various elemental and molecular components will be when in contact with various bodily fluids.

The WTC towers, built between 1966 and 1973, contained an enormous amount and variety of building materials (*8*). Some of the major components used in the construction are summarized in Table I. Much of the material listed in Table I was pulverized during the collapse of the towers and was turned into the fine dust that is described in this study. The makeup and composition of the dust is extremely complicated because of the enormous variety of materials that made up the WTC buildings. There are, however, three major components that make up the majority of the fine inorganic material in these dusts: concrete, gypsum, and glass fibers. These components probably originated primarily from lightweight concrete floors, wallboard, and spray-on fireproofing and ceiling tiles, respectively. In addition to the major components, the dust contained a large variety of other materials including metal-rich particles, particles of window glass, cellulose from wood and paper products, and asbestos. Optical micrographs of a typical sample are shown in Figure 1.

This chapter will present an overview of the components of the WTC dust and, when possible, suggest potential sources for these components within the buildings that were destroyed. This study primarily addresses the inorganic components of the dust. Organic particles and aerosols generated by the collapse of the WTC are discussed elsewhere in this volume.

Table I. Partial List of Building Materials Used for Construction of the Twin Towers

Material	*Amount*	*Unit*
Masonary	6,000,000	Square ft.
Painted surfaces/drywall	5,000,000	Square ft.
Wire	1,520	Miles
Conduit	400	Miles
Lighting fixtures	200,000	Each
Acoustical ceiling tile	7,000,000	Square ft.
Flooring	7,000,000	Square ft.
Window glass	600,000	Square ft.
Lightweight concrete floors	3,000,000	Cubic ft.
Structural steel	200,000	Tons

NOTE: Also included was spray-on fireproofing for steel, some containing asbestos.

SOURCE: Data summarized from Reference 6. Copyright 2002 New American Library

Analytical Methods

All but one of the samples analyzed in this study were collected by the two-member field team from the U. S. Geological Survey (USGS), Gregg Swayze and Todd Hoefen, as part of the calibration effort for the Airborne Visual and Infrared Imaging Spectrometer (AVIRIS) analysis of the WTC site described in Chapter 4 (*3, 9*). Dust samples were collected on September 17 and 18, 2001, primarily from horizontal surfaces that appeared to be undisturbed. Thirty-six samples were collected from outdoor locations. These samples had been subjected to one rainstorm on September 14th. One sample was collected from an indoor location across the street from the WTC plaza. Two samples were also collected from the coating on steel beams. One additional indoor sample was acquired from inside a 30th floor apartment two blocks southwest of the WTC plaza approximately two weeks after the event. The sample locations are described in Chapter 3. All samples were scooped into double zip-lock bags, identified by number and street location, and returned to the USGS analytical laboratories in Lakewood, Colorado.

Representative sample splits for analyses were obtained by the cone and quarter method and prepared according to each analytical method described

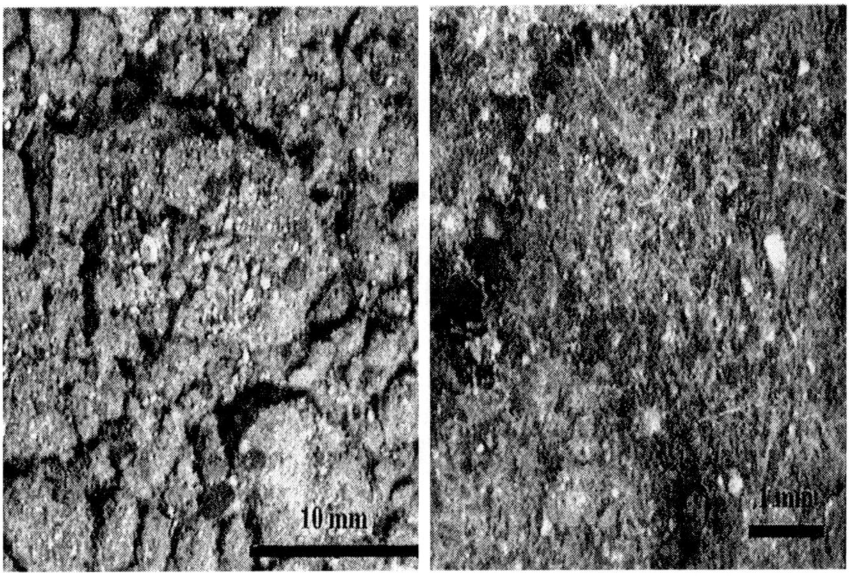

Figure 1. Optical micrographs of a typical dust sample at two magnifications. Fine-grain material is held together in clumps by abundant microscopic glass fibers (left). Glass fibers are visible in the higher magnification image (right).

below. This chapter will discuss results from x-ray diffraction (XRD) and electron and x-ray microanalytical analysis of the samples. Bulk chemistry and leaching studies are addressed in Chapter 12. Results of the airborne visual and infrared measurements, including the distribution of various dust components, are discussed in Chapter 4 (*9*).

X-ray Diffraction

X-ray powder diffractometry is used for the identification and sometimes quantification of crystalline compounds by interpreting their diffraction patterns. Because every crystalline material gives a unique x-ray diffraction pattern, the study of diffraction patterns of unknown phases offers a powerful means of qualitative identification.

The samples were split to obtain about a 3-gram specimen that was representative of the bulk sample. Each specimen was then dry pulverized with a mortar and pestle to an average particle size of about 50-60 micrometers. About 1 gram of the specimen was then packed in an aluminum sample holder and analyzed with an automated diffractometer (Scintag[a] X-1) fitted with a spinning sample holder using copper (Cu) K-alpha radiation. The sample was run at a power setting of 45 kilovolts (kV) and 35 milliamps (mA) at a stepping size of 0.02 degrees 2-theta with a one second counting time from 4 degrees 2-theta to 60 degrees 2-theta.

Scanning Electron Microscopy and Energy Dispersive Spectroscopy

Scanning electron microscopy (SEM) was performed using a JEOL 5800LV microscope operating in high-vacuum mode. Energy dispersive x-ray analysis (EDS) was performed using an Oxford ISIS EDS system equipped with an ultra-thin-window detector. Analytical conditions were: 15 KeV accelerating voltage, 0.5-3 nA beam current (cup), and approximately 30 % x-ray detector dead time. X-ray data reduction was performed using the Oxford ISIS standardless analysis package using the ZAF option.

Most SEM samples were prepared by distributing fine dust of selected samples onto double-stick conductive carbon tape. Splits of samples 19 and 18 were size separated dry using stainless steel sieves in an ultrasonic sieve system. Selected size separates were analyzed.

Quantitative chemical analysis was performed by EDS. The matrix corrections for atomic number, fluorescence and absorption used in these EDS analyses do not account for particle geometry. However, it has been shown that at the operating conditions used such errors are can be low (*10*). Our errors, in relative wt. %, estimated from analysis of 0.5 - 10 μm diameter particles of USGS BIR1-G basalt glass reference material (*11*) are approximately ± 13 % (1σ) for Na_2O, 4 % for MgO and CaO, 3 % for Al_2O_3, 2 % for SiO_2, and 7 % for FeO. Detection limits are 0.2 wt. % for most elements using EDS.

Electron Probe Microanalysis

Polished grain mounts were prepared from representative samples for quantitative analysis of single particles by electron probe microanalysis (EPMA). Quantitative EPMA of polished samples was performed using a five-spectrometer, fully automated, JEOL 8900 scanning electron microprobe. Analytical conditions were: 15 KeV accelerating voltage, 20 nA beam current (cup), point to 20 micrometer beam diameter, and 20 second peak and 10 second background counting time. Calibration was performed using well-characterized

---

[a] The use of product names is for descriptive purposes only and does not imply endorsement by the U.S. Geological Survey.

silicate and oxide standards. Analytical precision based on replicate analysis of standards was better than ± 2 % relative concentration for major and minor elements and equal to counting statistics for trace (< 1 %) elements. Matrix corrections were performed with the JEOL 8900 ZAF software.

Results

Anlysis by X-ray diffraction identified three major components in all dust samples: concrete phases, gypsum, and undifferentiated amorphous materials. These components appear to occur in roughly equal proportions in most samples based on visual observation. The concrete components include quartz, Ca-Si-rich phases, and lesser amounts of rock and mineral particles typical of concrete aggregates. Table II lists many of these components identified in the dust samples.

Table II. Phases and Minerals Detected by X-ray Diffraction in the WTC Dust Samples

Mineral/Phase	*Formula*
crystalline quartz	SiO_2
gypsum	$CaSO_4 \cdot 2(H_2O)$
calcite	$CaCO_3$
anhydrite	$CaSO_4$
amorphous material (includes glass fibers)	
muscovite	$KAl_3Si_3O_{10}(OH)_2$
portlandite	$Ca(OH)_2$
feldspar (various forms)	
amphibole (various forms)	
clay (various forms)	
lizardite (non-asbestiform serpentine)	$Mg_3Si_2O_5(OH)_4$
chrysotile	$Mg_3Si_2O_5(OH)_4$
dolomite	$CaMg(CO_3)_2$
calcium sulfate hydrate	$CaSO_4 \cdot nH_2O$
bassanite	$CaSO_4(0.5H_2O)$
calcium sulfate (various forms)	
thaumasite	$Ca_3Si(OH)_6[CO_3][SO_4] \cdot 12H_2O$
calcium silicate (various forms)	
larnite	Ca_2SiO_4

Gypsum, bassanite, and anhydrite are components of wallboard. These minerals are abundant in the WTC dust. Minor phases compatible with formation in gypsum deposits were also identified. Amorphous materials, specifically identified by optical microscopy and SEM, included glass and cellulose, generally in the form of wood, paper, and organic fibers.

Glass Fibers and Fragments

Glass fragments and man-made vitreous fibers (MMVF) were abundant in all samples analyzed. The electron micrographs shown in Figure 2 demonstrate the variety of glass textures and particle shapes. Glass fibers range in diameter from > 50 µm to < 1 µm with lengths up to several hundred micrometers. Surprisingly, the majority (> 85 %) of the disseminated microscopic glass fibers are very similar in composition. Table III lists representative EPMA analyses of MMVF found in our samples. The best compositional match for the majority of the WTC glass fibers is slag wool (SW), a by-product of pig iron production (*12*). In 1990, 70 percent of the slag wool produced in the United States was used in the manufacture of ceiling tiles (*12*). The glass fibers found in the steel beam coatings are also very similar in composition suggesting, that the beam coating products also contained slag wool as opposed to a more iron-rich glass fiber, such as rock wool (RW).

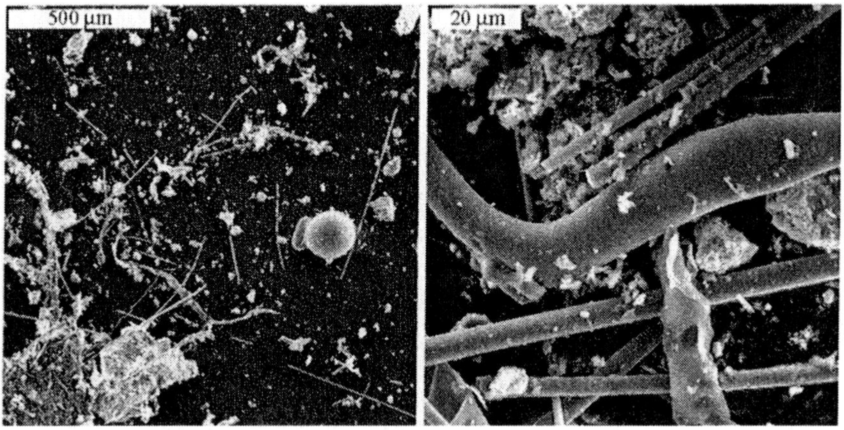

Figure 2. Typical examples of glass fibers and a glass sphere (left) of slag wool composition.

Table III. Representative Electron Probe Microanalysis Results of Typical Glass Fibers and Fragments in WTC Dust Samples

Species	*SW*	*SW*	*SWS*	*RW*	*HSF*	*SLG*	*SLG*	*SLF*
F	bdl	bdl	0.5	bdl	bdl	bdl	bdl	bdl
Na_2O	0.25	1.55	0.22	2.10	10.2	14.3	14.3	16
MgO	12.2	9.07	12.5	10.6	3.36	4.06	2.04	4
Al_2O_3	11.5	12.4	11.6	12.1	3.67	0.11	0.08	3
SiO_2	40.0	41.8	40.1	44.1	73.0	71.2	69.9	69
K_2O	0.62	1.44	0.59	1.15	0.75	0.01	0.01	1
CaO	32.5	32.2	33.0	23.5	7.86	8.21	10.8	7
TiO_2	0.62	0.51	0.58	0.83	0.05	0.00	0.01	bdl
MnO	0.35	0.35	0.30	0.09	0.02	0.04	0.00	bdl
FeO	0.77	0.51	0.57	5.28	0.57	0.12	0.37	bdl
Total	99.81	99.83	99.96	99.75	99.48	98.1	97.5	100
% of dust	>85		0.5–1	<<1	<<1	1–15		<<1

NOTE: Units are percent by weight; bdl= below detection limit. SW is slag wool (the most abundant fiber). SWS is slag wool glass spheres. RW is fibrous rock wool. HSF is high soda fiber. SLG is soda-lime glass (probably window glass). SLF is fibrous soda-lime glass.

NOTE: SLF values were determined by SEM/EDS analysis. The last row gives the percent of each glass type in the fine dust.

Pieces of yellow thermal insulation were widely scattered around the disaster site. The composition of this material, plotted in Figure 3, is similar to soda-lime glass (SLG) (*13*). Very few fibers of this composition were found in the fine microscopic portion of the dust.

Glass shards, fragments, and spheres are present in the dust samples as centimeter-size pieces to micrometer-size particles. The microscopic glass shards and fragments are much less abundant than the ubiquitous slag wool fibers in the fine dust. The composition of the particles are shown in Figure 3. Most of the glass fragments fall within the ranges for soda lime glass, a common type used as window glass (*13*). Glass spheres are generally less than 500 μm in diameter. At least 90 % of these glass spheres are of slag wool composition. Such spheres are byproducts of slag wool production (*12*) and did not form during the disaster. A few glass spheres have compositions outside the slag wool field, as shown in Figure 3.

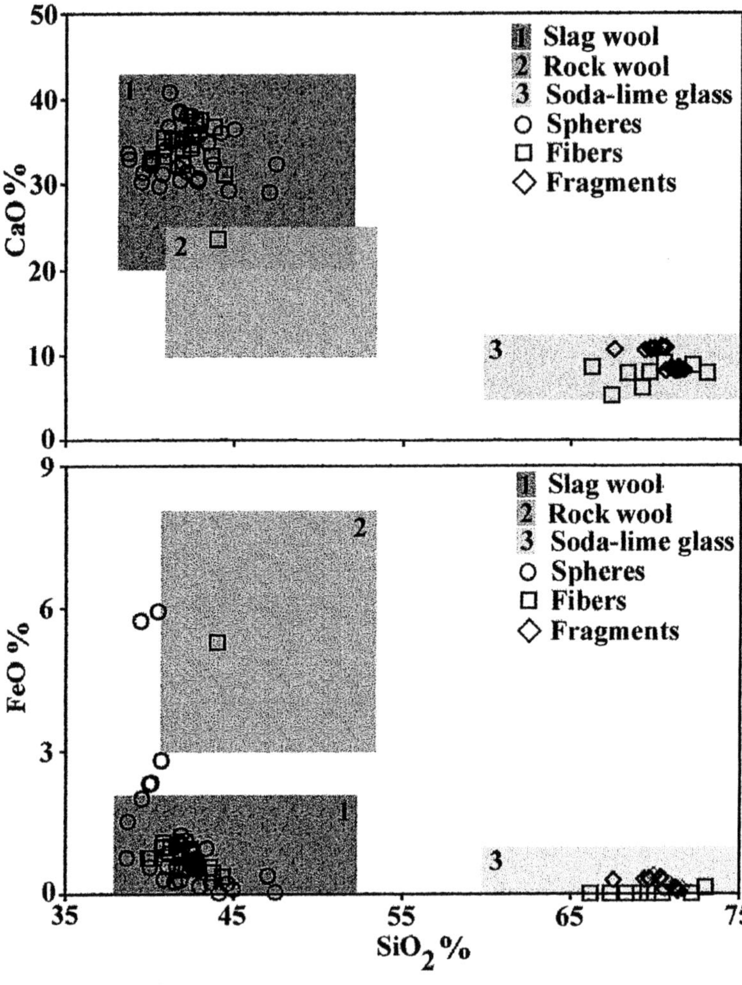

Figure 3. Compositions of man-made vitreous fibers (*MMVF) and glass fragments found in WTC dust samples. Most of the fibers and spheres fall into the slag wool field. Glass fragments show compositions compatible with soda-lime glass.*

Concrete

The lightweight concrete floors used in the twin towers must have disintegrated to a large degree into fine and very fine dust (*14*). Lightweight concrete is composed of aggregate, sand, and Portland cement (*15*). The aggregate material in our WTC concrete sample appears to be expanded shale. The sand is primarily quartz, but can contain feldspar, iron and titanium oxides micas, and other rock-forming minerals. Portland cement hydrates to form a large variety of Ca-rich compounds including calcium silicate hydrate, calcium aluminum hydrate, calcium aluminum iron hydrate, and portlandite (*15*). Minor gypsum is added to control the set of the concrete. The XRD results in Table II identified many of these components in the dust samples.

Figure 4 shows two examples of WTC concrete. Both examples are composed of multiple phases as indicated by the gray levels in the backscattered electron images. Larger grains of quartz sand are also shown in the images. Our analyses found two health-related components in the dust from concrete, portlandite ($Ca(OH)_2$) and crystalline silica. Portlandite is a concrete component that develops at varying concentrations during hydration of Portland cement, generally in the range of a few weight percent (*15*). Portlandite is highly caustic but will react rapidly with CO_2 in water to produce calcium carbonate via the reaction $Ca(OH)_2 + CO_2 \rightarrow CaCO_3 + H_2O$. Therefore, fine-grained portlandite

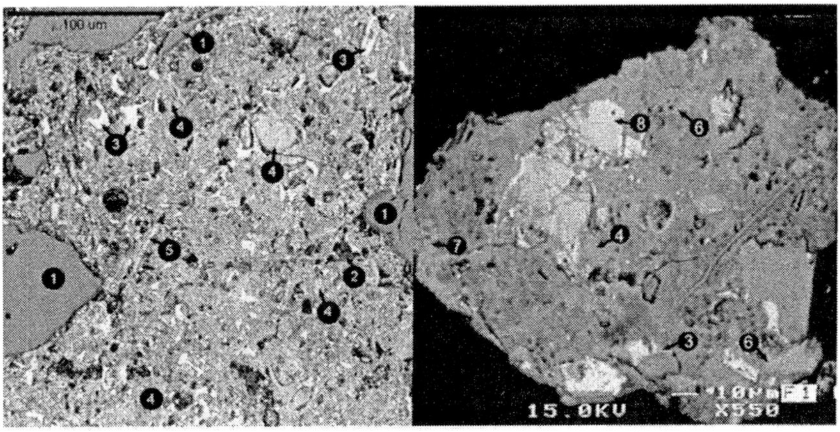

Figure 4. Backscattered electron images of polished specimens of WTC concrete showing a portion of a centimeter-sized piece (left), and a particle approximately 160 μm in diameter (right). Numbers show phases identified by Electron Probe Microanalysis : 1=quartz, 2=$Mg(OH)_2$, 3=Fe-rich Ca aluminate, 4=hydrated Ca silicate, 5=Ca-Al-Si glass, 6=portlandite, 7=calcite, and 8=tricalcium silicate.

in the outdoor dusts may be converted quickly to calcite by reactions with with atmospheric carbon dioxide (CO_2) and moisture. Sand-size crystalline silica particles, generally α-quartz, are a normal component of concrete. The amount of respirable quartz (< 3 μm diameter) should occur at much lower levels, on a weight percent basis, than the x-ray diffraction analysis would suggest. We have analyzed a portion of the < 20 μm diameter fraction of sample 19 using acid dissolution to remove concrete phases and gypsum. The sample was then pipetted onto a filter and analyzed in the SEM to determine the approximate amount of respirable quartz. The results suggest that approximately 0.8 weight percent of the < 20 μm dust is composed of quartz particles < 3 μm in diameter.

Gypsum and Anhydrite

Gypsum ($CaSO_4 \bullet 2H_2O$) is the major component of wallboard (drywall); minor and trace components can include dolomite, anhydrite, bassanite, $CaSO_4 \bullet 0.15H_2O$, and quartz. These minerals are abundant in the dust samples. Gypsum and possibly anhydrite and bassanite often occur in acicular (needle-like) particles or crystals. This morphological type is common in the WTC dust. The Ca-sulfate particles also occur in more equant and rounded morphologies. These sulfates are extremely soft and friable minerals and therefore would be expected to disintegrate into extremely fine particles. Strontium occurs at elevated levels in the dust (see Chapter 12); and appears to occur predominately as celestine ($SrSO_4$). Celestine is a likely trace component of wallboard, because it is a natural accessory phase in gypsum deposits (*16*).

Asbestos and Vermiculite

One of the primary concerns of health specialists following the WTC collapse was the potential for high levels of asbestos in the dust. Indeed, the primary reason the U.S. Geological Survey was asked to initiate a study of the dust was the Agency's expertise in the identification and mapping, by remote sensing, of asbestos and asbestos-containing materials (*3*). Scanning Electron Microscopy and Energy Dispersive Spectroscopy (SEM/EDS) were used for rapid survey analysis of representative dust samples. Within one day after receiving the samples, we found chrysotile asbestos was present in all samples except one at levels of approximately 0.1 to 1.0 volume precent (*3*). These results were confirmed by XRD. The one exception was sample 8, one of the steel beam fireproof-coating samples that contained 10 to 20 weight percent chrysotile asbestos. Figure 5 is a typical chrysotile bundle found in sample 8. Our studies did not find amphibole asbestos. However, Chatfield and Kominsky (*17*), using heavy liquid separation, found a small amount of amphibole asbestos, including amosite and richterite. The latter type appeared to be compatitable with amphibole types found to occur in the Vermiculite Mountain vermiculite deposit once mined near Libby, Montana (*18*).

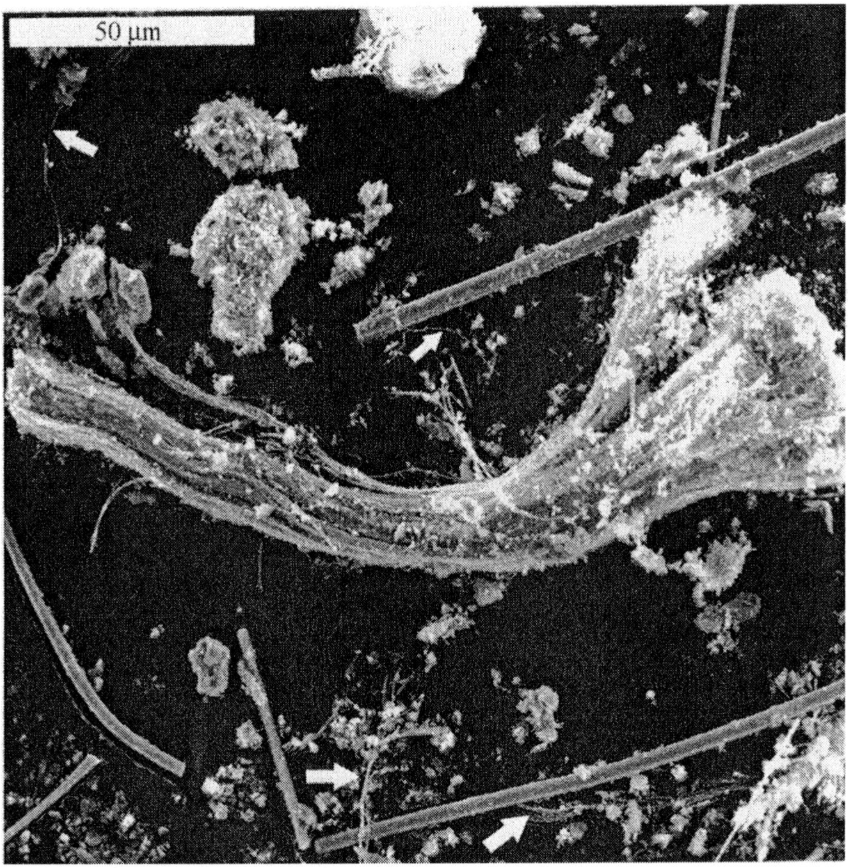

Figure 5. Electron micrograph of a chrysotile bundle from sample 8. Similar bundles were found in most samples analyzed, although, in much less abundance than sample 8. All samples analyzed by SEM contained small strands of chrysotile such as those shown by arrows.

The Libby vermiculite deposit supplied as much as 70 % of the world's vermiculite between 1920 and 1990, when the mine closed. Reports indicate that Libby vermiculite was used in construction of the WTC (*19*). Vermiculitefrom the Libby deposit has been shown to be contaminated by fibrous and asbestiform amphiboles (*18*) and recent studies link these minerals to significantly elevated incidences of lung abnormalities in those exposed (*20*). It is therefore possible that some of the vermiculite used in the construction of the WTC came from the Libby deposit, and might contain amphibole asbestos.

We have analyzed hand-picked vermiculite grains from samples 2, 15, 19, and 36 by EPMA and compared the results to known compositions from several vermiculite deposits and districts. As shown in Figure 6, vermiculite from the mine near Libby appears to have been used in the WTC; however, several other sources are also indicated, including Louisa, Virginia, Enoree District, South Carolina, and Palabora, South Africa. None of these other sources have been shown to produce a vermiculite product containing significant amounts of fibrous amphibole or amphibole asbestos.

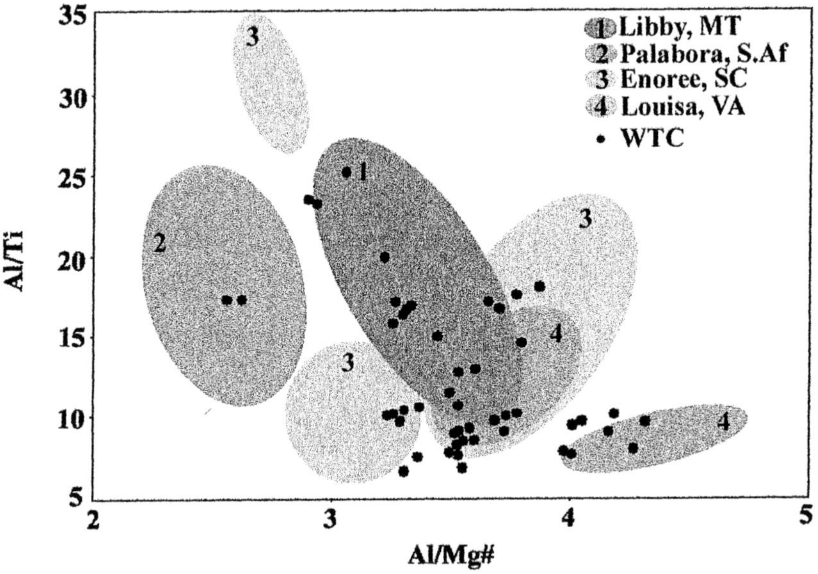

Figure 6. Electron Probe Microanalysis results from hand-picked vermiculite grains from several WTC samples. Cation ratios are plotted against fields defined from USGS analyses of commercial vermiculite samples.

Metals, Metalloids, and Metal-rich Particles

Many metals and metalloids are recognized health hazards, and the degree to which they are bioaccessible and hazardous depends on the reactivity of the phases in which they reside (*21*). Bulk values for metals and metalloids for representative WTC dust samples are given in Chapter 12. These results show that iron (Fe), zinc (Zn), strontium (Sr), molybdenum (Mo), and others are at elevated levels when compared to typical Eastern U.S. soils.

In this study we used backscattered electron imaging to identify particles containing the higher atomic number elements followed by EDS to identify the elements or compounds present in the particles. Knowledge of the compounds in which the metals reside can aid in understanding of the bioaccessability of a particular element. Hazardous elements of low atomic number, such as Be, are not easily detected by this method. It would be impossible to identify all of the types of particles present in the dust. Instead, we have tried to point out the types of particles that appear most abundant and also mention some of the more unusual particle types and compositions that we encountered.

The most abundant particles of higher atomic number contain high concentrations of Fe or Zn, usually in the form of oxides or possibly hydroxides. These particles are likely derived from sources such as rust on steel beams and rebar (Fe-rich) and galvanized coatings on air-handling shafts (Zn-rich). Typical Fe- and Zn-rich particles are shown in Figure 7. A few Fe metal particles were observed, but these particles are rare and may be abraded or smear metal from steel beams, rebar, or other Fe construction components. Spheres of Fe-rich phases, primairily oxides were also detected. These spheres were likely produced during welding or cutting of steel during construction or rescue operations.

Metallic lead (Pb) particles were detected in several samples. These lead particles were often found as groups of small (< 1μm) particles, or as single particles adhering to or embedded in other phases as shown in Figure 7.

Particle Size Analysis

Samples 18 and 19 were coned and quartered to obtain approximately 37 grams of sample for sieve analysis. The samples were ultrasonically dry-sieved in 3 inch diameter stainless steel sieves. The size fractions obtained were < 20 μm, 20 - 75 μm, 75 - 150 μm, 150 – 425 μm and, on sample 19 only, 425 to 850 μm. The ultrasonic sieve was operated at 20-minute intervals at a variety of amplitudes, ranging from 0 to 1.5 mm (vibration height is twice this number). Table IV summarizes the weights obtained for each size fraction. It was not

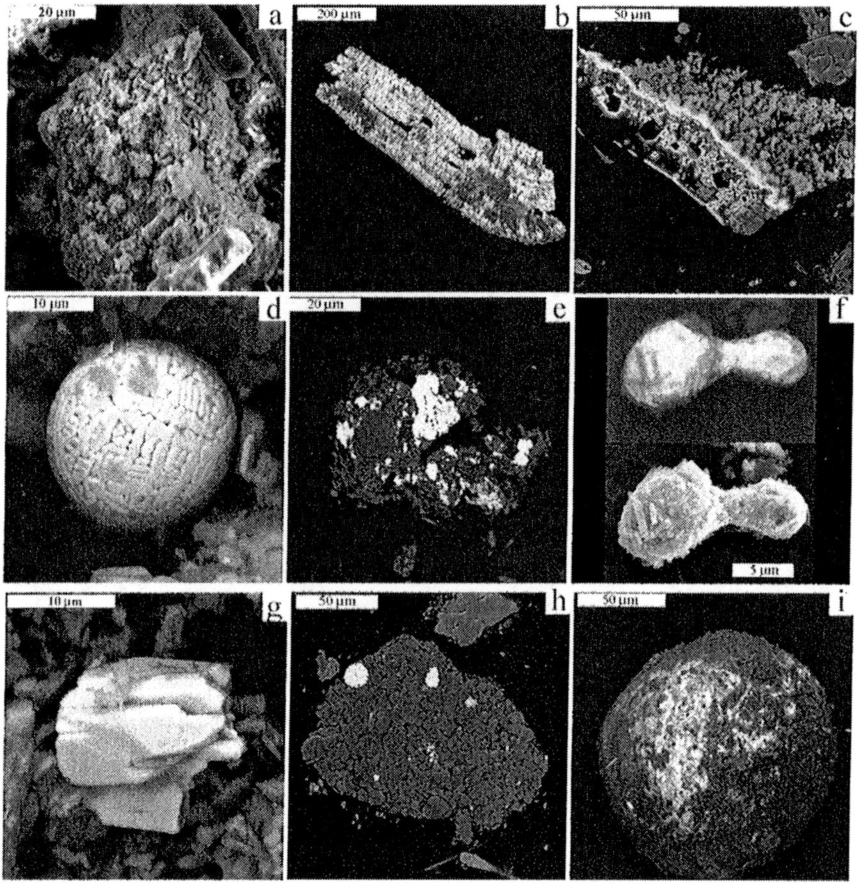

Figure 7. Examples of particles containing heavier elements. a) Fe oxide or hydroxide particle, possibly rust. b) Fe metal particle with silicate minerals coating portions of the surface (darker gray). c) Polished cross section of a Zn metal and Zn oxide particle, brighter areas are more Zn rich. d) Fe oxide sphere possibly produced by welding or cutting of steel beams. e) Carbonate-rich particle containing Pb (bright areas). f) Bi-rich particle coated with gypsum crystals shown in backscattered electron image (top) and secondary electron image (bottom). g) Celestine ($SrSO_4$) grain, likely from wall board. h) Carbonate-rich particle with FeS_2 framboids produced by bacterial action in a natural marine environment. This material is likely a fragment of decorative stone. i) Sphere composed primarily of Al, O and F. The bright areas on the surface are Pb-rich

Table IV. Particle Size Analysis Sample Weights Obtained for Each Size Fraction of WTC Samples 18 and 19

Size	*Mesh*	*Weight (18)*	*Fraction (18)*	*Weight (19)*	*Fraction (19)*
>850	20			22.09	0.61
850 – 425	40	2.55	0.15	3.57	0.10
425 – 150	60	3.09	0.19	4.68	0.13
150 – 75	100	3.75	0.23	2.23	0.06
75 – 38	200	3.76	0.23	1.74	0.05
38 – 20	400	3.08	0.19	1.34	0.04
<20	625	>0.40	0.02	>0.28	0.01
Total		16.63	1.00	35.93	1.00

NOTE: Units for sizes are in μm. Units for weights are in grams.

possible to remove all of the fine particles from the larger size fractions without rinsing. Rinsing was not done in order to preserve the sieved samples for further analysis as described in Chapter 12. Therefore, the weight shown for the < 20 μm fractions are minimum weights.

X-ray diffraction and SEM/EDS were used to evaluate the phase distribution as a function of particle size. X-ray diffraction was used to analyze the particle size fractions less than 150 μm in diameter. These analyses were performed without crushing. The results of the x-ray diffraction analysis showed no significant segregation of particle types by size. Each size range of both samples contained greater than 25 wt % gypsum and calcite, 5 to 25 wt % anhydrite, and less than 5 wt % quartz, bassanite, and portlandite.

The other method used to evaluate possible differences in the phase distribution as a function of particle size was SEM with elemental mapping by EDS. This method is more qualitative than the XRD analysis. Using the EDS x-ray images, the area percent for each element was obtained using image processing software. As above, no significant differences were observed between samples and size fractions.

Conclusions

The dust generated by the collapse of the WTC is composed primarily of disintegrated lightweight concrete from floors, gypsum from wall board and amorphous materials, including pulverized window glass and major amounts of slag wool fibers from fireproofing and probably ceiling tiles. All of the 36 samples collected by the USGS team had these materials as the three major inorganic components of the dust. The dust in all samples also contained chrysotile asbestos in amounts ranging from approximately 0.1 to 1.0 weight percent as determined by XRD. The primary heavy metals in all samples were Fe oxide and hydroxide phases and Zn oxides. These metals were likely derived from rust on steel construction components and galvanized metals used for building ventilation, respectively. Other heavy metals, such as lead, bismuth, copper, molybdenum, chromium, and nickel, were present at much lower levels occurring in a variety of chemical forms far too numerous to completely identify. Size separated fractions of the dust showed little if any chemical or mineralogical differences between the size fractions analyzed.

Acknowledgments

This paper was improved by discussions, comments and suggestions from Bradley Van Gosen, US Geological Survey, D.K. Hurcomb, U.S. Bureau of Reclamation, and an anonymous reviewer. We would like to thank Roger Morse for providing reference samples of WTC concrete. This work was funded by the U.S. Geological Survey Minerals Program.

References

1. Stephen, H. G.; Haykal-Coates, N.; Highfill, J. W.; Ledbetter, A. D.; Chen, L. C.; Cohen, M. D.; Harkema, J. R.; Wagner, J. G.; Costa, D. L. *Environ. Health Perspect.* **2003,** *111*, 981-991.
2. Prezant, D. J.; Weiden, M.; Banauch, G. I.; McGuinness, G.; Rom, W. N.; Aldrich, T. K.; Kelly, K. J. *N. Engl. J. Med.* **2002,** *347*, 806-815.
3. Clark, R. N.; Green, R. O.; Swayze, G. A.; Meeker, G.; Sutley, S.; Hoefen, T. M.; Livo, K. E.; Plumlee, G.; Pavri, B.; Sarture, C.; Wilson, S.; Hageman, P.; Lamothe, P.; Vance, J. S.; Boardman, J.; Brownfield, I.; Gent, C.; Morath, L. C.; Taggart, J.; Theodorakos, P. M.; Adams, M. *U.S. Geol. Survey OFR-01-0429*, 2001.
4. Lioy, P. J.; Weisel, C. P.; Millette, J. R.; Eisenreich, S.; Vallero, D.; Offenberg, J.; Buckley, B.; Turpin, B.; Zhong, M.; Cohen, M. D.; Prophete,

C.; Yang, I.; Stiles, R.; Chee, G.; Johnson, W.; Porcja, R.; Alimokhtari, S.; Hale, R. C.; Weschler, C.; Chen, L. C. *Environ. Health Prospect.* **2002,** *110,* 703-714.

5. Millette, J. R.; Boltin, R.; Few, P.; Turner, W. *Microscope* **2002,** *50,* 29-35.
6. McGee, J.K.; Chen L.C.; Cohen, M.D.; Chee G.R.; Prophete, C.M.; Haykal-Coates, N. *Environ. Health Perspect.* **2003,** *111,* 972-980.
7. Lioy P.J.; Weisel, C.P.; Millette, J.R.; Eisenreich, S.; Vallero, D.; Offenberg, J. *Environ. Health Perspect.* **2003,** *110,* 703-714.
8. Gillespie, A.K. *Twin Towers-The Life of New York City's World Trade Center*; New American Library: New York, 2002.
9. Clark, R.N.; Swayze, G.; Hofen, T.; Livo, E.; Sutley, S.; Meeker, G.; Plumlee, G.; Brownfield, I.; Lamothe, P.J.; Gent, C.; Morath, L.; Taggart, J.; Theodorakos, P.; Adams, M.; Green, R.; Pavri, B.; Sarture, C.; Vance, S.;' Boardman, Paper presented in ACS Symposium: *Urban Aerosols and their Impact: Lessons Learned from The World Trade Center Tragedy*, New York City, Sept. 10, 2003. *ACS Division of Environmental Chemistry Preprints of Extended Abstracts, paper 110,* **2003,** *43 (2)*, 1324.
10. Small, J.; Armstrong, J. T. In *Proceedings of Microscopy and Microanalysis*; Bailey, G. W., Ed.; Microscopy and Microanalysis, Volume 6 Supplement 2; Springer: New York, 2000; pp. 924-925.
11. Meeker, G. P.; Taggart, J. E.; Wilson, S. A. In *Proceedings of Microscopy and Microanalysis;* Bailey, G. W., Ed.; Microscopy and Microanalysis, Volume 4 Supplement 2; Springer: New York, 2000, pp. 240-241.
12. *Nomenclature of Man-Made Vitreous Fibers*; Eastes, W., Ed.; TIMA, Inc.: 1991.
13. *Hand Book of Glass Manufacture Volume 1*; Tooley, Fay V., Ed.; Ogden Publishing: New York, 3rd printing, 1961.
14. *World Trade Center Building Performance Study-Data Collection, Preliminary Observations, and Recommendations;* McAllister, T., Editor; FEMA 403; FEMA: New York, 2002.
15. Chandra, S.; Berntsson, L. *Lightweight Aggregate Concrete-Science, Technology, and Applications*; Noyes Publications: Norwich, NY, 2003.
16. Deer, W. A.; Howie, R. A.; Zussman, J. *Rock forming minerals; Volume 2 Chain Silicates;* Longmans, Green and Co. Ltd: London, 1963; 379 p.
17. Chatfield, E. J.; Kominsky, J. R. *Characterization of Particulate Found in Apartments after Destruction of the World Trade Center*; Chatfield Technical Consulting Limited, Mississauga, Ontario, Canada; 2002.
18. Meeker, G. P.; Bern, A. M.; Brownfield, I. K.; Lowers, H. A.; Sutley, S. J.; Hoefen, T. M.; Vance, J. S. *Am. Min.* **2003,** *88,* 1955-1969.
19. Schneider, A.; McCumber, D. *An Air That Kills;* G. P. Putman's Sons: New York, 2004.

20. Peipins, L. A.; Lewin, M.; Campolucci, S.; Lybarger, J. A.; Miller, A.; Middleton, D.; Weis, C.; Spence, M.; Black, B.; Kapil, V. *Environ. Health Perspect.* **2003,** *111*, 1753-1759.
21. Plumlee, G. S.; Ziegler, T. L. *The Medical Geochemistry of Dusts, Soils, and Other Earth Materials;* In *Treatise on Geochemistry*, Volume 9, Chapter 7; Sherwood Lollar, B., Ed.; Elsevier: Amsterdam, 2003.

Chapter 6

Persistent Organic Pollutants in Dusts That Settled at Indoor and Outdoor Locations in Lower Manhattan after September 11, 2001

John H. Offenberg[1,7], Steven J. Eisenreich[1,2], Cari L. Gigliotti[3], Lung Chi Chen[4], Mitch D. Cohen[4], Glenn R. Chee[4], Colette M. Prophete[4], Judy Q. Xiong[4], Chunli Quan[4], Xiaopeng Lou[4], Mianhua Zhong[4], John Gorczynski[4], Lih-Ming Yiin[4], Vito Illacqua[4], Clifford P. Weisel[5,6], and Paul J. Lioy[5,6]

[1]Department of Environmental Science, Rutgers, The State University of New Jersey, New Brunswick, NJ 08901
[2]Institute for Environment and Sustainability, Joint Research Centre, I–21020 Ispra, Italy
[3]Brookdale College, Lincroft, NJ 07738
[4]Nelson Institute of Environmental Medicine, New York University School of Medicine, Tuxedo, NY 10987
[5]Environmental and Occupational Health Sciences Institute, Robert Wood Johnson Medical School–UMDNJ and Rutgers, The State University of New Jersey, Piscataway, NJ 08854
[6]Department of Environmental and Occupational Medicine, Robert Wood Johnson Medical School–UMDNJ and Rutgers, The State University of New Jersey, Piscataway, NJ 08854
[7]Current address: U.S. Environmental Protection Agency, National Exposure Research Laboratory, Human Exposure Atmospheric Sciences Division, Research Triangle Park, NC 27711

During the initial days that followed the explosion and collapse of the World Trade Center (WTC) on September 11, 2001, fourteen bulk samples of settled dusts were collected at locations surrounding the epicenter of the disaster, and analyzed for persistent organic pollutants, including polycyclic aromatic hydrocarbons (PAHs), polychlorinated biphenyls (PCBs), and select organo-chlorine pesticides. The PCBs comprised less than 0.001% by mass in three outdoor samples analyzed, indicating that PCBs were of limited significance in the total settled dust across lower Manhattan. Likewise, organo-chlorine pesticides, were found at low concentrations in the bulk samples. Conversely, the PAHs comprised up to nearly 0.04% by mass of the settled outdoor dust in the six samples. Further size segregation indicated that the PAHs were found in higher concentrations on relatively large particles (10-53 μm). Significant concentrations were also found on fine particles (<2.5 μm), often accounting for ~0.005 % by mass. Twelve bulk samples of the settled dust were also collected at indoor locations surrounding the epicenter of the disaster. Concentrations of PCBs comprised less than one ppm by mass in the two indoor dust samples. The organo-chlorine pesticides were found at even lower concentrations in the indoor samples. The PAHs comprised up to 0.04% by mass of the indoor dust in the eleven WTC impacted indoor samples. Comparison of PAH concentration patterns shows that the dusts that settled indoors are chemically similar to dusts found at outdoor locations. Analysis of one sample of indoor dusts collected from a vacuum cleaner of a rehabilitated home shows markedly lower PAH concentrations (< 0.0005 mass %), as well as differing relative contributions for individual compounds. These PAH analyses may be used in identifying dusts of WTC origin at indoor locations, along with ascertaining further needs for cleaning.

Introduction

The attack on the World Trade Center (WTC) on September 11, 2001 resulted in an intense fire and the subsequent complete collapse of both structures and damage to adjacent buildings. A consequence of the burning and pulverization of buildings was the development of a large plume of dust and smoke that released both particles and gases into the atmosphere. The initial plume was dispersed in all directions from ground zero to many outdoor and indoor locations downwind. For the first 12 to 18 hours after the collapse, winds transported the plume to the east and then to the southeast toward Brooklyn (*1*).

To begin the process of assessing the possible impacts of the dust and smoke in lower Manhattan during the first few days after the disaster, samples of particles that initially settled in downtown New York City were taken from thirteen locations around the WTC site. These outdoor samples were collected during the first few days after the disaster in order to to determine the chemical and physical characteristics of the material that was present in the initial plume and to determine the absence or presence of contaminants that could affect acute or long-term human and ecosystem health. The compounds and materials present in the plume were expected to be similar to those found in building fires or in collapsed buildings. The primary differences were the intensity of the fire (as compared with normal building fires), the occurrence of a building collapse in conjunction with these fires, and the extremely large mass of material reduced to dust and smoke (*1, 2*). A summary of the potential hazards (*3*) as well as a list of the initial steps taken to assess human health exposure during and after the disaster are outlined elsewhere (*1, 4*). Initial measurements focused on the general composition of the dust and smoke, with a primary concern being asbestos (*3*). The approach employed for the initial characterizations focused upon analyzing a few of the dust samples utilizing a wide battery of tests and methodologies. Subsequent chemical analysis focused upon the compounds of greatest concern (*1*). Additional samples were collected at later dates in order to assess the dusts that had settled at indoor locations. Twelve bulk samples of the settled dust were collected at indoor locations surrounding the epicenter of the disaster, including one sample from a residence that had been cleansed and was once again occupied. Additionally, one sample was collected from just outside a fifth story window on the sill. The concentrations of organic compounds measured in these analyses are presented in this chapter.

Experimental Methods

Sample Collection

Details of the sample collection procedures can be found elsewhere (*1, 5-7*). All of the outdoor samples and all but two of the indoor dust samples were collected by investigators from New York University (NYU) and The Environmental and Occupational Health Sciences Institute (EOHSI) as part of an effort organized and completed by EOHSI and NYU. All samples obtained by NYU or EOHSI personnel, were collected using the protocols established for surface soil collection in studies of the dispersal of chromium laden hazardous waste in Jersey City, New Jersey (*8*), and the National Human Exposure

Assessment Survey (*9*). After collection, all samples were stored at 4° C prior to delivery of sub-fractions to the individual collaborators' laboratories for analysis.

Sample Characterization

Analyses were designed to provide a general, broad characterization of the content of the samples using a combination of techniques. This work provided an opportunity to classify the general morphology and focus the chemical analyses subsequently performed on each sample. Particle size distributions were determined by first sieving to remove large particles (> 53 µm) and then re-suspending the sample in a stream of air to collect size fractionated particles between 53 and 10 µm and < 2.5 µm (*7*). The amount of sample in the 2.5- to 10-µm fraction was very small and therefore not practical to analyze. The sum of the size fraction percentages does not total 100% of the original sample (< 53 μm) because of sample loss during fractionation.

The inorganic analyses included trace and toxic elements, ionic species, and functional groups. The organic chemical analyses included polycyclic aromatic hydrocarbons (PAHs), polychlorinated biphenyls (PCBs), select organo-chlorine (OC) pesticides, polychlorinated dibenzodioxins, polychlorinated dibenzofurans, phthalates, and hydrocarbons. Only the findings from the PAH, PCB, and pesticide analyses are presented here. A summary of the results of inorganic analyses, additional organic analyses, as well as asbestos, alpha and Beta radio-nuclide activity, and morphology are presented elsewhere (*1*).

Quantitative Chemical Analysis

Polychlorinated Biphenyls and Organo-Chlorine Pesticides

Approximately 0.7 g aliquots of the settled dusts were ultrasonically extracted in 30 mL dichloromethane, reduced in volume, and fractionated on a column of 3% water deactivated alumina. Samples were again reduced in volume and injected with a solution containing 50 ng of PCB congeners 30 and 204 (2,4,6-trichlorobiphenyl and 2,2',3,4,4',5,6,6'octachlorobiphenyl, respectively) as internal standards prior to analysis by gas chromatography with electron capture detection (GC/ECD) on a Hewlett Packard 6890 chromatograph for quantification of 68 PCB congeners, hexachlorobenzene, DDTs (4,4'-DDE, 2,4'-DDT, and 4,4'-DDT) and Mirex according to the procedures of Brunciak, *et al.* (*10*). All compounds were identified and quantified against known concentrations of authentic standards as described by Mullins (*11*). The NIST Standard Reference Material (SRM) 1649a – Urban Dust, Organics (*12*) was procedural blank samples (i.e. containing 0 g of dust) were also processed in parallel with the samples in order to quantify the overall operational limits of detection for the procedures. This PCB & OC pesticide analysis was performed on only a select subset of the samples.

Polycyclic Aromatic Hydrocarbons and Chlordanes

Approximately 0.7 g of each settled dust sample was weighed, ultrasonically extracted in 30 mL dichloromethane, reduced in volume under a gentle stream of clean, dry nitrogen and injected with a solution containing 100 ng each of four per-deuterated internal standards. These samples were analyzed by gas chromatography with mass selective detection (GC/MS) on a Hewlett Packard 6890/5973 for 37 individual PAHs and six chlordane species utilizing well established methods (*13, 14*). The mass selective detector was operated in selective ion monitoring mode with an electron impact ionization energy of 70 eV. The 37 PAHs reported herein are as follows: naphthalene, acenapthylene, acenapthene, fluorene, 1-methylfluorene dibenzothiophene, phenanthrene, anthracene, methylphenanthrenes + methylanthracenes (the sum of 1-methylphenanthrene, 2-methylphenanthrene, 1-methylanthracene, 2-methylanthracene, and 9-methylanthracene), 4,5-methylenephenanthrene, 3,6-dimethylphenanthrene, 9,10-dimethylanthracene, fluoranthene, pyrene, benzo[*a*]flourene, retene, benzo[*b*]fluroene, cyclopenta[*cd*]pyrene, benz[*a*]anthracene, chrysene + triphenylene, naphthacene, benzo[*b*]naptho[2,1-*d*]thiophene, benzo[*b*] fluoranthene + benzo[*k*]fluoranthene, benzo[*e*]pyrene, benzo[*a*]pyrene, perylene, indeno[1,2,3-*cd*]pyrene, dibenzo[*ah*]anthracene + dibenzo[*ac*]anthracene, benzo[*g,h,i*]perylene, and coronene. The chlordane species reported herein as the sum of chlordanes (Σ-chlordanes) include: *oxy*-chlordane, *trans*-chlordane, *cis*-chlordane, *trans*-nonachlor, *cis*-nonachlor and MC5. All compounds were identified and quantified against standard solutions containing known concentrations of authentic compounds. In the few cases where sample mass was extremely limited, the entire mass in each size segregated sample was utilized in a single analysis regardless of whether or not it amounted to 0.7 grams.

Quality Assurance

Several procedures were used in order to verify the accuracy and precision in the measurements presented here. Per-deuterated PAHs or non-industrially produced PCB congeners were spiked into each sample to track potential losses.

Surrogate recoveries indicated that there were limited sample losses during processing and handling. The concentrations reported herein have not been corrected to account for the minimal losses indicated by the recoveries of surrogate compounds that differed from 100%. Another step taken was the analysis of procedural blanks in parallel with the samples. The concentrations measured in the samples are several orders of magnitude greater than those observed in the procedural blanks for PCBs and PAHs, thereby indicating that the contribution of sample handling and preparation to the measured values was negligible. Finally, chemical analyses of standard reference materials in parallel with the samples allowed for direct assessment of the extraction, handling, and measurement methods utilized in this work. The concentrations measured in the NIST urban atmospheric particulates standard reference material (SRM 1649a), generally fall within the reported uncertainties for the certified concentrations (*12*). Thus, based upon the analysis of procedural blanks, surrogate recoveries and the analyses of SRM 1649a, we conclude that the sample handling and analysis techniques used here represent accurate measurements of the concentrations in both indoor and outdoor dusts, and furthermore do not introduce significant biases into the reported results. For further details we refer the reader to Offenberg *et al.* (*5*).

Results and Discussion

Analysis of Bulk Dust Samples

Concentrations of the sum of sixty-eight polychlorinated biphenyl congeners (Σ_{68}-PCBs) in the three bulk outdoor samples averaged 667 ± 113 ng/g and ranged from 562 ± 63 at Cherry Street and Market Street, to 723 ± 77 ng/g at Market Street and Water Street. Concentrations of the Σ_{68}-PCBs in the two bulk indoor samples collected at 114 Liberty Street and 90 & 100 Trinity Place averaged 537 ± 35 and 662 ± 33 ng/g, respectively (*6*). Furthermore, these concentrations on WTC related dusts that settled at both indoor and outdoor locations are not different from the measured concentrations of PCBs on ambient aerosols obtained in Jersey City, New Jersey, which averaged 890 ± 760 ng/g (*n*= 78) over the period 7/5/98 – 1/19/01. These measurements taken at a station located in Liberty State Park, 3.5 km west/southwest of the WTC (see Chapter 4, Figure 2), were made as part of the New Jersey Atmospheric Deposition Network (NJADN), and are summarized by Brunciak *et al.* (*10*). The Σ_{86}-PCB concentrations, comprising less than one ppm by mass of the bulk in the samples analyzed, indicate that PCBs were of limited significance in the dust that settled at indoor and outdoor locations across lower Manhattan.

Similar to the PCBs, concentrations of the organo-chlorine pesticides, hexachlorobenzene, Heptachlor, 4,4'-DDE, 2,4'-DDT, 4,4'-DDT and Mirex and the chlordanes were extremely low in the dusts measured at both indoor and outdoor locations. These organochlorine pesticide concentrations are much lower than those that were measured in Jersey City, New Jersey through the NJADN program (159 ± 111 ng/g, n=30 over period 7/5/98 – 3/30/99) (*10*), and indicated that these compounds were of limited significance in the dust that settled at indoor and outdoor locations across lower Manhattan.

Concentrations of the sum of 37 individual polycyclic aromatic hydrocarbons (Σ_{37}-PAHs) on dusts that settled at outdoor locations range from 42000 ng/g to 383000 ng/g (*5*). Measured concentrations of the PAHs in these bulk dust samples are high, especially in light of the extremely large masses of dust generated in this disaster. The total concentrations of 37 individual PAH's represented up to nearly 0.04% by mass of the bulk settled dusts measured at outdoor locations. The levels of individual PAHs ranged from hundreds of pg /g to > 40 μg/g. Further size segregation of three initial bulk samples and seven additional samples indicated that Σ_{37}-PAHs were found in higher concentrations on relatively large particles (10-53 μm), representing up to 0.04% of the total dust mass. Notable concentrations were also found on fine particles (<2.5 μm), often accounting for ~0.005 % by mass. The large variety of fires associated with the disaster would be expected to have burned at different temperatures, thus leading to a wide range of unburned and partially burned hydrocarbons derived from the diverse mix of plastics, metals, woods, and other synthetic products in the collapsed buildings.

Concentrations of the sum of 37 individual polycyclic aromatic hydrocarbons (Σ_{37}-PAHs) ranged from 59000 ± 9400 ng/g to 357000 ± 25000 ng/g in the indoor bulk dust samples (*6*). The lone dust sample that represents post rehabilitation conditions, and was analyzed as part of this work, contained an average concentration of 2390 ± 380 ng/g, nearly a factor of 25 lower than the lowest concentration measured from WTC impacted indoor dust samples. A single sample of dusts collected from a fifth story outdoor windowsill contained an average concentration of 189000 ± 23000 ng/g which is similar to that found in the outdoor dust samples. There is strong similarity in PAH content in both the indoor and outdoor settled dusts (see Figure 1). In contrast, concentrations of the same PAH species on airborne particulate matter at a long term monitoring site in Jersey City, New Jersey average 107000 ± 112000 ng/g of total suspended particulate matter (TSP) over the period spanning 7/5/98 to 1/19/01 (*15*). For a complete discussion of all methods and results pertaining to the outdoor dust samples, we refer the reader to Lioy *et al.* (*1*) and Offenberg *et al.* (*5*).

The PAHs associated with the bulk dusts that settled indoors were dominated by phenanthrene, fluoranthene, pyrene benz[*a*]anthracene and benzo[*b*+*k*]fluoranthene. The observed pattern is nearly identical to that

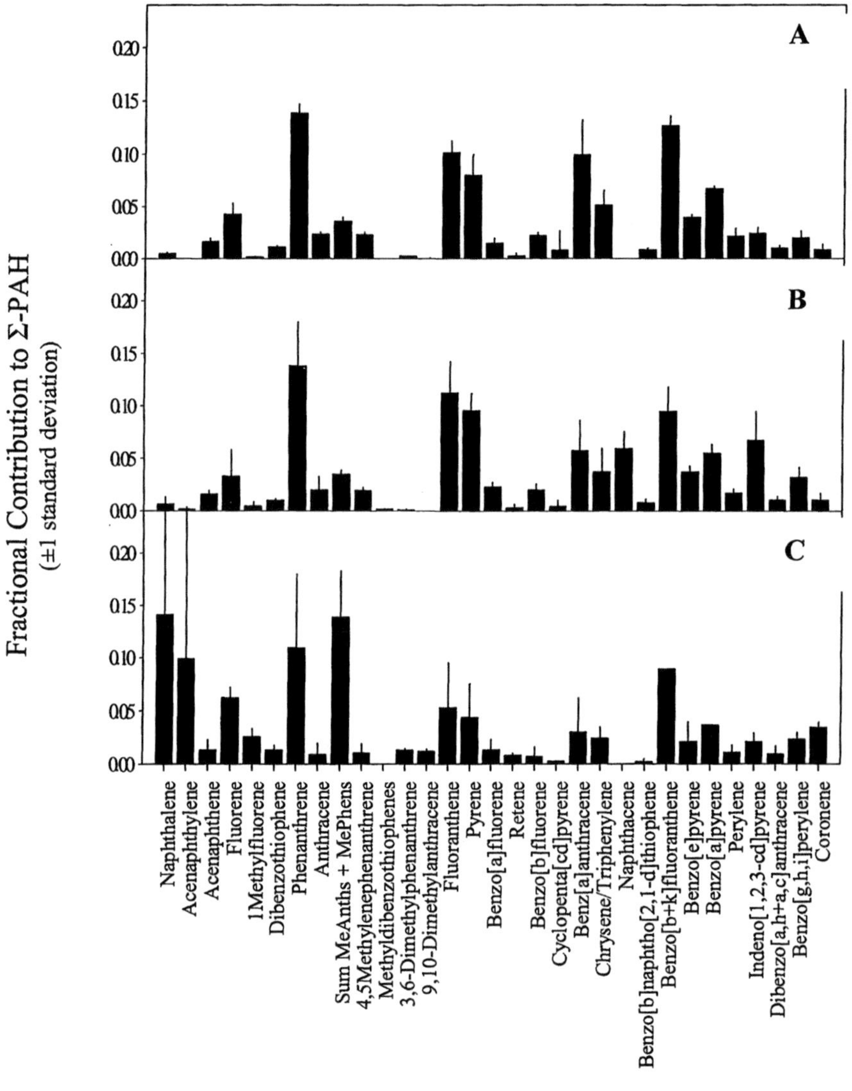

Figure 1. PAH compound fingerprints representing the average fractional contribution to (A) indoor WTC related settled dusts (n=11), (B) outdoor WTC related settled dusts (n=7) (5, 6) and (C) post rehabilitation dust from a vacuum cleaner at an indoor location (triplicate analysis of 1 sample) (6).

observed on dusts which were collected from outdoor locations. The similarity in both concentrations and the relative contributions of the several compounds analyzed is indicative of the direct linkage between the dusts generated outdoors and those that were carried into the indoor locations.

The total concentrations of 37 individual PAHs represented up to nearly 0.04% by mass of the bulk settled dusts measured at indoor locations. The levels of individual PAHs in indoor dusts ranged from hundreds of pg/g to > 50 μg/g. Concentrations in samples from within one building varied by almost a factor of three, while two samples from another building varied by almost a factor of four. The large variety of concentrations in dusts generated by the disaster may be a result of the wide variety of unburned and partially burned hydrocarbons derived from the diverse mix of plastics, metals, woods, and other synthetic products in the collapsed buildings. Additionally, varying levels of penetration of the outdoor dusts into the numerous areas of each building may have also influenced composition and thus the observed concentrations on the indoor dusts.

Although the total concentrations of PAHs were not the same among the samples collected, the relative pattern varied little, indicating that the PAHs in the settled dusts, which spread into many indoor locations across lower Manhattan, were relatively consistent among the samples, with variations in concentration but not in the source fingerprint. The lone sample that represents dusts collected from a home after rehabilitation (vacuum cleaner contents) shows a fingerprint that differs from the others. Greater relative proportions of naphthalene, acenaphthene, methylphenanthrenes and methylanthracenes, and coronene occur in the vacuum cleaner contents as compared with both the WTC impacted indoor and outdoor samples (*6*). In combination with the significantly lower concentrations, these differences in fingerprint patterns could be helpful in determining the extent of cleaning completed, or whether further cleaning of dusts should be necessary prior to re-occupation of contaminated buildings.

Conclusions

The Σ_{37}-PAHs comprised up to nearly 0.04% (<0.005 to 0.039%) by mass of the bulk settled dust in the six bulk outdoor dust samples. Further size segregation of three initial bulk samples and seven additional samples indicated that Σ_{37}-PAHs were found in higher concentrations on relatively large particles (10-53 μm), representing up to 0.04% of the total dust mass. Notable concentrations were also found on fine particles (<2.5 μm), often accounting for ~0.005 % by mass. Additionally, Σ_{37}-PAHs comprised up to 0.04% (<0.005 to

0.036%) by mass of the bulk indoor settled dust in the eleven WTC impacted indoor samples. Comparison of PAH concentration patterns (i.e. chemical fingerprints) shows that the dusts that settled indoors are chemically similar to previously measured WTC dusts found at outdoor locations. Analysis of one sample of indoor dusts collected from a vacuum cleaner of a rehabilitated home shows markedly lower PAH concentrations (< 0.0005 mass %), as well as differing relative contributions for individual compounds. These PAH analyses may be used in identifying dusts of WTC origin at indoor locations, along with ascertaining further needs for cleaning.

Acknowledgements

The authors wish to thank the NIEHS and Dr. Ken Olden for the supplemental funds provided to the NIEHS Centers at EOHSI (ES05022-12), and at the NYU Institute of Environmental Medicine (ES00260) to complete these analyses, as well as the EPA University Partnership with the National Exposure Research Laboratory (CR827033), which supports Dr. Lioy. We would also like to thank Dr. Dan Vallero, US EPA, NERL, for obtaining the necessary approvals to gain access to the sampling sites surrounding Ground Zero. The United States Environmental Protection Agency through its Office of Research and Development partially funded the research described here. This work has been subjected to peer-review and has been cleared for publication. Mention of trade names or commercial products does not constitute an endorsement or recommendation for use. Additionally, we would like to thank The Henry and Camille Dreyfus Foundation for financial support during the completion of this work. This is a contribution of the NJ Agricultural Experiment Station of Rutgers University. Finally, we would like to express our sympathy and continuing concern for the survivors and for the families of the victims of September 11, 2001.

References

1. Lioy, P. J.; Chen, L. C.; Weisel, C.; Millette, J.; Vallero, D.; Eisenreich, S.; Offenberg, J.; Buckley, B.; Turpin, B.; Zhong, M.; Cohen, M. D.; Yang, I.; Stiles, R.; Johnson, W.; Alimokhtari, S. *Environ. Health Perspect.* **2002,** *110*, 703-714.
2. Claudio, L. *Environ. Health Perspect.*, **2001,** *109*, A528-537.
3. Landrigan, P. K. *Environ. Health Perspect.* **2001,** *109*, A514-515.
4. Manuel, J. S. *Environ. Health Perspect.* **2001,** *109*, A526-527.

5. Offenberg, J. H.; Eisenreich, S. J.; Chen, L. C.; Cohen, M. D.; Chee, G.; Prophete, C.; Weisel, C.; Lioy, P. J. *Environ. Sci. Technol.* **2003,** *37*, 502-508.
6. Offenberg, J.; Eisenreich, S.; Gigliotti, C.; Chen, L. C.; Xiong, J.; Quan, C.; Lou, X.; Zhong, M; Gorczynski, J.; Yiin, L.-M.; Illacqua, V.; Lioy, P. J. *J. Exposure Anal. Environ. Epi,* **2004,** *14*, 164-172.
7. McGee, J. K.; Chen, L. C.; Cohen, M. D.; Chee, G. R.; Prophete, C. M.; Haykal-Coates, N.; Wasson, S. J.; Conner, T. L.; Costa, D. L.; Gavett, S. H. *Environ. Health Perspect.* **2003,** *111*, 972-980.
8. Kitsa, V.; Lioy, P. .J; Chow, J. C.; Watson, J. G.; Shupack, S.; Howell, T.; Sanders, P. *Aerosol Sci. Technol.* **1992,** *17*, 213-229.
9. Pellizzari, E.; Lioy, P. J.; Quackenboss, J.; Whitmore, R.; Clayton, A.; Freeman, N.; Waldman, J.; Thomas, K.; Rodes, C.; Wilcosky, T.; *J. Exposure Anal. Environ. Epi.* **1995,** *5*, 327-358.
10. Brunciak, P. A; Dachs, J.; Gigliotti, C. L.; Nelson, E. D.; Eisenreich, S. J. *Atmos. Environ.* **2001,** *35*, 3325-3339.
11. Mullins, M. D. 1985. PCB workshop, U.S. Environmental Protection Agency Large Lakes Research Station, Gross Ile, MI.
12. Wise, S. A.; Sander, L. C.; Schantz, M. M; Hays, M. J.; Benner, B. A. *Polycyclic Aromatic Compounds*, **2000,** *13*, 419-456.
13. Offenberg, J. H.; Baker, J. E. *Enviro. Sci. Technol.* **1999,** *33*, 3324-3331.
14. Naumova, Y. Y.; Eisenreich, S. J.; Turpin, B. J.; Weisel, C. P.; Morandi, M. T.; Colome, S. D.; Totten, L. A.; Stock, T. H.; Winer, A. M.; Alimokhtari, S.; Kwon, J.; Shendell, D.; Jones, J.; Maberti, S.; Wall, S. J. *Environ. Sci. Technol.* **2002,** *36,* 2552-2559.
15. Gigliotti, C. L.; Dachs, J.; Nelson, E. D.; Brunciak, P. A; Eisenreich, S. J. *Environ. Sci. Technol.* **2000,** *34,* 3547-3554.

Chapter 7

Characterization of Size-Fractionated World Trade Center Dust and Estimation of Relative Dust Contribution to Ambient Particulate Concentrations

Polina B. Maciejczyk[1], Rolf L. Zeisler[2], Jing-Shiang Hwang[3], George D. Thurston[1], and Lung Chi Chen[1]

[1]Department of Environmental Medicine, New York University School of Medicine, Tuxedo, NY 10950
[2]National Institute of Standards and Technology, Gaithersburg, MD 20899
[3]Institute of Statistical Science, Academia Sinica, Taiwan

After the collapse of the World Trade Center (WTC), bulk samples of settled dust were collected close to the WTC. Each sample was separated into four fractions according to aerodynamic particle size. All samples <2.5 µm fraction were analyzed by X-ray Fluorescence, and some 2.5 – 10 µm and 10 – 53 µm fractions were analyzed by Instrumental Neutron Activation Analysis. The 10 – 53 µm fraction contained the highest concentrations of elements used in construction materials (Al, Ca, Mg, Ti, Fe, Zn), while the <2.5 µm fraction also contained combustion related Cl and Sb. Factor analysis of ambient samples (collected five blocks east of Ground Zero) was used to predict the factor scores for the 28 dust samples of the <2.5 µm fraction. A new combined dust factor was calculated and applied to predict the relative contribution of WTC dust to ambient PM2.5 concentrations in the three months after the WTC collapse.

Introduction

The collapse of the World Trade Center (WTC) on September 11, 2001 was an unprecedented event. It released millions of tons of material into the air from pulverized and incinerated building materials, furniture, office equipment, and unburned jet fuel. The debris plume from the collapse covered a large area around the WTC, and penetrated into many buildings in downtown Manhattan. Additional pollutants were released by the ensuing fires, which persisted until December 20, and by the recovery and clean up processes that followed. Thousands of survivors, residents and commuters, along with policemen, firemen, healthcare workers, and civilian volunteers, were exposed to very high concentrations of these pollutants immediately after the collapse. Many residents, who returned to their community a few weeks after the disaster were also exposed and have complained of persistent respiratory symptoms. Thus, it is important to characterize the exposure environment with respect to the physicochemical nature of the particles that existed immediately after the collapse and to examine the air quality around the WTC area. The findings from these studies will facilitate early detection of potential health effects.

Our efforts have focused on the characterization of the ambient particulate matter and fallout dust. With the intense fire caused by the jet fuel and subsequent pulverization of the large mass of WTC ($>10 \times 10^6$ tons), the physical and chemical composition of these is expected to be consistent with that seen during collapsed buildings and fires. The types of materials potentially present have been reported previously (*1*) and include asbestos from concrete, polychlorinated biphenyls (PCBs) from electrical wiring, dioxins from jet fuel combustion, particulate matter from pulverized concrete and other building materials, and lead and other metals from computer monitors. Inhalation of these materials, especially at the very high concentrations as those encountered by survivors and rescue workers, could potentially produce adverse health effects. Furthermore, since large quantities of the pulverized building materials were dispersed in a wide area, these dusts could penetrate into the air intakes of many office and apartment buildings and infiltrate through the windows and doorways of residential homes. Re-suspension of these dusts could pose further health risks for workers returned to their offices and residents who remained in these areas.

The analyses reported here include a general characterization of the percent distribution of various materials present in dust samples collected at outdoor (n=15) and indoor (n=13) locations close to the WTC and a detailed measurement of the inorganic components of the dust mass. We also sought in this analysis to determine how much of the atmospheric fine particle mass (PM2.5) measured in lower Manhattan after the collapse was associated with the WTC dust.

Experimental Methods

Sample Collection

Bulk samples of fallout dust that initially settled in downtown New York City (NYC) were taken from several undisturbed protected locations around the WTC site. Thirteen samples were taken on September 12, 2001 and additional six samples were collected on September 13, 2001. The locations of these sampling sites are shown in Plate 1. The samples were collected from protected external ledges and window sills around buildings surrounding the WTC complex and from thick deposits on the top of cars. Indoor samples were collected on November 19 at various floors of the apartment buildings located at Liberty Street and Trinity Place. In addition, several samples were sent to our laboratory by residents on the 2nd and 3rd floors of an apartment building located at Nassau Street. After collection, all samples were stored at room temperature in sealed bottles before being fractionated by size. These processed samples were then sent to individual laboratories for analysis. Chain of custody procedures were maintained throughout sample transferal and analyses.

Between September 14 and December 31, 2001, ambient fine particulate samples (PM2.5) were collected daily (sometimes twice daily) at New York University Downtown Hospital (NYUDT), located approximately five blocks east of Ground Zero (see Plate 1). Since we could not attend to the sampling site daily due to the security precautions around Ground Zero, we employed a new NYU-designed multichannel automated sequential sampler (*2*) to collect PM2.5 samples on Teflon filters (Gelman "Teflo", 37 mm, 0.2 μm pore size). Filter samples were stored at constant temperature and humidity (21 ± 0.5°C, 40 ± 5%RH) until analyzed.

Size Fractionation of the WTC Settled Dust Samples

The cardinal feature that influences the toxicity of inhaled particles is their size distribution, as described in Chapter 1. Very large particles, even with high intrinsic toxicity, would not be inhaled or penetrate deeply into the respiratory tract and are not likely to cause effects in small airways. Therefore, the analytical plan was to first size separate the WTC settled dust bulk samples into several

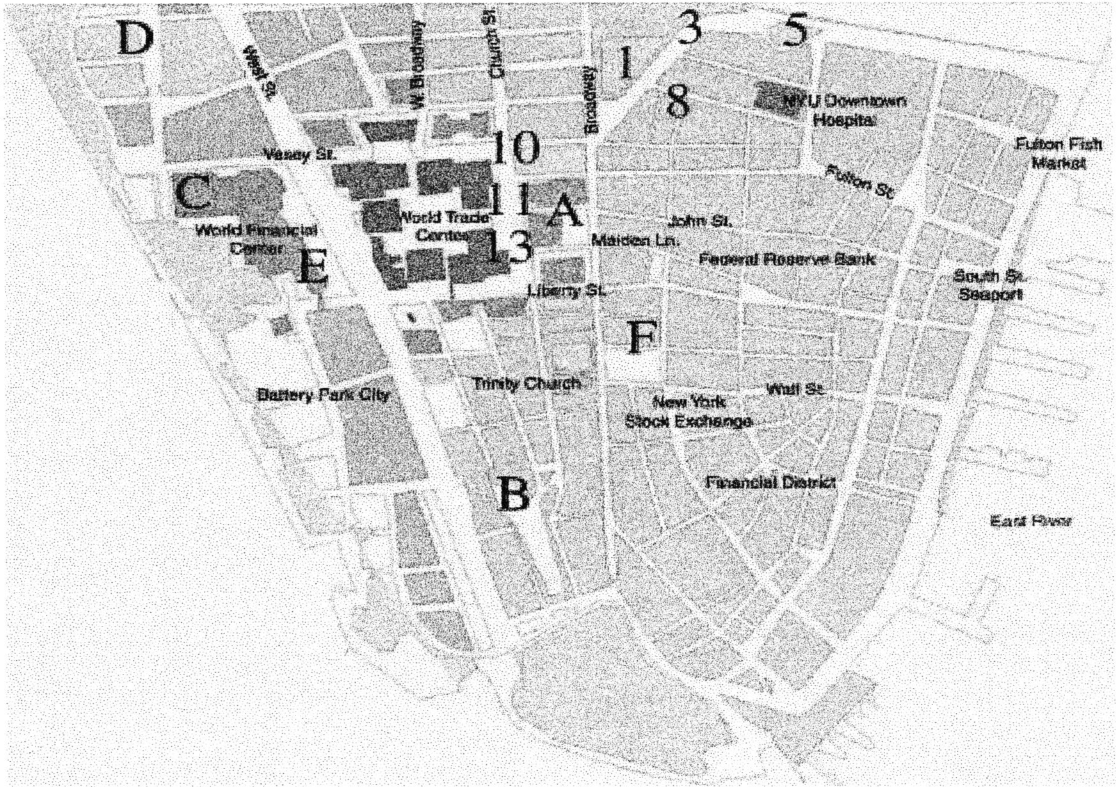

Plate 1. *Sites where WTC dust samples were collected on September 12, 2001 (numbers) and September 13, 2001 (letters). (See page 9 of color inserts.)*

size classes before qualitative and quantitative analyses to identify components that may cause adverse health effects.

The size fractionation method used for dust samples performed at NYU has been described previously (*3*). Briefly, dusts were first mechanically separated using a sieve with a mesh size of 53 µm. The fractions of particles below 53 µm were further separated, aerodynamically, into three size fractions (10 – 53 µm, 2.5 – 10 µm, and < 2.5 µm). Particles were resuspended by a jet of filtered air, passing through an inlet (Wedding Inlet, 10 µm cut size, Andersen Instrument Co., Fultonville, NY) before entering a cyclone with a size cut of 2.5 µm (BGI, Inc., Walthman, MA). Particles between 10 – 2.5 µm were collected by the cyclone while particles smaller than 2.5 µm, which penetrated through the cyclone, were collected on Zefluor filters (Pall, 47 mm). Filters were weighed before and after sample collection using a microbalance (Model MT5, Mettler-Toledo) in a temperature and humidity controlled weighing room. Particles collected in the Wedding Inlet and cyclone were swept into plastic test tubes and weighed. The average recovery rate was 89.4 ± 4.2%

Analytical Methods

Size fractionation was followed by a visual characterization of the asbestos contents and other components of the WTC dusts using a polarized light microscope at 400-450X magnification (4-mm objective) and was performed by Ambient Group Inc. (New York, NY). Chemical analyses were then performed on each sample. Due to the limited quantities of material available, only a subset of the samples was analyzed. This report focused on the inorganic analyses of trace and toxic elements. The organic analyses for polycyclic aromatic hydrocarbons, polychlorinated biphenyls, pesticides, dioxins, furans, phthalates, and general hydrocarbons are reported elsewhere (*4, 5*). We used the ambient particulate PM2.5 samples in our factor analysis model, and a comprehensive discussion of the elemental composition analysis and concentration trends are also presented elsewhere (*6, 7*).

Analyses for pH and Soluble Ionic Species

Aliquots of the size fractionated samples below 53 µm (53-10 µm, 10-2.5 µm, and <2.5 µm) were placed in test tubes and distilled and deionized water was added to obtain a concentration of approximately 30 mg/ml. The tubes were inverted several times and then sonicated for 15 minutes. The samples were then centrifuged for 10 minutes. The supernatants were removed to new tubes and stored at 4°C before analysis. An aliquot of the supernatant was used for pH measurement on a digital pH meter (Model 601, Orion Research Inc., Cambridge, MA) on the day of extraction. A separate aliquot was used to measure soluble ionic species by ion chromatography (Dionex DX 500 system). An IonPac AS14 column, ASRS Ultra-Suppressor, and 3.5mM Na_2CO_3/1.0mM $NaHCO_3$ eluent were used for anion analysis while an IonPac CS12A, CSRS II Ultra-Suppressor, and 20 mM methanesulfonic acid eluent were used for cation analysis.

Calibration curves were constructed using seven working standards prepared by diluting an NIST-traceable standard (Fisher) using Milli-Q water. Each of the working standards was subsequently run as a sample to verify the calibration curve. Samples were run once the calibration curve was verified. A control (usually the working standard #2) was run followed by a blank for every 10 samples. After all samples and their duplicates were analyzed, these 7 working standards were analyzed again followed by two additional NIST traceable stock standards (Dionex). Samples that were originally off-scale were diluted with Milli-Q water and tested again.

X-Ray Fluorescence

Ambient PM2.5 samples (n=155) and dust samples of fraction <2.5 µm were analyzed for 34 elements by non-destructive energy dispersive X-Ray fluorescence (ED-XRF) spectroscopy (Model EX-6600–AF, Jordan Valley, Israel) using five secondary fluorescers (silicon, titanium, iron, germanium, and molybdenum). The ED-XRF instrument was calibrated with thin elemental layers deposited on Nuclepore film substrates (Micromatter Co., Eastsound, WA) for elements ranging from atomic number Z=11 (Na) to Z=82 (Pb) with deposited masses gravimetrically determined to ±5%, and dual element polymer films (U.S. EPA and ManTech Environmental Technology, Inc.). The gain and baseline of the instrument were checked before each batch of samples using copper and tin foils. Additionally, check standards and NIST SRM 2783 were run in the middle and at the end of each batch. Spectra were analyzed with the XRF spectral software package XRF2000v3.1 (U.S. EPA and ManTech Environmental Technology, Inc.).

Instrumental Neutron Activation Analysis

Instrumental Neutron Activation Analysis (INAA) was used to determine the elemental composition of the 2.5 – 10 µm, 10 – 53 µm fractions of 4 outdoor and 3 indoor samples. Samples of dust (11 – 64 mg) were packaged in polypropylene film (SPECTRO-FILM™, Somar International Inc., Tuckahoe,

NY), heat sealed and irradiated, together with element standards, control material NIST SRM 2702 and blank polypropylene at the NIST Center for Neutron Research reactor (RT-4 pneumatic tube, 3.4 × 10^{13} $cm^{-2}s^{-1}$ thermal neutron flux) in a sequence of irradiation and counting steps described as follows (*8*). Each sample was subjected to a short time irradiation (30 s) followed by 2 counts for elements with short-lived nuclides (counts A and B). After several days of decay, this cycle was followed by a simultaneous long-term irradiation of all samples, blanks, and controls, together with standards, for 4 h + 4 h with a 180 degree inversion of the irradiation capsule to insure flux homogeneity and 2 counts for elements with intermediate and long-lived nuclides (counts C and D). Counts A, B, and C, were made with high-resolution high-purity germanium detectors (25% to 32% relative efficiency) equipped for a high count-rate acquisition and an electronic loss-free counting correction (*9*), count D was made with a high-resolution detector (42% relative efficiency) with conventional dead time and pile-up correction. Counting conditions were as follows: A – 8 min decay, 5 min count @ 20 cm; B – 15 min decay, 10 min count @ 10 cm, C – 3 d to 4 d decay, 2 h @ 10 cm, D – 20 d to 25 d decay, 6 h count @ 5 cm. The spectral data files were evaluated with the Canberra neutron activation software package (Canberra Industries, Meriden, CT) using the comparator method to arrive at concentration values for each element included in the standards. The results obtained for SRM 2702 indicated statistical control of all measurements.

Results

Size Fractionation

The relative contribution of each aerodynamic size fraction of the WTC settled dusts is shown in Figure 1. More than 97% of the dust particles' mass was larger than 10 µm aerodynamic diameter. Only 0.43 ± 0.13% and 0.57 ± 0.15% of the outdoor and indoor dusts were fine particles less than 2.5 µm. Approximately equal percentages of the dust were particles 2.5–10 µm: 1.32 ± 0.40% and 1.62 ± 0.46% of outdoor and indoor samples, respectively. As expected, the fractions of particles larger than 53 µm were much less for indoor samples (39.9 ± 14.8%) as compared to other outdoor samples (53.9 ± 11.6%). Consequently, more 10–53 µm particles were found in the indoor samples (57.9 ± 14.3% compared to 44.3 ± 11.2% outdoors). However, the fractions <10 µm of the indoor samples (0.61%) were similar to those found outdoors (0.53%).

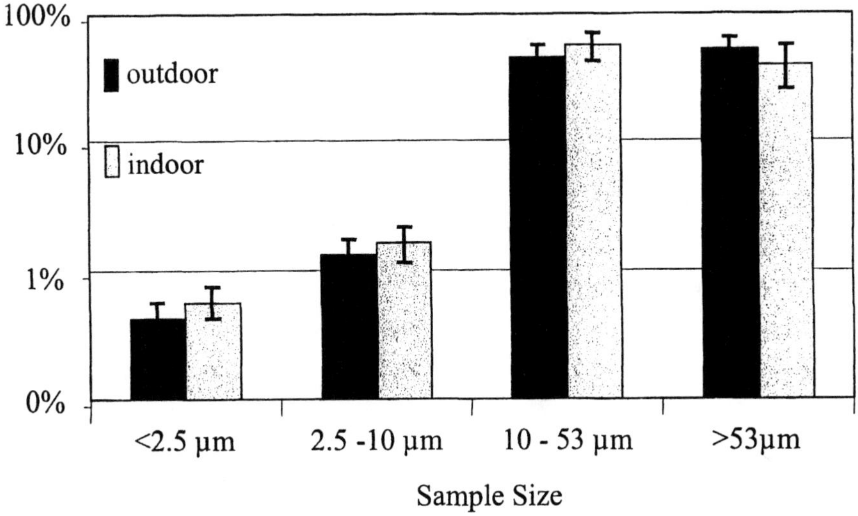

Figure 1. Contribution of each size fraction to the total dust.

Morphological Composition of WTC Settled Dusts

The compositions of the outdoor bulk dust samples as well as the sized fractionated settled dusts are shown in Table I. The content and distribution of material were indicative of the complex building debris and combustion plume formed. For the bulk samples, 40% of the material was classified as non-fibrous, consisting of cement, gravel and road dusts. Fiber glass was 30 to 40%, carbon particles were 10%, and the remaining material was cellulose. Only trace amounts (< 1%) of asbestos was found in these samples. The compositions of samples collected by other investigators were similar to those found around the WTC sites, even though the other sampling locations were approximately 1 km away from Ground Zero (*5*). The compositions of dust fractions <2.5 µm were very different from those of the bulk samples. Non-fibrous material contributed 80%, the remaining were carbon particles (15%), and cellulose (5%). No asbestos was found in this size class. The dusts collected at the apartment building were also different from the dusts collected around the WTC sites, except for one sample collected from the roof of the building (data not shown). More non-fibrous material, less fiberglass, and various amount of cellulose were found in these samples. Similar to other samples, only a trace or no asbestos was found in the indoor samples.

Soluble Ionic Species and pH

As shown in Figure 2, the pH of the majority of the suspensions of the bulk WTC settled dusts was greater than 10. The pH decreased with decreasing

Table I. Morphological Composition of Selected Dust Samples

Sample	*Size*	*Chrysotile*	*Amosite*	*Cellulose*	*Fiberglass*	*Non-fibrous*	*Carbon Particles*
10	Bulk	0.3%	ND	10%	40%	40%	10%
11	Bulk	0.3%	ND	15%	35%	40%	10%
13	Bulk	0.3%	ND	20%	30%	40%	10%
1	>53 µm	ND	ND	40%	20%	20%	20%
1	53-10 µm	Tr	ND	5%	40%	40%	15%
1	<2.5 µm	ND	ND	5%	none	80%	15%
NS	Bulk	0-0.5%	ND	13±10%	30%	57±10%	—

NOTE: Sample numbers refer to Plate 1. Results reported for Nassau Street (NS) are the arithmetic means and standard deviations of 6 indoor samples.

NOTE: ND is none detected, Tr is trace found, — indicates not done.

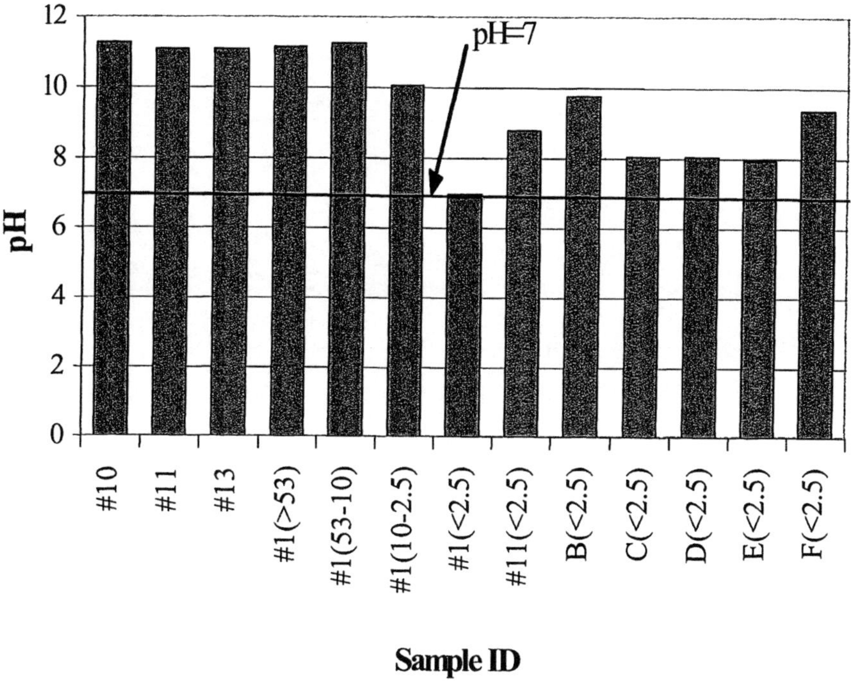

Figure 2. Sample pH of the suspensions of selected dust samples. Sample size fraction is indicated in parenthesis. Where size fraction is not indicated, the samples are bulk dust.

particle size, with particles less than 2.5 µm closer to neutral pH. Since larger particles are likely to be caught in the upper airway, their alkalinity was partially responsible for the reported "WTC cough" (*10*).

The concentrations of soluble ions measured via ion chromatography are shown in Figure 3. The concentrations of ions in all dust fractions varied to some degree, indicating uneven spatial distribution of the collapsed building materials or heterogeneity in these samples. The bulk samples contain, on average, 8.2 ± 0.6 and 14.5 ± 1.2 µg/g of Ca^{2+} and SO_4^{2-}, respectively. The concentrations of these highly correlated ions in the bulk samples (r^2= 0.996) reflect the content of gypsum in the fraction larger than 53 µm. Both soluble calcium and sulfate ion concentrations increased with decreasing particle size and remained correlated. Their mean concentrations in the dust fraction < 2.5µm were 150 ± 33 and 385 ± 85 µg/g, respectively. Concentrations of other soluble ions (Na^+,

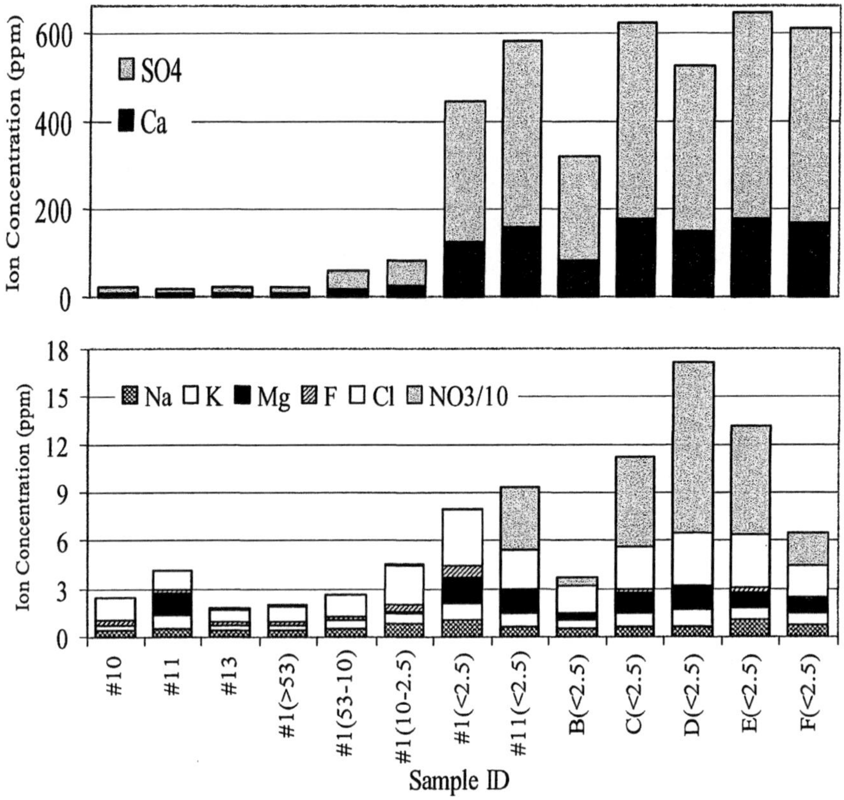

Figure 3. Soluble ion concentrations(μg/g) in selected dust samples. Sample size fraction is indicated in parenthese. Where size fraction is not indicated, the samples are bulk dust.

K^+, Mg^{2+}, F^-, CL^- and NO_3^-) are shown in Figure 3. Interestingly, chloride is the dominant species (other than calcium and sulfate) in all size classes. The average molar ratio of CL^-/Na^+ in <2.5 μm fractions was 2.31 ± 0.55 with a poor correlation (r^2 = 0.286). However, this ratio was 1.49 ± 0.42 in average bulk samples with a higher correlation (r^2 = 0.952). These data suggest that other chlorinated compounds, perhaps produced by the fire, were also present in the fine dusts in addition to coarse sea salt (see Chapter 9). Nitrate ion was detected only in fine fractions of dusts except for sample #1, and was highly correlated (r^2 = 0.984) with ammonium ion (data not shown).

Elemental Concentrations

Mass fraction element concentrations as determined via XRF and INAA analyses are shown in Plate 2 and listed in Table II. Not all elements could be determined by both methods (e.g., silicon, sulfur, and lead are not measured by INAA, and our XRF was not calibrated for the rare-earths). All dust fractions contained substantial amounts of calcium (9.7 – 23%), aluminum (0.7 – 2.7%), magnesium (0.4 – 2.8%), and iron (0.2 – 1.0%). There are also elevated concentrations of titanium which presumably originated from paint. These elements were present in construction materials (gypsum boards, cement blocks, steel beams, etc.) of the collapsed buildings, and their concentrations were progressively higher in the coarser dust fractions. Other elements that followed this trend and were present in μg/g quantities are zinc, copper and manganese. Some of the elements (antimony and chlorine) were produced by the fires rather than by the mechanical grinding of the building collapse, and thus are much elevated in the fine fraction of dust samples. Traces of lead, chromium, vanadium, and nickel, elements important for toxicological considerations, were also present.

As shown in Plate 2, there appears to be a slight trend for the concentrations of the elements to be lower in the indoor than in the outdoor samples. Since the indoor samples were collected approximately two months after the collapse, these samples may contain slightly more materials not measured by either XRF or INAA (e.g., carbon as the result of fires, and pre-existing house dust).

Factor Analysis

Factor analysis of ambient samples (PM2.5) was used to predict the factor scores of the < 2.5 μm fraction for the 28 dust samples. First, multivariate factor analysis and varimax rotation strategy were applied to 26 elements in the 155 ambient PM2.5 samples to extract 3 factors. We then used the model estimates to predict the factor scores for the < 2.5 μm fraction of the 28 dust samples. The predicted scores for the dust samples were well separated from those of the ambient samples and showed a strong linear correlation with the highest variance between factors F2 and F3 (see Figure 4). This indicates that the dust source could be represented by the wnd direction as well as the regression line fitted on these 28 dust sample factor score points.

A new combined factor score (F23) was obtained by rotating the axes to the regression line so that the variances of the new combined factor scores are maximized. This combined factor score is described as;

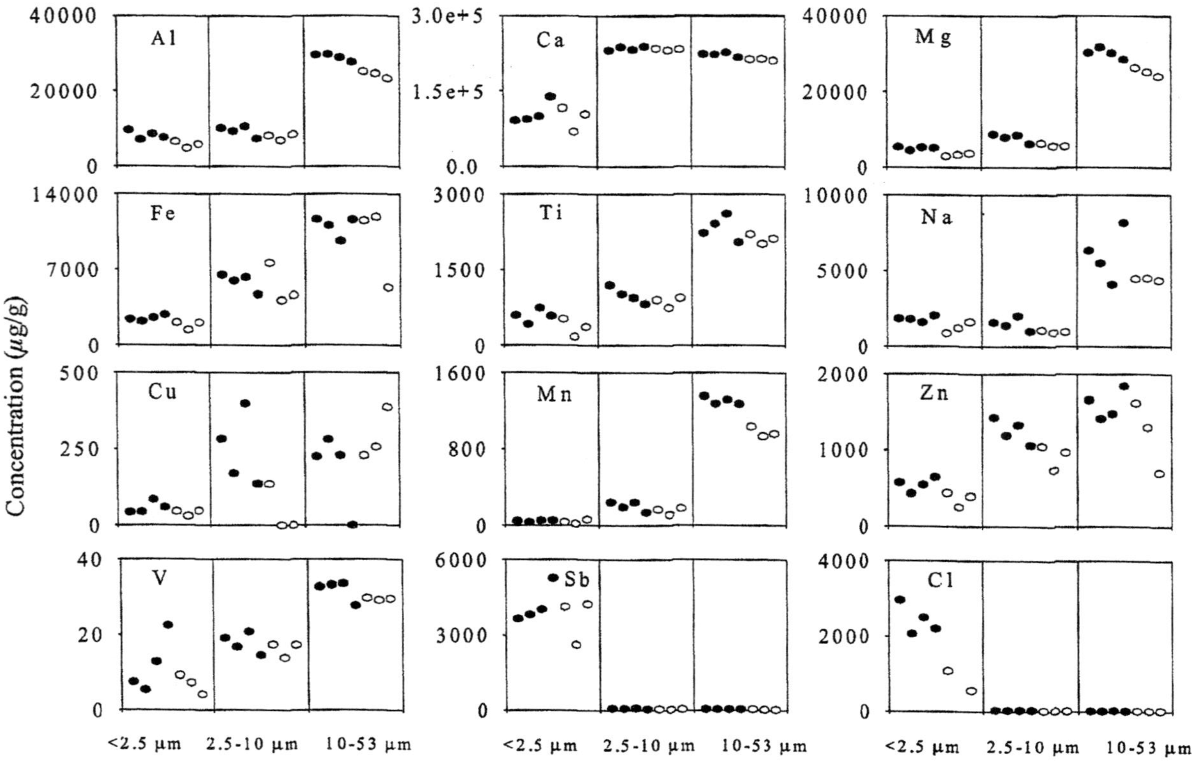

Plate 2. Elemental composition of selected outdoor (●) and indoor (○) dust samples by size classification. (See page 10 of color inserts.)

Table II. Mean Elemental Concentrations and in Selected Dust Samples

	10 - 53 μm		*2.5 - 10 μm*		*<2.5 μm*	
	mean	*sd*	*mean*	*sd*	*mean*	*sd*
Na	5,300	1,500	1,300	420	1,500	460
Mg	28,000	3,000	7,100	1,400	4,400	1,800
Al	27,000	2,600	8,800	1,400	7,200	2,200
Si					28,000	6,000
S					130,000	23,000
Cl	19	3.2	24	7.4	1,400	1,600
K	300	780		<DL	1,300	610
Ca	220,000	5,900	230,000	3,200	97,000	16,000
Ti	2,200	210	930	140	540	400
V	31	2.3	17	2.4	8.3	5.7
Cr	85	22	75	21	30	15
Mn	1,200	180	180	49	41	14
Fe	10,000	2,400	5,700	1,200	2,300	950
Co	4.4	1.0	3.1	0.60	6.0	4.6
Cu	230	120	160	140	64	35
Zn	1,400	370	1,100	230	550	290
As	2.7	0.8	2.0	0.4		<DL
Se					20	7.3
Br					68	24
Rb	16	3.3	12	1.8	32	10
Sr					210	37
Sb	40	14	54	18	3,800	630
Ba					27	37
Pb					65	27
La	38.0	12.0	9.3	3.6		
Sm	9.1	2.5	1.5	0.3		
Ag	1.6	0.7		<DL		
Ce	80	22	17	7.1		
Hf	3.0	0.7	0.6	0.1		
Sc	6.2	1.6	1.5	0.3		
Th	7.2	1.9	1.4	0.3		

NOTE: Units ar μg/g of sample, n=7 for 10 - 53 μm and 2.5 - 10 μm, n=28 for <2.5 μm. K was only detected in one sample of size 10 – 53 μm. <DL is below the detection limit, sd is standard deviation for the measurement.

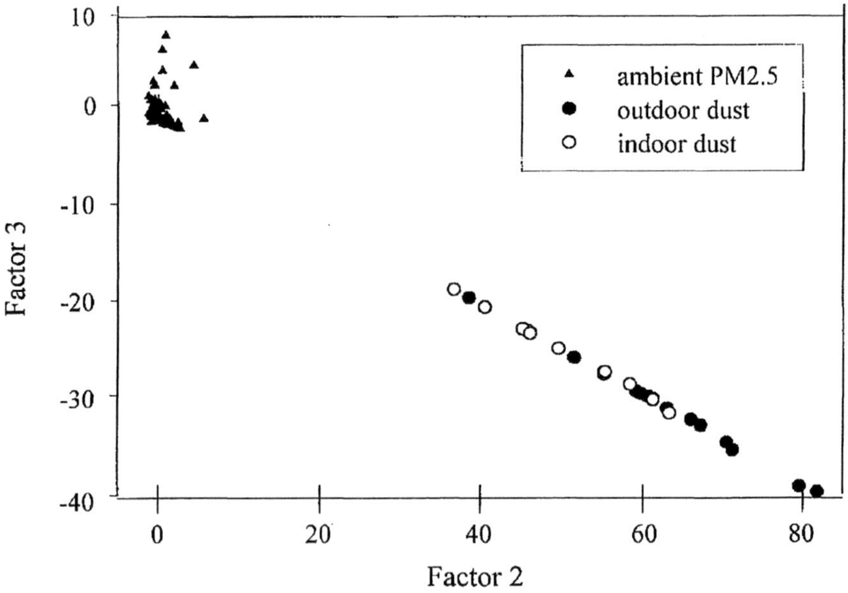

Figure 4. Factor scores of the ambient PM2.5 samples and the < 2.5 μm fraction of the outdoor and indoor dust samples.

$$F23 = \cos(\theta)F2 + \sin(\theta)F3 \qquad (1)$$

where θ is the angle between the axis for Factor 2 and the regression line for the dust samples. The scores for factor F1 and the new combined factor F23 were then plotted and the regression line was fitted for the 28 dust samples. Similarly, we rotated the axes so that the axis of the F23 factor matched the regression line. This new factor, in which the dust samples have the highest variance, was used to represent the dust source, while the axis perpendicular to the new dust axis represents another possible source, presumably combustion. The new factor scores thus were calculated as;

$$F_{dust} = \cos(\varphi)\ F23 + \sin(\varphi) \qquad (2)$$

$$F_{other} = \cos(\varphi + \pi/2)\ F23 + \sin(\varphi + \pi/2)\ F1 \qquad (3)$$

where φ is the angle between the axis for the combined factor and the regression line. These two recombined factor scores were then used to recreate the z-score matrix for the ambient PM2.5 samples that allowed for the calculations of the factor loadings. As shown in Figure 5, the factor loadings for elements of combustion origin are clearly separated from the crustal elements. Sulfur was associated with both dust and combustion sources, and additionally with regional sulfate aerosol which is inclusive in the category of other sources. Several elevated sulfur levels were observed at other New York state localtions during regional pollution episodes in September and October 2001 (6, 7).

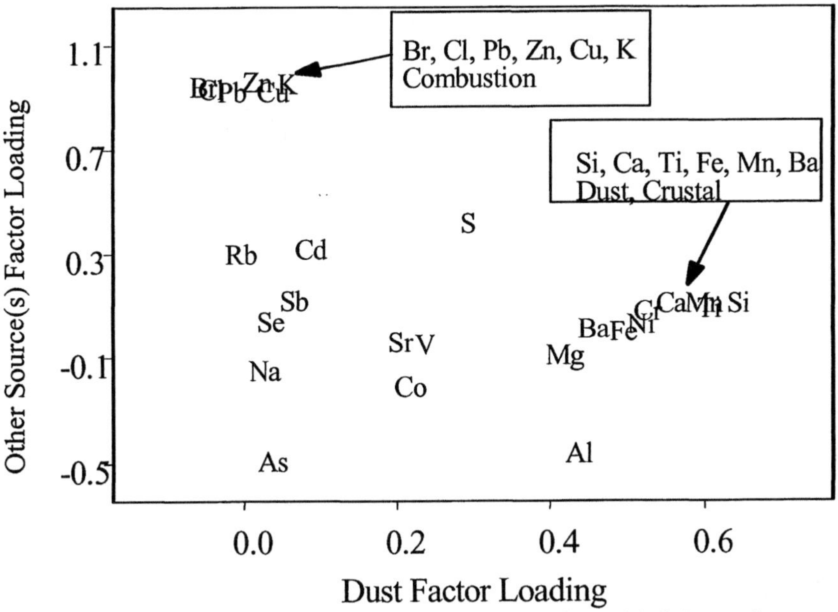

Figure 5. Recalculated factor loadings for ambient PM2.5 samples.

Following the regression to the PM2.5 mass concentrations, we were able to predict the relative contribution of the WTC dust to the ambient PM2.5 concentrations three months after the WTC collapse. As shown in Figure 6, fire impacts (labeled as "other sources" since it could also include local and transported regional pollution) largely ended in mid-October, but the WTC dust contribution increased in mid-October and diminished by mid-November. These modeling results compare favorably with the Positive Matrix Factorization model (*7*).

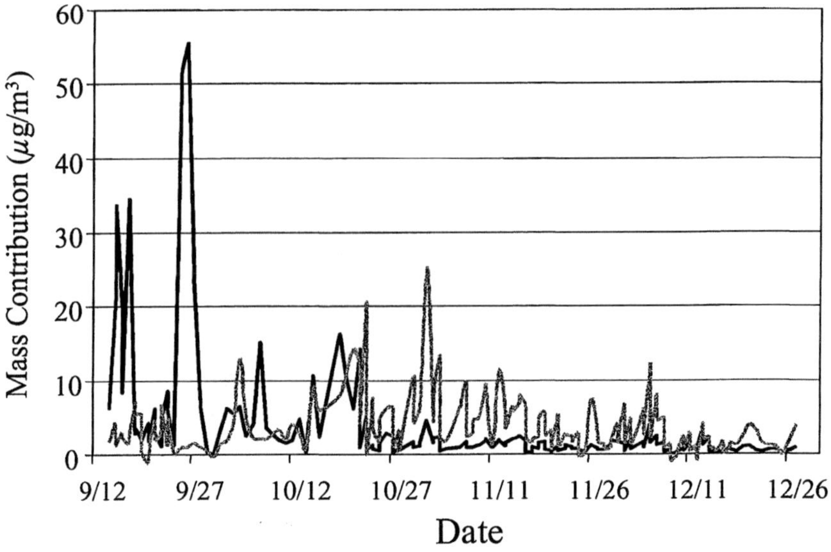

Figure 6 . Factor analysis model-predicted relative mass contributions of WTC WTC dus t (---) and other sources (—) to ambient PM2.5 from 9/14 – 12/30/01.

Acknowledgements

Certain commercial equipment, instruments, or materials are identified in this paper in order to specify the experimental procedure adequately. Such identification is not intended to imply recommendation or endorsement by the sponsoring organizations, nor is it intended to imply that the materials or equipment identified are necessarily the best available for the purpose.

The authors wish to thank the NIEHS and Dr. Kenneth Olden for the supplemental funds provided our NIEHS Center, at EOHSI (ES05022-12), and the NIEHS Center at the NYU School of Medicine (ES00260) to complete these analyses. NYU is also funded in part by a U.S. EPA PM Center Grant (R827351). Finally, the authors wish to express their deepest sympathy and continuing concern for the families of the victims, and survivors of September 11, 2001.

References

1. Claudio, L. *Environ. Health Perspect.* **2001,** *109*, A529-536.

2. Lippmann, M.; Gorczynski, J.; Chen, L. C. "Multichannel sequential sampler for $PM_{2.5}$ and PM_{10}." (in preparation).
3. McGee, J. K.; Chen, L. C.; Cohen, M. D.; Chee, G. R.; Prophete, C. M.; Haykal-Coates, N.; Wasson, S. J.; Conner, T. L.; Costa, D.L.;Gavett, S. H. *Environ. Health Perspect.* **2003,** *111*, 972-980.
4. Offenberg, J. H.; Eisenreich, S. J.; Chen, L. C.; Cohen, M. D.; Chee, G.; Prophete, C.; Weisel, C.; Lioy, P. J. *Environ. Sci. Technol.* **2003**, *37*, 502-508.
5. Lioy, P. J.; Weisel, C. P.; Millette, J. R.; Eisenreich, S.; Vallero, D., Offenberg, J.; Buckley, B.; Turpin, B.; Zhong, M.; Cohen, M. D.; Prophete, C.; Yang, I.; Stiles, R.; Chee, G.; Johnson, W.; Pocja, R.; Alimokhtari, S.; Hale, R. C.; Weschler, C.; Chen, L. C. *Environ. Health Perspect.* **2002,** *110*, 703-714.
6. Thurston, G.; Maciejczyk, P.; Lall, R.; Hwang, J.; Chen, L.C. *Epidemiology* **2003,** *14*, S87-88.
7. Thurston, G.; Maciejczyk, P.; Lall, R.; Hwang, J.; Chen, L.C. "Source signatures and impacts of World Trade Center fine particle air pollution in Lower Manhattan." *Science* **2004** (submitted).
8. Greenberg, R. R.; Fleming, R. F.; Zeisler, R. *Environm. Intern.* **1984**, *10*, 129-136.
9. Zeisler, R. *J. Radioanal. Nucl. Chem.* **2000,** *244*, 507-510.
10. Chen, L.C; Thurston, G. *Lance*t **2002,** *360*, S37-38.

World Trade Center Fine Particle and Volatile Organic Emissions

Chapter 8

Characterization of the Plumes Passing over Lower Manhattan after the World Trade Center Disaster

Robert Z. Leifer[1], Graham S. Bench[2], and Thomas A. Cahill[3]

[1]U.S. Department of Homeland Security, Environmental Measurements Laboratory, 201 Varick Street, New York, NY 10014–7447
[2]Center for Accelerator Mass Spectrometry, Lawrence Livermore National Laboratory, 7000 East Avenue, Livermore, CA 94550–9234
[3]DELTA Group, Department of Applied Science, University of California at Davis, One Shields Avenue, Davis, CA 95616

During the three-month period from October 1, 2001 to December 15, 2001 impactor measurements were carried out on the roof of the Environmental Measurements Laboratory (EML), approximately 45 m above the street level in New York City, to characterize plumes passing over the laboratory. Depending on the wind direction it was hoped to intercept aerosols originating from the disaster of the World Trade Center. More than 25 plumes were detected in our samples. With southwest winds on October 3, an aerosol plume, transported from the WTC, was observed over EML. The total particulates collected on all stages reached a peak mass concentration of 267 µg/m^3 at 08:15 with more than 39% of the mass in the 0.56 to 0.34 μm stage. The October 3rd sample showed that almost 75% of the total mass at the peak seen at EML was below 1.15 µm. The observed size distribution was bimodal and showed a strong temporal variation as the modes shifted from a dominant coarse to a fine mode. The detected plume lasted for more than 9 hours. It is this fine particulate mode that is of concern to human health exposure.

Introduction

In response to the disaster at the World Trade Center (WTC), the Department of Energy's (DOE) Environmental Measurement Laboratory (EML) at 201 Varick Street, initiated a sampling program on the roof of the laboratory approximately 45 m above street level, to characterize plumes passing over the building. We were hoping to intercept debris from Ground Zero, which is approximately 2 km south-southwest of EML. The sampler, an 8-stage Davis Rotating-drum Universal-size-cut Monitoring (DRUM) sampler (see Chapter 9) operated on a 42-day cycle and provided three sets of drums for chemical analysis during the period from October 2, 2001 to the end of December 2001. The DRUM impactor is a unique instrument that not only separates the atmospheric aerosols into 8 size fractions but also, because of the nature of the collection, allows for measurement of the mass and chemical composition in short time intervals needed to characterize atmospheric aerosol processes. This sampler, on loan from the DELTA (Detection and Evaluation of Long-Range Transport of Aerosols) Group at the University of California at Davis was used for collecting low volume environmental samples for analysis of mass, optical aerosol properties, trace element concentration, organics, and asbestos. The DRUM sampler contains eight stages with 50% aerodynamic cutoff diameters of 5 µm, 2.5 µm, 1.15 µm, 0.75 µm, 0.56 µm, 0.34 µm, 0.26 µm, and 0.09 µm. The inlet rain hat removes particles above 12 µm. The impactor was operated at a sampling rate of 10 L/min. Measurements of the ambient pressure, temperature, wind speed, wind direction and relative humidity were available from a meteorological system mounted on the roof of EML. Samples were returned to the University of California, Davis where they were analyzed for optical properties, size and morphology, mass, hydrogen (as an organic surrogate), elemental and organic concentrations. Mass was measured in vacuum by scanning transmission ion microscopy. Hydrogen was measured by proton elastic scattering analysis. Because the samples were exposed to vacuum during this analysis, the remaining hydrogen is primarily found in inorganic ions and organic matter. The concentration of elements from sodium through molybdenum and selected heavy elements was established by synchrotron x-ray fluorescence. Complete discussion of these analytical techniques can be found elsewhere (*1*).

In this chapter, we report only on the concentrations and mass distribution of the aerosols sampled at the EML site. Results of other analyses performed on these samples are reported in Chapter 9.

Results and Discussion

DRUM Impactor Mass Characterization

During the month of October 2001 more than 25 plumes were detected passing over EML, which is surrounded by apartment houses where home heating oil is burned. In addition, diesel trucks continually pass by the building, emitting large amounts of sulfur bearing aerosols in the fine particulate mode. These local sources, along with the complex wind patterns in the surrounding area, make for difficult plume source identification. These emissions, contributed to the 25 plumes observed at EML during October 2001.

Figure 1 provides a time history of the aerosol mass concentration for the 8 stages measured on the roof of EML. The mass concentration, in µg/m^3, is shown for each stage. The curves have been clipped and the scales have been set for the clearest presentation. The characteristic shape of each stage is quite unique and is related to various emission sources. During the period from October 1 through October 6 the metropolitan New York area was impacted by a regional plume. Monitoring sites located in the New York and New Jersey metropolitan area (Plate 1) all showed significant increases in the PM-2.5 concentration during this time period (Plate 2). This observation is discussed in more detail below. Superimposed on this regional plume were additional plumes most likely coming from local sources. These peaks can be seen in the various stages of Figure 1. On October 3, the most dramatic peak can be seen superimposed on the regional plume. The total particulates collected on all stages reached a peak mass concentration of 267µg/m^3 at 08:15 with more than 39% of the mass (103 µg/m^3) in the 0.56-0.34 µm stage. The daily average PM-2.5 seen on the roof of EML was 48 µg /m^3 within the EPA ambient air quality standard of 65 µg /m^3 for a 24-hour average (*3*). There are no EPA standards for fine particulates below 1 µm but the sample obtained on October 3rd showed that almost 75% of the total mass at the peak seen at EML was below 1.15 µm. These fine particulates, breathed by unprotected workers at Ground Zero, are known to produce various health problems, as described in Chapter 1 (*4*). The composition of this plume, which lasted approximately 10 hours, was dominated by sulfur (*1*) also a known health risk factor (*4*).

This type of plume is related to a very close source emitting particulates over a very long period of time. It is our contention that this peak seen on October 3rd has its origin from Ground Zero emissions. Further evidence will be forth coming to support this hypothesis. Additional peaks seen during the month may be from Ground Zero emissions but the mass data by itself cannot definitively provide the necessary evidence. Wind field characterization can help

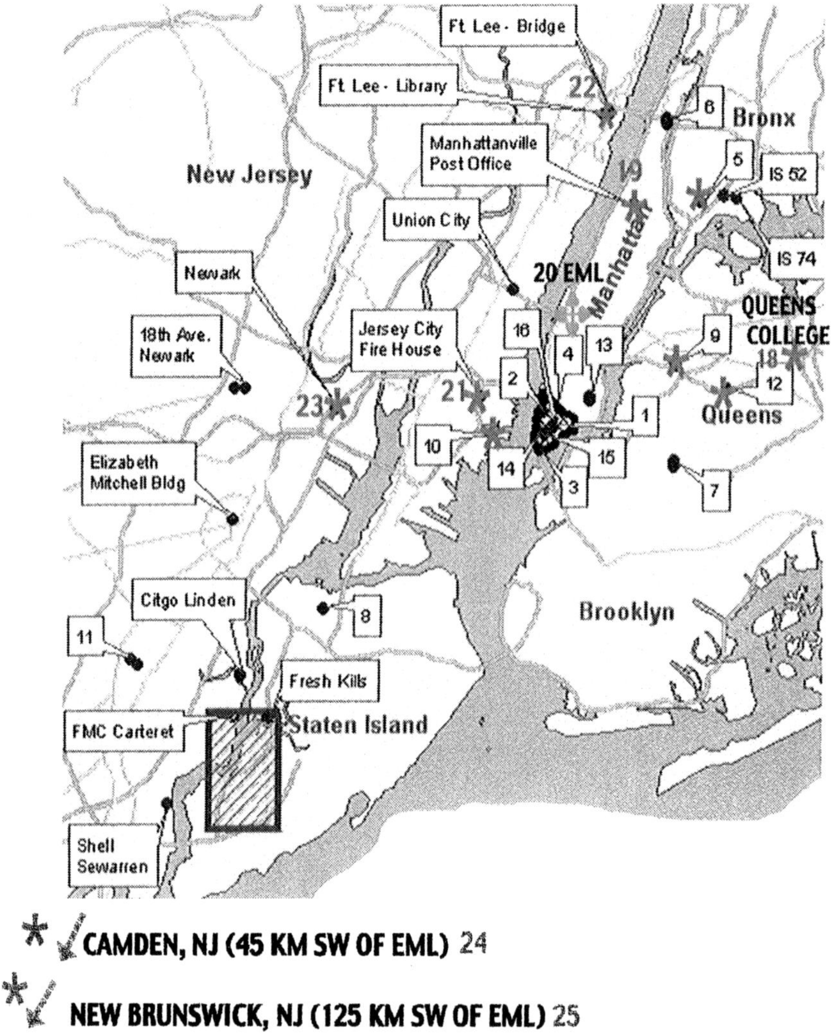

Plate 1. Monitoring site locations in the New York and New Jersey metropolitan areas. Data shown in Plate 1 are from (*): PS 154, Bronx, NY (5), PS 199, Queens, NY (9), Liberty State Park, NJ (10), Maspeth Library, Long Island, NY (12), Queens College, Queens, NY (18), Manhattanville Post Office, NY (19), Environmental Measurements Laboratory, NY (20), Jersey City, NJ (21), Fort Lee, NJ (22), Newark, NJ (23), Camden, NJ (24), and New Brunswick, NJ (25). (See page 11 of color inserts.)

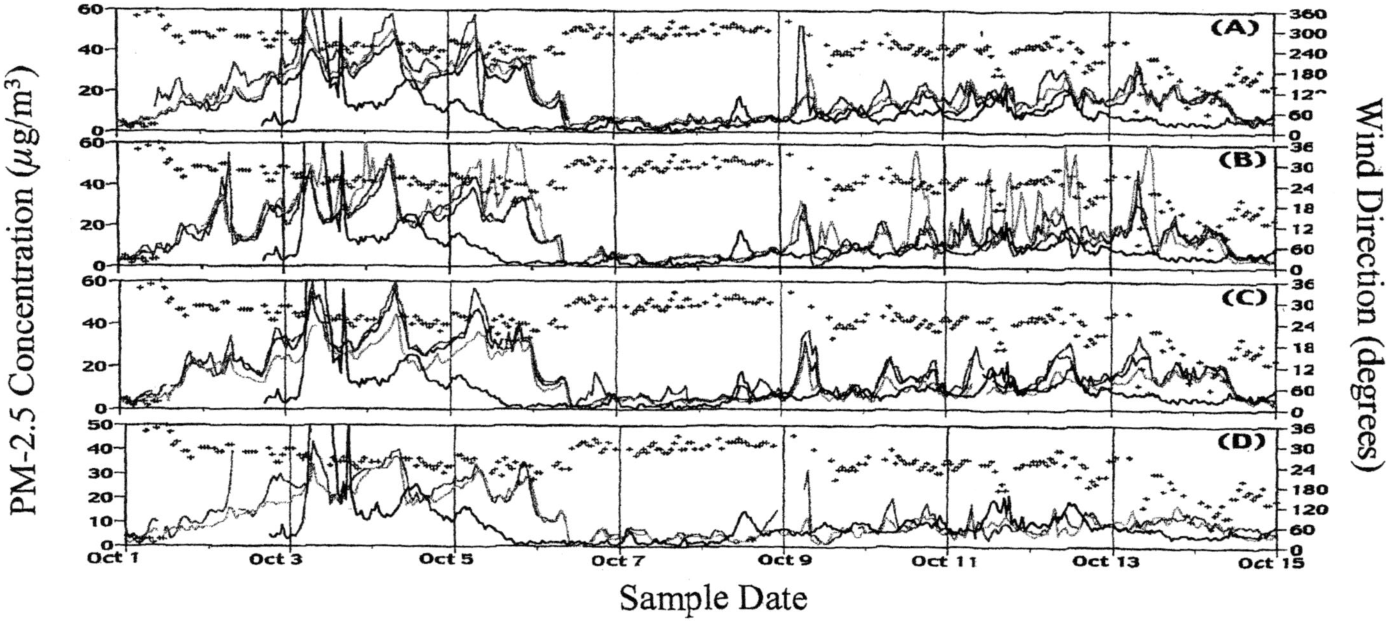

Plate 2. Comparison of measured PM-2.5 concentrations at EPA monitoring sites (colored lines) with the calculated PM-2.5 concentration at EML (black lines) along with wind directions measured in Central Park (black crosses). Sites north of EML; Fort Lee, NJ (blue), Manhattanville, NY (green), and PS 154 , Bronx, NY (red), are shown in (A). Sites west of EML; New Jersey City, NJ (blue), Newark, NJ (green), and Liberty State Park, NJ (red) are shown in (B). Sites East of EML; PS 199, Queens, NY (blue), Maspeth Library, NY (green) and Queens College, NY (red), are shown in (C). Sites south of EML; Camden, NJ (blue) and New Brunswick, NJ (red), are shown in (D). (See page 12 of color inserts.)

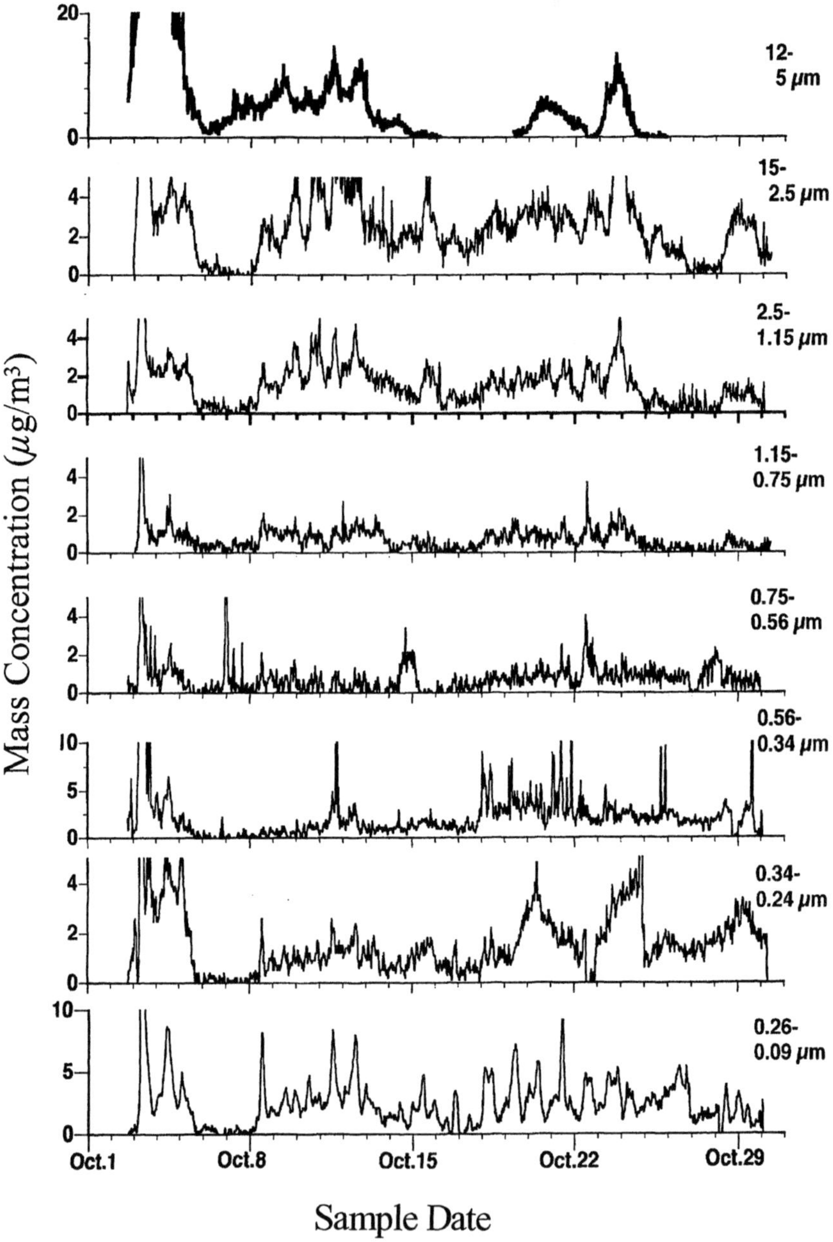

Figure 1. Particulate mass concentrations measured at EML with the 8-stage DRUM impactor for the month of October.

narrow down the identification process but the meteorology of New York City is quite complex. Some peaks are so sharp or of such short duration (Figure 1; 0.56-0.34 μm stage), that they may represent local sources close to EML and not from Ground Zero. The identification of the origin of plumes in lower Manhattan is therefore extremely difficult. No canyon transport models are presently available for New York City. Measurements of wind directions obtained from LaGuardia, Newark, JFK and Tetterborough airports, Central Park and the roof of EML for the period October 1 to October 11 are shown in Plate 3. The observed NOAA wind directions show large differences, approaching 50° for the same time period. Wind directions observed at EML are influenced by local building structures and must be cautiously interpreted. The effects of tall buildings on the wind field make it difficult, except for only a few cases, to say for certain that the observed plumes measured at EML came from Ground Zero when the dominant winds were from the southerly quadrants.

Comparison with Metropolitan Area Measurements

Because of the nature of the impactor stage cuts we are able to sum the last 6 stages of the DRUM sampler to calculate the total PM-2.5 mass concentration at EML. The PM-2.5 data obtained at the EPA monitoring sites surrounding EML (Table I) provide additional evidence on the transport of pollutants past the EML site. With southerly winds blowing past the Ground Zero emissions, one would expect that the transport of the plume past EML would also impact sites north of EML. To help characterize the plumes seen at EML, we have chosen PM-2.5 monitoring sites north, south, east and west of the EML site. The sites in New Jersey are maintained by the New Jersey Department of Environmental Protection, Bureau of Monitoring Sites, while those in New York are maintained by the New York Department of Environmental Conservation, Air Resources Division. Data from these sites can provide further clarification of the source of the plumes passing over EML.

Plate 2 shows a comparison of PM-2.5 concentrations measured at EML with the sites north, west, east, and south of the EML site. We have only included the data for the period of October 1 to October 15 to allow a clear comparison of the sites. Every site shows characteristic peaks throughout the month. As stated earlier, the metropolitan New York City area was under the influence of a regional plume during the first of the month that lasted for at least six days. Every site shows the impact of this plume. On October 3rd every site shows an additional peak superimposed on the broad regional plume. The sites south of EML show this characteristic additional peak appearing a few hours

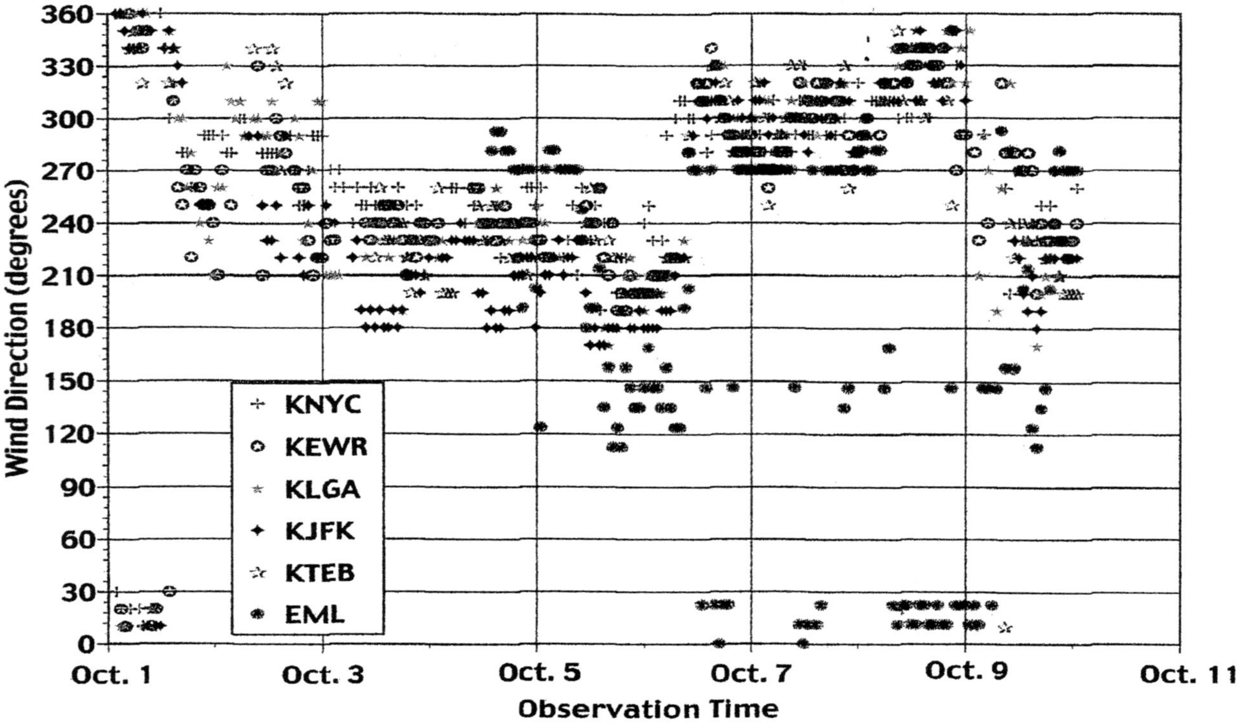

Plate 3. Comparison of wind directions from LaGuardia (KLGA), Newark (KEWR), JFK (KJFK) and Tetterborough (KTEB) airports, Central Park (KNYC) and the roof of EML (EML) for early October 2001.
(See page 13 of color inserts.)

Table I. PM-2.5 Monitoring Sites Surrounding the EML Sampling Site.

Direction From EML	*Site Number*	*Site Name*
North	5	PS 154, Bronx, NY
	19	Manhattanville Post Office, NY
	22	Fort Lee, NJ
West	10	Liberty State Park, NJ
	21	Jersey City, NJ
	23	Newark, NJ
South	24	Camden, NJ
	25	New Brunswick, NJ
East	9	PS199, Queens, NY
	12	Maspeth Library, Long Island, NY
	18	Queens College, Queens, NY

NOTE: Site numbers correspond to Plate 1.

earlier than at EML because of the distance separating the sites. The fact that both the Camden (Plate 1, site 24) and New Brunswick (Plate 1, site 25) sampling sites show the peak at the same time probably reflects that this additional peak is part of the regional plume impacting the area. The sites surrounding EML reported peak concentrations ranging from 30 to 60 $\mu g/m^3$ during this time, similar to measurements taken at Camden and New Brunswick. There is no available chemical data for these sites to clearly identify the source of the plume, but it is known that combustion or secondary aerosols dominate the fine particulate mode.

The unique feature of the data obtained at EML on October 3rd is the magnitude of the concentrations, especially in the fine particulate stages (see Figure 1). It appears to be a coincidence that the Ground Zero emissions reaching EML are superimposed on this regional plume. If this were the only evidence for a Ground Zero influence, one would have to be cautious about associating the plume measured at EML with emissions from Ground Zero. The last stage of the impactor, which collects fine particulates in the size range below 0.2 µm (Figure 1, bottom panel), shows this same characteristic peak and probably reflects a nearby source. In a later section, a detailed size distribution will provide supporting evidence for the origin of the large peak seen at EML on October 3, 2001. During the periods when the winds are southerly it is difficult to say that other peaks seen at EML, as compared to the other monitoring sites, are associated with Ground Zero emissions. Except for October 12, 2001, the sites north of EML show no temporal correlation with peaks observed at EML when the wind is from the southerly direction. The winds during the period of October 10 to October 13 show a southwesterly direction. The only other site that shows any correlation with the EML site during southerly flow is the Liberty State Park site (Plate 1, site 10), which is in close proximity to Ground Zero and just west across the Hudson River. The dates where similar peaks occur at Libeerty State Park and EML are October 11 and October 12. At no other times in October can any of the peaks seen at EML be easily correlated with the data from these other monitoring sites.

Comparison with 2002 Data

Many PM-2.5 plumes can often be seen impacting the metropolitan New York City area. During the month of October in the year 2002, regular patterns of plumes were observed at the Manhattanville Post Office (Plate 1, site 19) and PS 154 (Plate 1, site 5) monitoring sites. Data obtained at these sites in October 2001 and October 2002 are compared with the calculated PM-2.5 concentrations obtained at EML (Plate 1, site 20) for October 2001 in Plate 4. Table II provides a comparison of the monthly average PM-2.5 concentrations for October 2001 at these sites along with their standard deviations. The PM-2.5 averages and standard deviations in Table II are comparable for all sites except for EML (site 20). If the major peak occurring on October 3rd is removed from the EML monthly average, the average value drops to 8.4 $\pm$ 5 $\mu g/m^3$ but the standard deviation becomes closer to that seen at the other sites.

Table II. October Monthly Average PM-2.5 concentrations.

Site Number	*Year*	*Concentration ($\mu g/m^3$)*	*Standard Deviation ($\mu g/m^3$)*
19	2001	13.5	8
19	2002	12.6	9
5	2001	14.6	9
5	2001	12.2	8
20	2001	9.9	15

NOTE: Site numbers correspond to Plate 1.

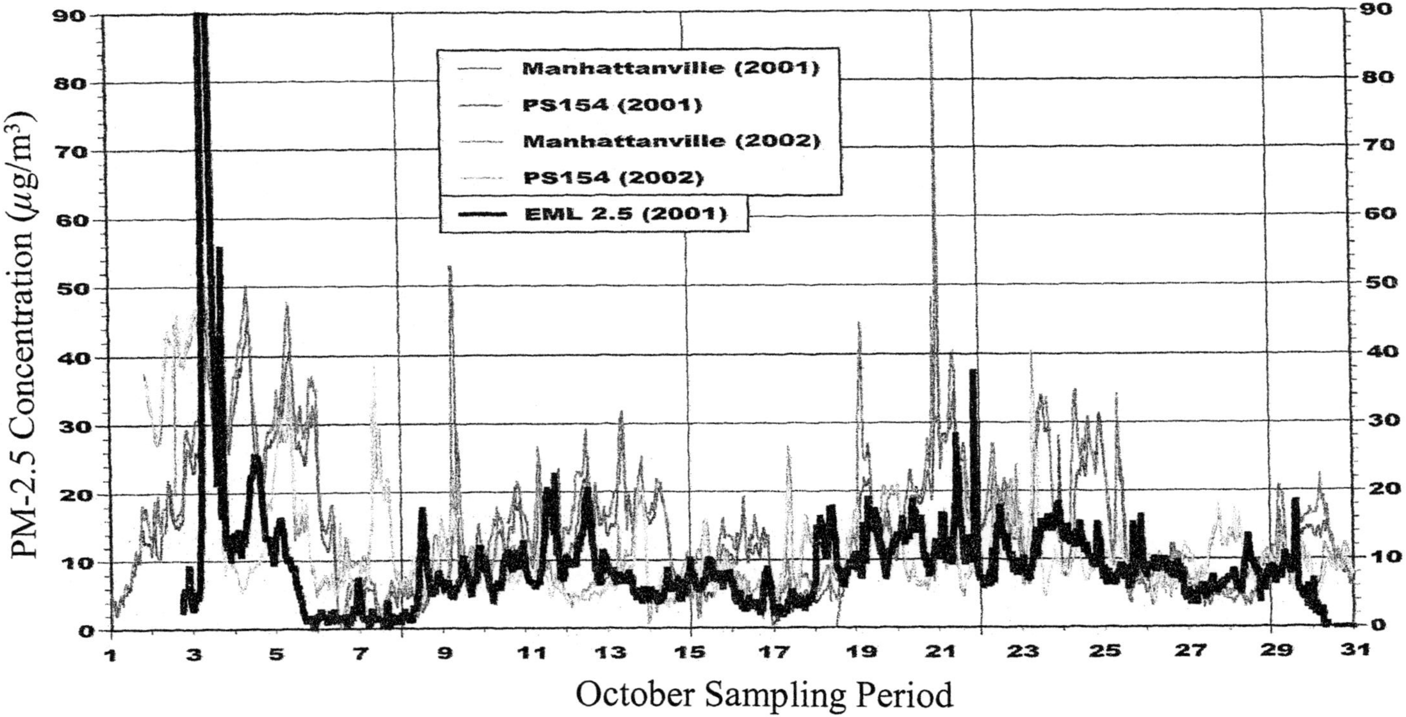

Plate 4. Comparison of October 2001 and 2002 PM-2.5 concentrations measured at the EPA monitoring sites at Manhattanville Post Office (Figure 2, site 19) and PS 154 (Figure 2, site 5) with the calculated PM-2.5 concentration at EML for October 2001. (See page 14 of color inserts.)

In summary, the aerosol plumes seen at EML are common throughout the metropolitan New York City area and their presence doesn't necessarily indicate impacts from emissions at Ground Zero when the winds are blowing from the southern sector.

Aerosol Size Distributions

Background Size Distributions

To help identify the source of plumes passing over EML, we analyzed the aerosol mass size distribution during the times of peak PM-2.5 concentrations, as well as during periods when the concentrations were the lowest, especially during a rain event. It is well known that the modes in the particle size distribution provide insight into the atmospheric mechanisms of aerosol formation (Chapter 1, Figure 1) (*5*). It is these modes that play a key role in exposure and risk assessment. The data obtained on October 17 show some of the lowest concentrations of the month. For comparison purposes, these low periods could be considered as background or clean events. The aerosol size distributions of this background event, shown in Figure 2 (▲), are bimodal with the fine particle mode occurring at a larger particle diameter than observed during the October 3rd peak also seen in Figure 2 (■).

October 3rd Size Distributions

Plate 5 (a-d) provides a series of plots of the normalized mass size distribution for the period 12 AM to 7:30 PM, October 3, 2001. The early morning distributions (12 AM to 3:00 AM) are quite uniform and dominated by large aerosols (Plate 5a). Beginning with the next sampling period (3:45 AM), a transition is seen from the large aerosols to finer aerosols (Plate 5b). The region between the 0.34 µm and 0.56 µm shows fine aerosol peaks systemically growing with time, reflecting source emissions changes. In a similar fashion the coarse fraction decreases with time. This transition is dominated by a combustion source from Ground Zero. What also can be seen is the transition back to the coarse dominated aerosol mode (Plate 5c) in the early afternoon. By 2:15 PM the systematic transition is not evident (Plate 5d). An elevated peak in the region of 0.34 µm and 0.56 µm at 5:15 PM is seen in Plate 5d. This aerosol distribution reflects the sudden change in source strength that can occur at Ground Zero. The combined effect of Ground Zero emission changes and the regional plume impact is evident in this data set. What can be clearly seen from these plots is that the October 3rd plume could only have come from a local

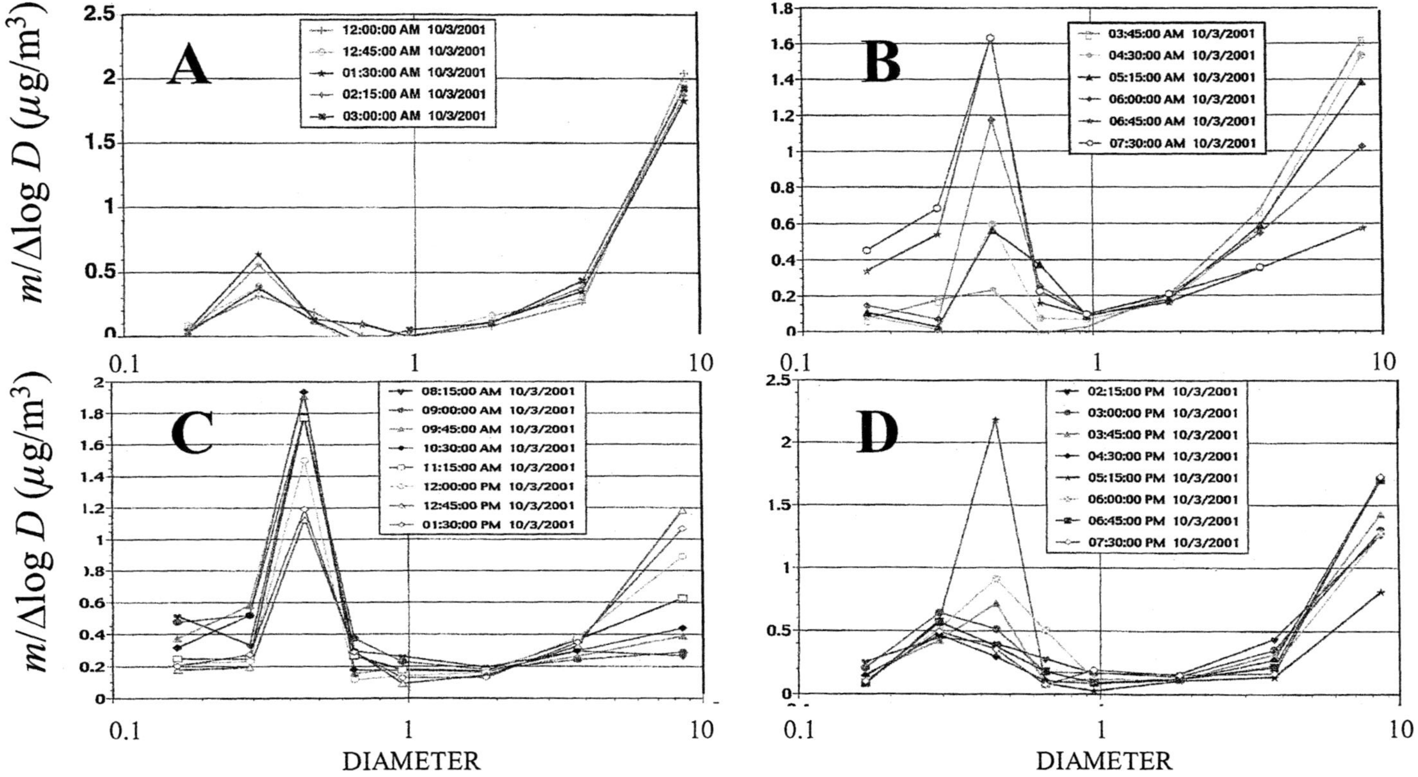

Plate 5. Normalized aerosol mass size distribution observed on October 3 when EML intercepted a plume from Ground Zero. (A) 12:00 am to 03:00 am. (B) 03:45 am to 07:30 am (C) 8:15 am to 1:30 pm (D) 2:15 pm to 7:30 pm.
(See page 15 of color inserts.)

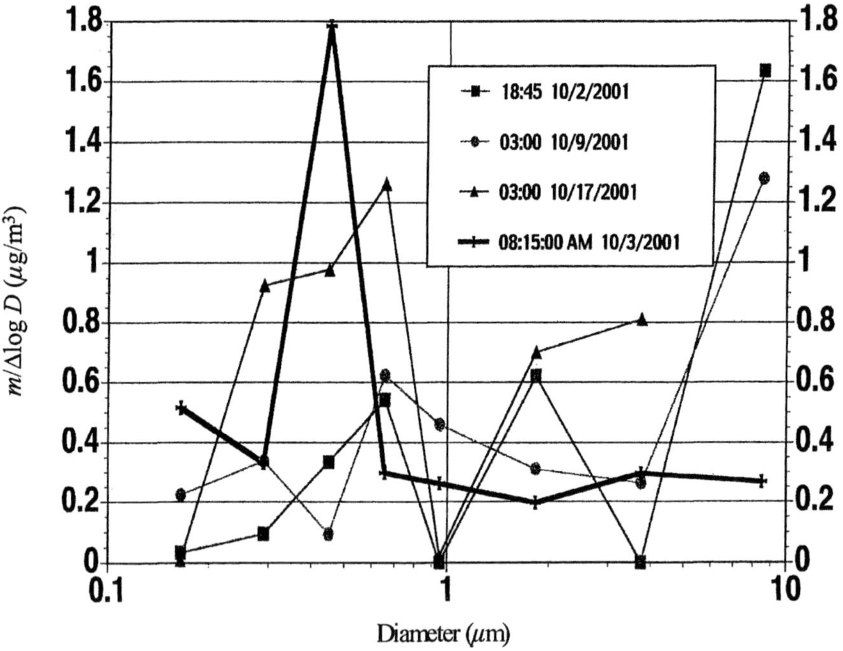

Figure 2. Normalized aerosol mass size distribution obtained during low concentration background events compared to the mass size distribution obtained at EML on October 3, 2001.

source and not from a regional plume. The magnitude of the concentrations measured at EML leads us to believe that the plume had its origin at Ground Zero. The chemistry of this plume has been discussed in detail elsewhere (*1*). The aerosol size distributions can be useful in risk assessment analyses.

Other Size Distributions during October

The regional aerosol plume lasted until October 5th. There were 25 plumes investigated during October and only three plumes had size distributions similar to that of October 3rd. These three events are shown in Figure 3 compared to the October 3rd plume. The distributions selected are based on the peak PM-2.5

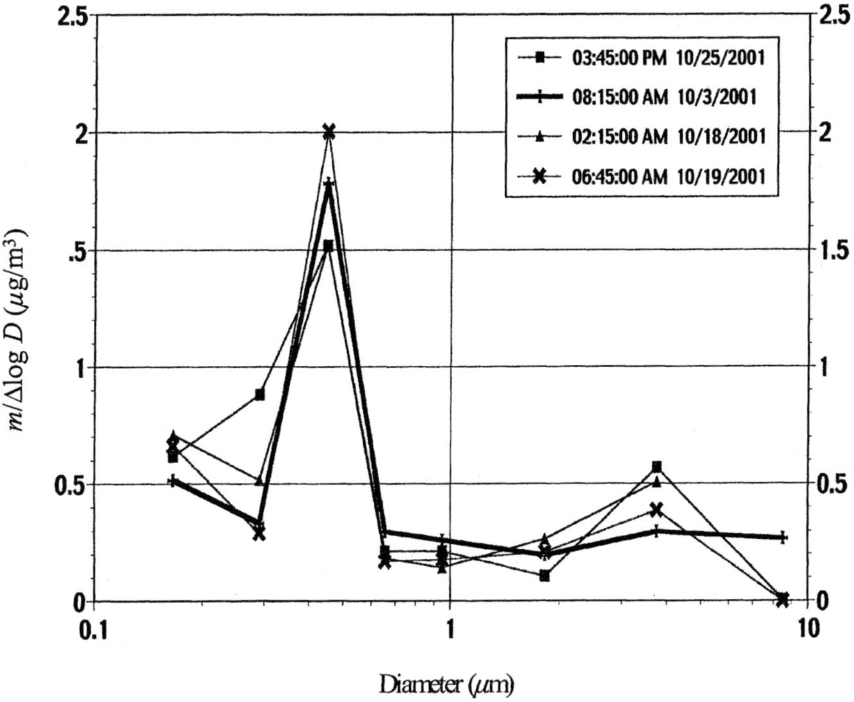

Figure 3. Normalized aerosol mass size distribution obtained at EML on October 3, 18, 19, and 25, 2001.

concentrations during each event. The temporal transformation from coarse to fine particle sizes did not occur for these three dates. Also, none of the monitoring sites north of EML showed peaks occurring approximately at the same time as was seen at EML for any of these dates. Of these, the only date that had winds from a southwesterly direction was October 19, but no monitoring site north of EML showed a similar concentration pattern on that date. In future publications, the analysis of the aerosol chemical composition combined with the mass and size distribution data of these plumes will help to further clarify the origin of each of the plumes.

Conclusions

The Environmental Measurements Laboratory is surrounded by apartment houses where home heating oil is used. In addition, diesel trucks continually pass by the building, emitting large amounts of sulfur bearing aerosols. These local sources, along with the complex wind patterns in the surrounding area, make for difficult plume source identification. These emissions, contributed to the 25 plumes observed at EML during October 2001. Analysis of the data leads us to believe that one plume, on October 3rd, can be identified with emissions from Ground Zero. The October 3rd plume, at the peak concentration, had more than 39% of the mass in a very narrow size range, between 0.34 µm and 0.56µm. Investigation of the size distribution over a 20-hour period provided evidence of the dynamic processes occurring at Ground Zero. No sources in the local area, other than from Ground Zero emissions, could show a shifting of the bimodal distribution as observed in this data set. The distribution was first dominated by coarse particles, and then changed to one dominated by fine particles over the time frame examined. Comparison of the EML data with data from surrounding monitoring sites provided additional evidence and insight into the possible transport of material from Ground Zero to the surrounding community. It is this comparison that provided the necessary information to determine what was happening on October 3rd in the NYC metropolitan area as well as over EML.

A more extensive network of DRUM type samplers located in a region, especially at upwind and downwind locations, can provide the necessary database for aerosol characterization. The DRUM sampler used at EML was the only sampler available for high-resolution aerosol measurements, including mass and chemical composition. These results could not be compared with any other data set.

Acknowledgement

This research was carried out during the employment of Robert Leifer at the Environmental Measurements Laboratory, U.S. Department of Energy (Now U.S. Department of Homeland Security), New York, NY 10014. Robert Leifer is currently retired. Present Address: 14 Holland Lane, Monsey, NY 10952.

We would like to thank Peter Roiz, Rienzie Perera, and Fabian Raccah for their help in setting up the equipment at EML. The publication of this paper is sponsored by EML.

References

1. Cahill T. A.; Cliff, S.S.; Perry, K.P.; Jimenez-Cruz, M.; Bench, G.; Grant, P.; Ueda, D.; Shackelford, J. F.; Dunlap, M.; Meier, M.; Kelly, P.B.; Riddle, S.; Selco, J; Leifer, R. *Aerosol Science and Technology*, **2004,** *38*,165.
2. "EPA Response to 9-11: Fixed Air Monitoring Locations" [http://www.epa.gov/wtc/monstations.htm#regional] US. Environmental Protection Agency, 2004.
3. National Ambient Air Quality Standards (NAAQS) for particulate matter (PM) set in 1997 (Chapters 1 through 5, EPA/600/P-99/002aD) and Volume II (Chapters 6 through 9, EPA/600/P-99/002bD).
4. Lippmann, M.; Frampton, M.; Schwartz, J.; Dockery, D.; Schlesinger, R.; Koutrakis, P.; Froines, J.; Nel, A.; Finkelstein, J.; Godleski, J.; Kaufman, J.; Koenig, J.; Arson, T.; Luchtel, D.; Liu, L-J.S.;OBerdörster, G.; Peters, A.; Sarnat, J.; Sioutas, C.; Suh, H.; Sullivan, J.; Utell, M.; Wichmann, E.; Zelikoff, J. The U.S. Environmental Protection Agency Particulate Matter Health Effects Research Centers Program: A Midcourse Report of Status, Progress, and Plans, Environmental Health Perspectives, June 2003, III #8, 1024.
5. Morawska, L.; Thomas, S.; Jamriskka, M.; Johnson G. *Atmospheric Environment*, **1999**, *33*, 4401.

Chapter 9

Very Fine Aerosols from the World Trade Center Collapse Piles: Anaerobic Incineration?

Thomas A. Cahill[1], Steven S. Cliff[1], James F. Shackelford[1], Michael L. Meier[1], Michael R. Dunlap[1], Kevin D. Perry[2], Graham S. Bench[3], and Robert Z. Leifer[4]

[1]DELTA Group, Department of Applied Science, University of California at Davis, One Shields Avenue, Davis, CA 95616
[2]Department of Meteorology, University of Utah, Salt Lake City, UT 84112–0110
[3]Center for Accelerator Mass Spectrometry, Lawrence Livermore National Laboratory, 7000 East Avenue, Livermore, CA 94550–9234
[4]U.S. Department of Homeland Security, Environmental Measurements Laboratory, 201 Varick Street, New York, NY 10014–7447

By September 14, three days after the initial World Trade Center collapse, efforts at fire suppression and heavy rainfall had extinguished the immediate surface fires. From then until roughly mid-December, the collapse piles continuously emitted an acrid smoke and fume in the smoldering phase of the event. A knowledge of the sources, nature, and concentration of these aerosols is important for evaluation and alleviation of the health effects on workers and nearby residents. Here we build on our earlier work to ascribe these aerosols to similar processes that occur in urban incinerators. The simultaneous presence of finely powdered (circa 5 µm) and highly basic (pH 11 to 12) cement dust and high levels of very fine (< 0.25 µm) sulfuric acid fumes helps explain the observed health impacts, as do the high levels of very fine mass and its constituents. The unprecedented levels of several metals in the very fine mode aerosols can be tied to the liberation of those metals that are both present in elevated concentrations in the debris and have depressed volatility temperatures caused by the presence of organic materials and chlorine under anaerobic conditions. Health concerns focus on the workers at the site, as plume lofting protected most of New York City.

Introduction

The collapse of the World Trade Center structures (South Tower, North Tower, and WTC 7) presented two very different types of air pollution events as outlined in Chapter 2. These include the initial fires and collapse-derived "dust storm" and the continuing emissions from the debris piles. By September 14, three days after the initial World Trade Center collapse, efforts at fire suppression and heavy rainfall had extinguished the immediate surface fires. From then until roughly mid-December, the collapse piles continuously emitted an acrid smoke and fume in the smoldering phase of the event. The presence of fuels, including diesel and electrical insulating oils and combustible materials in the WTC buildings, and lack of oxygen, were two factors that allowed the smoldering fires to burn for over three months. While this chapter focuses on the second air pollution event, the three month smolder phase, both situations shared the unusual aspect of a massive ground level source of particulate matter in a highly populated area with potential health impacts.

The U.S. EPA has summarized five causal factors most likely to explain the statistically solid data connecting fine PM-2.5 aerosols and human health (*1, 2*).

1. Biological aerosols (bacteria, molds, viruses, etc.).
2. Acidic aerosols.
3. Very fine/ultra fine (< 0.25 μm) insoluble aerosols.
4. Fine transition metals.
5. High temperature organics.

A knowledge of the sources, nature, and concentration of the WTC aerosols is important for the evaluation and alleviation of the health effects on workers and nearby residents. We especially needed to ascertain if these potentially causal factors of aerosol health impacts were present in the WTC plumes.

Experimental Methods

In order to better understand the WTC aerosols a slotted Davis Rotating-drum Universal-size-cut Monitoring (DRUM) impactor, developed by the Detection and Evaluation of Long-Range Transport of Aerosols (DELTA) Group at the University of California at Davis (*3-8*) was shipped to New York and set up on the roof of the Department of Energy's (DOE) Environmental Measurement Laboratory (EML) at 201 Varick Street, 45 m above ground level and roughly 1.5 km north-northeast of the World Trade Center site. The DRUM operated until late December, after the last surface fires had been extinguished (*9*).

The samples were collected in 8 size modes from the inlet (circa 12 μm) to 5.0 μm, 5-2.5 μm, 2.5-1.15 μm, 1.15-0.75 μm, 0.75-0.56 μm, 0.56-0.34 μm, 0.34-0.26 μm, and 0.26-0.09 μm aerodynamic diameter. Samples were analyzed by a suite of non-destructive spatially resolved beam based techniques (*7, 10, 11*) for mass, hydrogen (as an organic surrogate), elements from sodium to molybdenum, some heavy elements by synchrotron x-ray fluorescence (S-XRF), and particle size by scanning electron microscopy. Time resolution ranged from 1 ½ hr to 3 hr depending on the size of the exciting beam.

The presence of the WTC plume at the Varick Street site depended on both the source emission rate and the meteorology (*12*), with both wind direction and plume lofting from the hot collapse piles as the key parameters. In addition, the nature of the aerosols themselves, especially as they deviated in concentration, size, and composition from upwind and regional aerosols, also provided information identifying the WTC impacts at the site. We chose six parameters as tracers of the WTC plume. These parameters deviated by an order of magnitude or more from typical regional and urban values, and all of which appeared in time defined plumes of typically 3 to 6 hours duration. These parameters, listed in Table I, were used to establishing the influence of the WTC plume at the Varick Street sampling site.

Five events observed at Varrick Street met all 6 criteria, four more events met 5 of the 6 criteria, and these 9 events are viewed as "highly probable events". Three events met only one or two of the parameters and these are assumed to be "non-WTC plumes" or uncertain sources, including a local power plant. Six of the events met 3 or 4 of the parameters and these are labeled "probable WTC influence". The analysis that follows uses only data from the 9 highly probable events.

In terms of the WTC plume impact on New York City, on most days the plumes lofted above NYC, to roughly 500 m (*13*), so that only individuals on the collapse piles or very near the WTC site were subject to these aerosols.

Table I. Criteria for Establishing the WTC Plume Influence at the Varick Street Sampling Site.

Parameter type	*Parameter definition*	*Criterion for WTC influence*	*# events meeting the criterion*
Very fine Aerosols	0.26 - 0.09 μm	> 3 μg/m^3	18
Wind direction	HYSPLIT regional trajectories	SSW quadrant	14 + 1 calm
Very fine organics	Organic matter by hydrogen (OMH)	> 1.0 μg/m^3	16
Fine Cement dust	5.0 - 2.5 μm	Si: EF > 2.5	16
Coarse Sulfate	5.0 - 2.5 μm	> 0.3 μg/m^3	15
Ground based haze	LaGuardia Airport visibility	L_v < 15 km	5

NOTE: EF is enrichment factor. L_v is limit of visibility.

Results

Table II shows very fine particle concentrations for six of the highly probable events, plus background data from October 7, a non-WTC plume event. We also include for comparison, data from earlier studies of two highly impacted sites (*2, 14, 15*) with similar characteristics, namely a maximum 3 hr plume in a study of at least 4 weeks duration and using the same size and analytical protocols.

Coarse particles at the sampling site were similar to the initial collapse aerosols (cement, dry wall, glass, etc.) (*16*), but also had chemicals and soot from the ongoing combustion (*17, 18*). The presence of unprecedented levels of very fine particles (0.26 - 0.09 μm) by mass and number in narrow plumes was more typical of an industrial source. Upwind sources were a very minor contribution as shown by direct comparison with upwind aerosol sites (*9, 19, 20*) and the size distributions were grossly different from typical ambient aerosols as described in Chapter 8 (*21*).

Data taken within 200 m of the site by the EPA (*19*) and meteorological analysis (*9*) indicate the mean 24 hr PM-2.5 aerosol mass at the site was in excess of 200 μg/m^3 from mid September through early October, much of it in the very fine and presumably ultra-fine modes, together with heavy PM-10 dusts. The very fine particle mass fell off sharply in October (*9*) and tests done near the

Table II. Results of Very Fine Aerosol (0.26 - 0.09 µm) Sampling at Varick Street, October, 2001, Compared to Samples Obtained from Kuwait, July, 1991 (*14*) and Beijing during ACE – Asia, March, 2001(*15*).

Date	*WTC impact*	Mass	*Organics*	SiO_2	H_2SO_4	*V*	*Ni*
Oct. 7	No	0.5	0.04	0.02	0.1	0.1	0.1
Oct. 3	Yes	50.6	9.3	1.4	17.1	115	22.7
Oct. 4	Yes	9.3	3.2	2.6	3.7	61	8.0
Oct. 5	Yes	4.7	1.5	0.9	1.9	3	1.3
Oct.12	Yes	8.2	3.0	2.9	2.2	24	7.1
Oct.15	Yes	4.8	1.6	1.3	1.1	3	17.3
Oct.24	Yes	3.8	1.0	0.2	1.6	5	6.7
Oct.29	No	2.4	1.2	0.07	0.9	2	1.6
Kuwait	-	Na	Na	0.6	5.5	Na	5.0
Beijing	-	Na	Na	1.1	6.7	0.8	1.8

NOTE: Units are $\mu g/m^3$.Units for V and Ni are ng/m^3. Na is not available.

WTC site for very fine particles in May, 2002, were generally < 10% of the October, 2001 24 hr average for plume impact days at Varick Street (*22*).

Discussion

Plate 1 shows an example of typical crustal elements (silicon, calcium, iron, and aluminum) in the very fine (0.26 - 0.09 µm) mode from October 2 to October 30, 2001. There are several important points to note in this plate, First, the crustal elements aluminum, calcium, and iron seen so abundantly in the coarser particles are essentially absent in this very fine mode. Thus, the signal is not due to a sampling artifact, as is also shown by the low silicon values in the sub-micron sizes, and the aerosol formation mechanisms are not simply mechanical grinding. Second, the silicon only appears when the wind is from the WTC site. Third, the peak values trend downward in October, and even the sample color grows fainter. October 24 was by all criteria (meteorology, coarse calcium, etc.) a very strong plume impact period, but by then the silicon has gone. The same pattern is seen in some other anthropogenic elements, notably vanadium, in both the coarse and fine modes.

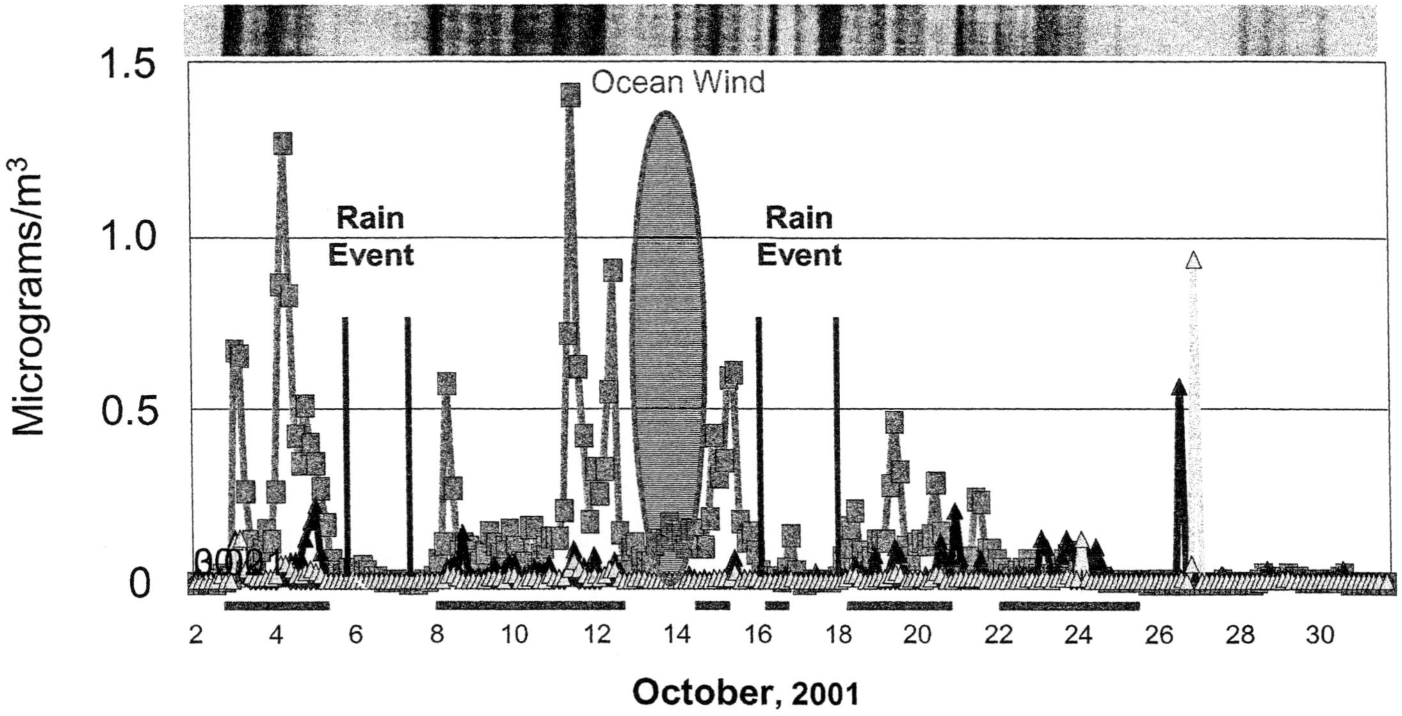

Plate 1. Typical crustal elements; silicon (—), calcium (—), iron ((—), and aluminum (—) in the very fine particle mode (0.26 - 0.09 μm) of samples obtained at Varick Strees during October, 2001. At top is shown the reflected light picture of this DRUM stage. Periods of favorable wind direction that would bring the WTC plume to Varick Street are indicated below the plot (—), Two rain events and an eastern ocean wind are also indicated.

These data then pose problems in determining the aerosol sources. We see aerosols typical of a high temperature industrial source, yet the temperatures in the near sub-surface collapse piles were not very high. We see some elements in the aerosols vastly enhanced over typical earth crustal values and others, equally present in the dust, absent in the aerosols. We propose a model derived from the theory and data describing municipal incinerators. It was discovered that the presence of organic matter and especially chlorine in the wastes liberated metals during incineration by greatly reducing their volatility temperatures (*23*).

In Table III, we present the concentrations in the WTC bulk dusts compared to their volatility temperatures in the presence of 10% chlorine for EPA criterion metals and a few other species of interest. These data are ordered by their volatility temperature (the point at which the vapor pressure reaches 10^{-6} atmospheres). The concentrations of many metals are greatly enhanced over typical Earth crustal levels, as to be expected in the smoking wreckage of highly computerized building. The top elements, chromium, beryllium, and barium, all have volatility temperatures that are probably higher than most parts

Table III. Concentration in the WTC Bulk Dust and Volatility Temperatures of Elements in the Presence of 10% Chlorine (*23*).

Element	*Boiling Point*	*Earth Crustal*	*Dust (24)*	*Dust (16)*	*Volatility Temp*	*Principal Species*
Chromium	2639	102	71.5	165	1594	CrO_2/O_3
Beryllium	1280	2.8	1.7	3.2	1042	$Be(OH)_2$
Barium	1634	425	195	381	895	$BaCl_2$
Nickel	2834	84	15.5	43.5	686	$NiCl_2$
Antimony	697	0.2	Na	Na	653	Sb_2O_3
Silver	2190	0.004	4.9	2.3	620	AgCl
Selenium	Na	0.05	< 1	Na	315	SeO_2
Cadmium	761	0.15	3.8	7.2	211	Cd
Vanadium	3480	120	18.3	38.9	147	VCl_4
Thallium	1464	9.6	< 1	1.4	136	TlOH
Osmium	4224	0.0015	Na	Na	40	OsO_4
Arsenic	814	1.8	< 1	2.6	32	As_2O_3
Mercury	353	0.085	0.37	Na	25	Hg
Silicon	1725	9000	Na	Na	12	$SiCl_{414}$
Lead	1748	14	98	305	-12	$PbCl_4$

NOTE: Concentration units are ppm. Units for temperature are °C. Na is not available

SOURCE: Bulk dust compositions were obtained from the EPA (*24*) and Lioy *et al.* (*16*).

of the collapse piles, since steel was not melted but was observed to glow red when pulled from the piles. Nickel, antimony, and silver are marginal, but all the rest have volatility temperatures well below the documented temperature of the collapse piles during October.

The mechanism we propose for the enhanced elemental concentrations in the WTC aerosols is the formation of metal containing gasses in the oxygen-poor, organic and chlorine rich collapse piles, followed by emission through the rubble as a gas, and conversion at or near the surface into oxidized forms and thus very fine particles. The other factor determining aerosol composition is the concentrations in the WTC dusts. The elements beryllium (which we can not analyze by S-XRF), antimony, silver, selenium, cadmium, thallium, osmium, arsenic and mercury are present only at the ppm levels in the WTC dusts, and thus would be at low levels in aerosols no matter what their volatility temperatures were.

Combining these data into Table IV, we see that the elements actually observed in the aerosols share the characteristics of relatively high concentrations in the collapse pile debris along with low volatility temperatures. Other species, including most other crustal elements absent in the very fine mode, do not have depressed volatility temperatures and thus are not seen in the very fine aerosols. Chromium and barium, both with high volatility temperatures,

Table IV. Average Concentrations in the WTC Very Fine Aerosols Collected on October 3 (0.26 – 0.09 μm), in Background Air Collected on October 7, and in WTC Dust Samples, Compared to their Volatility Temperatures in the Presence of 10% Chlorine.

	Oct. 7 background	*Oct. 3 WTC plume*	*WTC dust*	*Volatility temperature*
Silicon	11	698	abundant	12
Vanadium	0.1	114	30	147
Lead	< 0.5	26	200	-12
Nickel	0.1	23	30	686
Chromium	<0.1	1.5	120	1594
Barium	<0.1	<0.5	290	895

NOTE: Units for aerosols are ng/m3. Units for dust are ppm. Temperature units are °C.

NOTE: Background air obtained on October 7, 2001 contained 0.53 $\mu g/m^3$ of fine aerosols with 0.04 $\mu g/m^3$ organics, and 0.04 $\mu g/m^3$ sulfur. The WTC plume obtained on October 3, 2001 contained 50.7 $\mu g/m^3$ of fine aerosols, with 9.3 $\mu g/m^3$ organics and 5.6 $\mu g/m^3$ sulfur.

although abundant in the dusts, are largely absent in the aerosols. Vanadium and nickel are equally present in the dust, but because of the much lower vanadium volatility temperature, vanadium has 5 times more mass in the aerosols than nickel.

Further evidence for the role of chlorine in releasing certain metals from the collapse piles is shown in Figure 1. Vanadium and chlorine are together in the very fine mode aerosols but at a ratio well below that of VCl_4. The chlorine also has very low levels of associated sodium and thus is not present as NaCl.

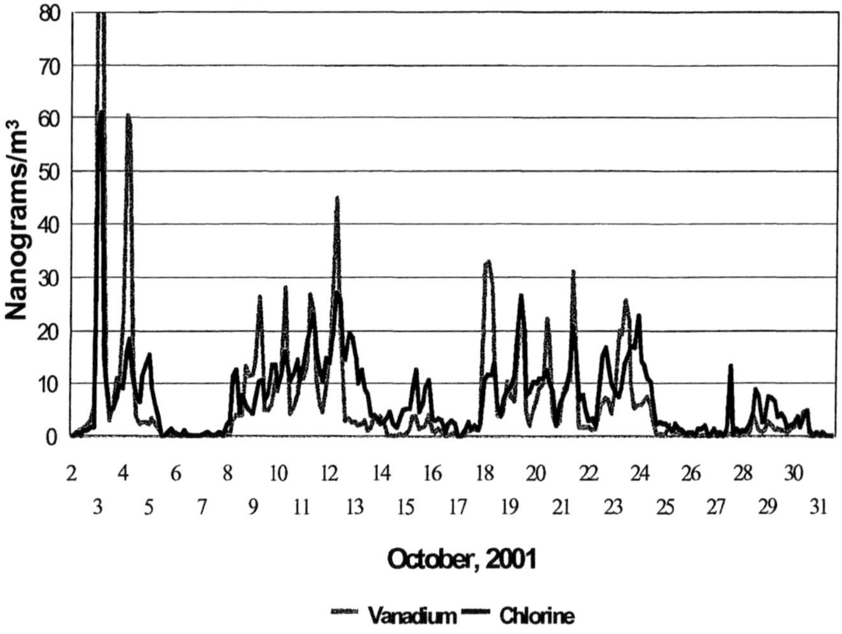

Figure 1. Vanadium and chlorine concentrations in the very fine (0.26 - 0.09 μm) aerosol mode at Varick Street during October, 2001.

Conclusions

By September 14, three days after the initial World Trade Center collapse, efforts at fire suppression and heavy rainfall had extinguished the immediate surface fires. From then until roughly mid-December, the collapse piles continuously emitted an acrid smoke and fume in the smoldering phase of the event. Knowledge of the sources, nature, and concentration of these aerosols is important for evaluation and alleviation of the health effects on workers and nearby residents. Data obtained by the U.S. EPA within 200 m of the site and meteorological analysis indicate the mean 24 hr PM-2.5 aerosol mass at the site was in excess of 200 $\mu g/m^3$ from mid September through early October, much of it in the very fine and presumably ultra-fine modes, together with heavy PM-10 dusts. Very fine particles fell off sharply in October and tests done near the WTC site in May, 2002, were generally < 10% of the October, 2001 plume impact days at Varick Street.

The simultaneous presence of high concentrations of very fine powdered and highly basic (pH 9 to 12) cement dust and high levels of very fine (< 0.25 µm) sulfuric acid fumes help explain observed health impacts. The unprecedented levels of several metals in the very fine mode can be tied to liberation of those metals that are both present in the debris and have depressed volatility temperatures caused by the presence of organic materials and chlorine in anaerobic, fuel rich conditions. These plumes have the potential for health impacts, as potentially causal health factors in very fine particles reached unprecedented ambient levels in the plumes from the WTC collapse piles. This would affect workers at the site who did not use adequate respirators, but because of plume lofting, would not have affected most New York residents well away from the site.

Acknowledgements

The authors would like to gratefully acknowledge the scores of people and numerous organizations that contributed their time and effort to this work, most on a volunteer basis. The work benefited greatly from sampler and analytical developments occasioned by the ACE-Asia study NSF ATM 0080225, support from the center for Accelerator Mass Spectrometry, LLNL, and resources Center for X-ray Optics of the Advanced Light Source, LBNL. This work was performed in part under the auspices of the U.S. Department of Energy by University of California, Lawrence Livermore National Laboratory under Contract No. W-7405-Eng-48. We would like to especially thank William

Wilson, U.S. EPA, for his careful review and thoughtful discussions which led to significant improvements in the paper, the very helpful analysis of Prof. Ian Kennedy (UCD Engineering). The authors gratefully acknowledge the NOAA Air Resources Laboratory (ARL) for the provision of the HYSPLIT transport and dispersion model and the ARL READY web site used in this publication, and the support of the American Lung Association for the May, 2002 study.

References

1. Devlin, R. "What we know and what we think we know about health effects of fine particles; Is it enough?" Paper presented at The American Association for Aerosol Research Annual Conference, Charlotte, NC, October 7-11, 2002.
2. EPA 600/P-95/001af/ *Vol. 1*, **1996**, 6-170 - 6-220.
3. Lundgren, D.A., *J. Air Poll. Cont. Assoc.* **1967,** *17*, 225.
4. Cahill, T. A.; Goodart, C.; Nelson, J. W; Eldre, R. A.; Nasstrom, J.S.; Feeney., P.J. Proceedings of International Symposium on Particulate and Multi-phase Processes. Ariman, T.; Veziroglu, T. N. Eds. Hemisphere Publishing Corporation, Washington, D.C. 2, 319.
5. Marple, V.; Rubow, K;, Anath, G.; Fissan, H. *J. Aerosol Sci.* **1986,** *17*, 489.
6. Raabe, O.; Braaten, G., D. A.; Axelbaum, R. L.; Teague, S. V.; Cahill, T. A. J. *Aerosol Sci.* **1998,** *19*, 183.
7. Cahill, T. A; Wakabayashi, P. *Measurement Challenges in Atmospheric Chemistry*; L. Newman, Ed. American Chemical Society: Washington, DC, 1993; Chapter 7, p 211.
8. Raabe, O. *private communication* **1997**.
9. Cahill, T. A.; Cliff, S. S.; Perry, K. D.; Jimenez-Cruz, M.; Bench, G.; Grant, P.; Ueda, D.; Shackelford, J. F.; Dunlap, M.; Meier, M.; Kelly, P. B.; Riddle, S.; Selco J.; Leifer, R. *Aerosol Sci. Technol.* **2004,** *38,* 165.
10. Bench, G.; Grant, P.G.; Ueda, D.;. Cliff, S.S; Perry, K.D.; Cahill, T.A. *Aerosol Sci.Technol.* **2002,** *36*, 642.
11. DQAP 7.02; UC Davis DELTA Group DRUM Quality Assurance Protocols, v. 7.02, 2002; [http://delta.ucdavis.edu/technology.htm].
12. "NOAA ARL Real-time Environmental Applications and Display sYstem Dispersion Models" NOAA ARL HYSPLIT Model; [http://www.arl.noaa.gov/ready/hysplit4.html] National Oceanic and Atmospheric Association.
13. North East States for Coordinated Air Use Management (NESCAUM) visibility photographs, New Jersey to WTC towers, 2001.
14. Reid, J. S.; Cahill, T. A.; Dunlap, M. R.; *Atmos. Environ.* **1994,** *28*, 2227.

15. Seinfeld, J.R., Carmichael, G. R., Arimoto, R., Conant, W. C., Brechtel, F. J., Bates; T. S.; Cahill, T. A; Clarke, A.D.; Flateau, P. J.; Huebert, B. J.; Kim, J. J.; Markowicz, K. M.; Masonis, S. J.; Quinn, P. K.; Russell, L. M.; Russell, P. B.; Shimizu, A.; Shinozuka, Y.; Song, C. H.; Tang, Y.; Uno, I.; Vogelmann, A. M.; Weber, R. J.; Woo, J-H.; Zhang, X. Y. ACE *Bull. Am. Met. Soc.* **2004,** *85,* 367.
16. Lioy, P.J;, Weisel, C.P;, Millette, J. R.; Eisenreich, S.; Vallero, D.; Offenberg, J.; Buckley, B;, Turpinm B.; Zhong, M.; Cohen, M. D.; Prophete, C.; Yang, I.; Stiles, R.; Chee, G.; Johnson, W.; Porcja, R.; Alimokhtari, S.; Hale, R. C;, Weschler, C; Chen, L. C. *Environ. Health Perspect.* **2002,** *110,* 703.
17. Natusch, D.F.S; Wallace, J.R.; Evans, C.A. *Science* **1974,** *183,* 202.
18. Natusch, D.F.S.; Wallace, J.R. *Science* **1974,** *186,* 695.
19. Pinto, J.; Grant, L.D.; Huber, A.H.; Vette, A.F. Paper presented in ACS Symposium: *Urban Aerosols and their Impact: Lessons Learned from The World Trade Center Tragedy,* New York City, Sept. 10, 2003. *ACS Division of Environmental Chemistry Preprints of Extended Abstracts, paper 1370,* **2003,** *43 (2),* 1370.
20. Cahill, T. A.; Wakabayashi, P.; James, T. *Nuclear Instruments and Methods in Physics Research B: Beam Interactions with Materials and Atoms,* **1996,** *109/110,* 542.
21. Leifer, R. Z.; Bench, G.; Cahill, T.A. Paper presented in ACS Symposium: *Urban Aerosols and their Impact: Lessons Learned from The World Trade Center Tragedy,* New York City, Sept. 10, 2003. *ACS Division of Environmental Chemistry Preprints of Extended Abstracts, paper 254,* **2003,** *43 (2),* 1393.
22. Cahill, T.; Cliff, S.S.; Perry, K.D.; Jimenez-Cru;, M. Final Report to the American Lung Association, New York City, 2003.
23. Barton, R.G; Clark, W. D.; Seeker, W.R. *Combust. Sci. & Technol.* **1990,** *74,* 327.
24. *"EPA Response to September 11"* [http//www.epa.gov/wtc] U. S. Environmental Protection Agency.

Chapter 10

Semivolatile Organic Acids and Levoglucosan in New York City Air Following September 11, 2001

Michael D. Hays[1], Leonard Stockburger[2], John D. Lee[3], Alan F. Vette[2], and Erick C. Swartz[2]

[1]National Risk Management Research Laboratory and [2]National Exposure Research Laboratory, U.S. Environmental Protection Agency, Research Triangle Park, NC 27711
[3]Arcadis, G&M Inc., P.O. Box 13109, Research Triangle Park, NC 27709

Organic acid compounds and levoglucosan, an important molecular marker of burning cellulose, were detected in New York City air samples collected between 9/26/01–10/24/01, 500 m from Ground Zero. The sampling of Ground Zero emissions at our site is commensurate with a southwesterly wind flow. Aerosol phase distributions captured with a High Capacity Integrated Organic Gas and Particle Sampler show many of the organic acids are semivolatile. However, underlying the phase equilibrium for certain acid molecules is a set of factors more complex than vapor pressure alone can explain. At our collection site, quartz filter acid concentrations by class follow predicted seasonal trends and are generally below the annual average and episodic concentrations determined earlier for urban Los Angeles. Here, analysis of distribution patterns of in-series acid homologues is consistent with a pyrogenic burden, the specifics of which are discussed. By evaluating the inorganic bulk chemistry of airborne PM-2.5 emissions from Ground Zero after 9/11/2001 with X-ray fluorescence, tentative assignments to sources of construction materials (flame-retardant, paint, cement, and polymer materials) are construed.

Introduction

In New York City (NYC) on 9/11/01, a coordinated terrorist attack reduced the two main structures of the World Trade Center (WTC) and nearby buildings to more than one million tons of debris and ignited an intense fire (> 1000 °C) that would persist for weeks. In tandem with the emissions from the combustion and pyrolysis of fallout debris, the unusually high degree of attrition of the construction and workplace materials produced a chemically complex aerosol plume, which became dispersed, mixed, and aged in the lower Manhattan atmosphere. Even with the search and rescue operations considered complete just weeks following the WTC collapse, smoldering fires and emission sources from debris removal operations at Ground Zero (the term assigned post 9/11 to the 16 acre WTC area impacted by the disaster) continued to release a potentially hazardous mixture of gas- and particle-phase matter to the urban NYC atmosphere. Though fine particulate matter (PM-2.5) may typify only a small mass percentage of the original dust plume and settled fallout debris resulting from the WTC disaster, there is particular interest in the chemical properties of the airborne fine aerosols originating from this site. This interest stems from the onset of acute respiratory symptoms in the workforce and general public at or near Ground Zero and the established link between the inhalation of fine aerosol particles into the deep lung and negative human health impacts (*1-3*).

Organic acid compounds and levoglucosan (an important molecular marker of burning cellulose) in ambient fine aerosols sampled between September 26 and October 24, 2001 in lower Manhattan are the particular focus of this work. To the best of our knowledge, these data are the first that identify and quantify individual polar constituents in aerosols putatively ascribed to Ground Zero emissions. Justifications for profiling the polar constituents include their candidacy for use in chemical mass balance models for source apportionment (*4*) and their important function in the comprehension of atmospheric conditions, chemical transformations, and regional air quality issues (*5*).

The fine aerosols are collected using a recently introduced High Capacity Integrated Organic Gas and Particle Sampler (HiC IOGAPS) system (*6*), which applies annular denuder and quartz fiber filter technology in series to capture and partition corresponding gas- and particle-phases. Swartz et al. have determined that the HiC IOGAPS efficiently collects and measures the partitioning of semi-volatile molecular constituents of the *n*-alkane and polycyclic aromatic hydrocarbon (PAH) compound classes (*7*). The efficacy of HiC IOGAPS for the field collection of organic acid species is investigated here for the first time.

The polar organic compounds were identified and quantified by conventional solvent extraction and gas chromatography/mass spectrometry (GC/MS) techniques. Additionally, ion chromatography and X-ray fluorescence methods are implemented to classify the water-soluble anion-cation and elemental constituents in the aerosols. These experiments further our understanding of the inorganic chemical properties of Ground Zero-related PM-2.5 emissions and indicate the degree to which polymer, mineral-like, and metal-containing construction materials may have affected PM-2.5 chemistry in the air shed near the WTC site following the disaster. Evidence of the presence of airborne concrete dust particles, for example, could be of particular significance due to their alkalinity ($CaOH_2$ and $CaCO_3$ form during cement hydration), which may lower airborne organic acid concentrations via an acid-base reaction pathway. Analysis by polarized light microscopy confirms a significant concrete component by mass in the settled dust (*8*).

On the temporal scale studied (9/26/01–10/24/01), the nature of the fine aerosols that formed the WTC plume immediately following the structural collapse of the WTC towers on 9/11 is not defined by this particular effort. Instead, our observations depict the chemical properties of the fine aerosols (from 9/11) left lingering in the atmosphere and the gradual effects on the atmosphere in the event aftermath and include emissions from the smoldering fires and removal operations at Ground Zero, the airborne resuspension of the settled dust due to wind turbulence and/or mechanical disturbance (*7*), and NYC air pollution as background (*9*).

Experimental Methods

Sampling

Detailed accounts of the sampling of the NYC aerosols, including instrumentation, collection site position, meteorology, protocol and operations, and sampling period are given elsewhere (*7*). Briefly, a High Capacity Integrated Organic Gas and Particle sampler (HiC IOGAP) (URG model 3000DB) with a 2.5 μm cyclone inlet for particle discrimination was used to collect semi-volatiles for subsequent polar organic compound speciation. Two sampling schemes were implemented. *Scheme 1* used two XAD-4 coated annular denuders, a quartz fiber filter (90 mm diameter), and three XAD-4-impregnated quartz fiber filters in series to study gas- and particle- phase partitioning. In this arrangement, the annular denuders captured the gas-phase, whereas the material collected on the downstream quartz filter series was assigned to the particle-phase. *Scheme 2* was identical to *Scheme 1* but did not make use of the denuder system and thus provided total compound concentrations without partitioning data. All quartz filters were prefired (540 °C, 12 h) before use.

The HiC IOGAP sampler was operated roughly 500 meters northeast of the World Trade Center on the 16th floor (about 50 meters above street level) of the EPA Federal Building at 290 Broadway, New York, NY. The volumetric flow rate of the HiC IOGAP sampler was set to 85 L/min, and temperature was controlled at 4 °C above ambient to prevent condensation of water. *Scheme 1* was operated from 9/26/01-10/05/01for 11.5 hours per day. *Scheme 2* was operated from 10/06/01-10/24/01 for 23 hours per day. Sampling typically began at 12:00 PM (EST). A street level dust sample was taken ~50 m north of the WTC on 9/24/01. Data from an on-site nephelometer (M903, Radiance Research) was used to infer particulate mass concentrations in real time. Field blanks of quartz and XAD-impregnated quartz filters were used to correct for background acid levels due to sample handling, transport, equipment, and set up. Field blanks showed aromatic carboxylic acids but contained few other targeted molecules from the polar organic compound set studied.

Analysis of the polar constituents was performed for the five sampling periods congruent to the companion effort on the organic speciation of neutral compounds (*7*); these were 9/26/01, 10/4/01, 10/06/01, 10/12/01, and 10/20/01. Prevailing wind directions corresponding to these periods were west-southwest (9/26), southwest (10/4, 10/12, and 10/20) and north (10/06). On 9/26, 10/4, 10/12, and 10/20, the emissions from Ground Zero impacted our sample collection site and relatively high sample mass loadings were observed. Conversely, the 10/6 period experienced a morning precipitation event coinciding with a northerly wind direction and was thus used as a background check for semivolatile polar constituents in the NYC atmosphere. Wind speed was variable, ranging from 2-10 m/s.

Sample Preparation

An exhaustive description of the preparation, extraction, and concentration procedures of the aerosol sample sets is also given elsewhere (*6*). In brief, the denuder and filter samples were extracted in a solvent mixture of equal volumes of hexane, dichloromethane and methanol (Hex:$MeCl_2$:MeOH). Denuders were sequentially extracted three times using a "rolling rinse" method. Quartz fiber filters and XAD-4-impregnated quartz filters were extracted twice by mild sonication. All extracts were passed through a Teflon membrane (0.45 μm pore size) to filter solid particles from the sample liquid. Sample filtrate volumes were reduced to 0.5 mL using a vacuum evaporator (Labconco, RapidVap), for which temperature, pressure, vortex speed, and time were programmed to prevent bumping and to ensure optimum conditions for solvent removal. At this stage, the concentrator tube walls were rinsed with 1 mL of equal parts $MeCl_2$:Hex, and the extract (1.5 mL) was transferred to a vial for analysis. During method optimization, 1.58 μg of a deuterated *n*-alkane surrogate (n-$C_{24}D_{50}$) was spiked into the extract mixture prior to the reduction and transfer steps and showed a recovery of 80-120% (*6*). Preceding the extraction, a 1 mL spike of a deuterated internal standard suite of polycyclic aromatic hydrocarbons was used to determine analyte losses due to sample processing; acceptance limits were set at 80-120%.

Sample Derivatization and GC/MS Analysis

To convert the organic acids to their less polar methyl ester analogs, a 100 μL aliquot of each WTC extract was methylated with freshly prepared diazomethane reagent trapped in vacuum-distilled benzene. (Note: Diazomethane is a potentially explosive and poisonous gas; caution is warranted when preparing it.) Prior to methylation, the extract aliquot was spiked with two internal standard suites (3 μL each) containing decanoic-d_{19}, heptadecanoic-d_{33}, and phthalic 3,4,5,6-d_4 acids and radiolabeled levoglucosan-$^{13}C_6$. Addition of these compounds to the extract at this stage precluded the determination of their recoveries but was required for determining target analyte concentrations using a calibration database populated with response factors (the internal standard method). For similar solvent extraction methods, an established recovery range from 80 to 120% was confirmed for these compounds (*10*). A portion of the methylated extract was silyated using a commercially available BFSTA reagent (Supelco, Bellefonte, PA). This reagent replaced any active hydrogen in the extract with a trimethylsilyl group and allowed polar, nonvolatile levoglucosan, a molecular marker of cellulose combustion, to be successfully chromatographed. For purposes of quality control, solvent blanks were methylated and silyated and subsequently examined for traces of polar organic analytes. No target compounds were detected in the solvent or derivatizing reagent mixtures.

The GC/MS analysis and quantification scheme for these polar target analytes in the derivatized extracts was described by Hays et al. (*10*). Briefly, the GC/MS (HP6890/5973; Agilent Technologies) analysis was performed using an automated injection system, split-splitless inlet operating in splitless mode, an ultra-low bleed capillary column (5MS, Agilent Technologies, 30 m length, 0.25 mm i.d., 0.25 μm film thickness), and oven temperature programming (65 °C for 10 min, increased to 300 °C at 10 °C/min, 300 °C for 41.5 min). The GC/MS transfer line temperature was held fixed (300 °C). The mass spectrometer (MS) was operated in scan mode (50-500 amu, 3 scans/sec). Enhanced Chemstation software (V.B.01.00; Agilent Technologies) was used for instrument control, data acquisition, and data analysis.

Quantification of the polar target analytes in the derivatized extracts was accomplished with a multi-level calibration database of concentration and response ratios for the certified authentic standards and isotopically labelled compounds in the internal standard suites. Table I lists the targeted analytes, their internal standard assignments, method detection limits (MDLs), and the linear correlation coefficient (r^2) of the multi-level calibration used for quantification. Method detection limits were determined using the approach outlined in EPA publication SW-846 (*11*). Acid concentrations at the picogram level generally defined the MDLs, and the r^2 values ranged from 0.992-1.000. Levoglucosan (r^2=0.997) was the only analyte targeted by the analysis not listed in the table. The compound classification scheme of Rogge et al. (*12*) was utilized here. Of the 49 compounds sought, 32 had authentic standards. For constituents without an authentic standard, their identification was confirmed individually by comparing the MS fragmentation pattern to the NIST library mass spectrum or the mass spectrum of a secondary standard. In the event of a valid match, an analogous compound was used for quantification.

Organic Carbon/Elemental Carbon Analysis

The organic carbon /elemental carbon (EC/OC) content of the particulate samples was measured using a carbon analyzer implementing a thermal-optical transmittance technique (Sunset Laboratories; Forest Grove, OR) and the National Institute for Occupational Safety and Health (NIOSH) Method 5040 (*13*). Birch and Cary offered the details pertaining to instrument operation (*14*). Calibration of the analyzer was verified on a 24 h basis. The control and data acquisition were achieved using software provided by the manufacturer and a personal computer running a Windows (Microsoft Corp.; Seattle, WA) operating system.

X-ray Fluorescence and Ion Chromatography Analyses

The elemental analysis (up to 49 elements) of fine particulate matter collected on low background Teflon filters (Gelman) was preformed by energy dispersive X-ray fluorescence (XRF) (Kevex EDX-771; Thermo Electron Corp, CA), which implemented extended counting times and multiple secondary fluorescers. The elements ranged from Al to Pb with detection limits between 0.1 and 1.0 ng/m^3 at 1σ. Only elements detected at two times the standard error were reported. Crystalline components of the settled dust were identified using a

Table I. Targeted Polar Analytes Listed in Order of Elution.

Compound	*MDL*	r^2
hexanoic acid[a]	0.09	0.998
butanedioic acid[a]	0.10	0.999
octanoic acid[a]	0.12	0.999
pentanedioic acid[a]	0.04	0.999
nonanoic acid[a]	*	*
hexanedioic acid[a]	0.03	0.999
decanoic acid[a]	0.09	0.999
heptanedioic acid[a]	0.12	0.999
pinonic acid[b]	0.08	1.000
undecanoic acid[a]	*	*
octanedioic acid[a]	0.16	0.998
1,2-benzenedicarboxylic acid[b]	0.06	1.000
1,4-benzenedicarboxylic acid[b]	0.04	1.000
1,3-benzenedicarboxylic acid[b]	0.09	1.000
dodecanoic acid[a]	0.03	0.999
nonanedioic acid[a]	0.05	0.996
1,2-benzenedicarboxylic acid, 4-methyl[b]	0.03	1.000
tridecanoic acid[a]	*	*
decanedioic acid[c]	0.06	0.999
tetradecanoic acid[c]	0.05	0.999
pentadecanoic acid	*	*
1,2,4-benzenetricarboxylic acid[b]	0.05	0.997
9-hexadecenoic acid[c]	0.04	1.000
hexadecanoic acid[c]	0.02	0.999
heptadecanoic acid	0.04	0.995
8,11-octadecadienoic acid[c]	0.02	0.999
9-octadecenoic acid[c]	0.02	0.997
9,12,15-octadecatrienoic acid[c]	0.03	1.000
octadecanoic acid[c]	0.06	0.995
benzenetetracarboxylic acid[b]	*	*
nonadecanoic acid[c]	*	*
pimaric acid[b,d]	*	*
sandaracopimaric acid[b,d]	0.03	1.000
eicosanoic acid[c]	*	*
isopimaric acid[b,d]	*	*
6,8,11,13-abietatetraen-18-oic acid[b,d]	*	*
dehydroabietic acid[b,d]	*	*
heneicosanoic acid[c]	*	*
abietic acid[b]	0.09	0.998

Table I. Continued.

Compound	*MDL (ng)*	r^2
docosanoic acid[c]	0.02	0.997
tricosanoic acid[c]	*	*
tetracosanoic acid[c]	0.02	0.992
pentacosanoic acid[c]	*	*
hexacosanoic acid[c]	*	*
heptacosanoic acid[c]	*	*
octacosanoic acid[c]	0.03	0.996
nonacosanoic acid[c]	*	*
triacontanoic acid[c]	0.04	0.998

NOTE: Internal standard assignments are ([a]) decanoic acid-d_{19}, ([b]) phthalic 3,4,5,6-d_4 acid, and ([c]) heptadecanoic-d_{33}. . Componds identified by ([d]) were confirmed using the retention time and mass spectra of secondary standards.* denotes compound quantified using an analogous primary standard. The C_{28}-C_{30} *n*-alkanoic acids were quantified using a quadratic equation.

NOTE: Units are ng. MDL is method detection limit, r^2 is the calibration curve linear correlation coefficient.

Siemens X-ray Diffractometer D-500 (Bruker AXS, Madison, WI). Water- soluble ions in the aerosols were analyzed by ion chromatography (IC) (DX-120, Dionex Corp.) using conductivity detection.

Results and Discussion

PM-2.5 Mass and Chemical Composition

In this work, PM-2.5 mass concentrations were inferred from the nephelometer, which continuously measures the scattering coefficient of light by particles suspended in air. Plate 1 shows the ambient PM-2.5 mass concentrations at the lower Manhattan monitoring site from September 24 to November 1, 2001. The dotted line at 65 $\mu g/m^3$ indicates the EPA revised primary 24-hr standard for PM-2.5. Although this value is sporadically exceeded at the site over almost the entire sampling period, the 24-hr mean PM-2.5 concentration is greater than 65 $\mu g/m^3$ only on October 3rd (66 $\mu g/m^3$) and 4th (90 $\mu g/m^3$). No clear diurnal trends are observed over these two consecutive days or for the periods selected for detailed speciation. This result may be symptomatic of the relatively consistent

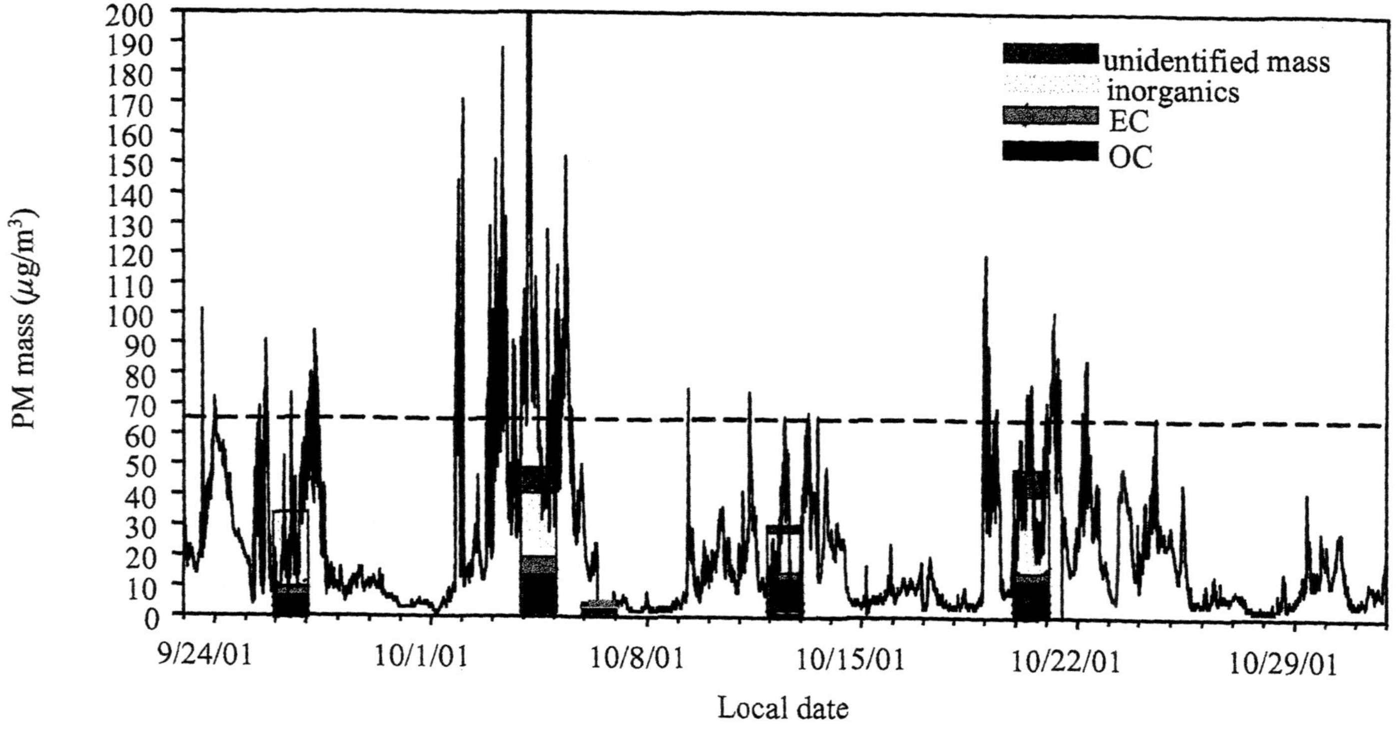

Plate 1. Chemical mass balance assignments for the sample sets obtained on 9/26, 10/4, 10/6, 10/12, and 10/20 superimposed on the real-time nephelometer data (solid line). The EPA revised primary 24-hr standard for PM-2.5 is shown by the dotted line at 65 g/μm^3. *(See page 17 of color inserts.)*

meteorological conditions and the limited chemical processing of the aerosols due to the short transport distance from the emissions source (WTC site) to the receptor site (at 290 Broadway). A maximum PM-2.5 concentration of 1049 μg/m^3 is observed at ~4 AM in the morning on October 4. This episode is anomalous and unsustained; lasting less than 1 hr before returning to lower concentrations (5-fold lower on average), and is likely due to the prevailing southwest wind direction that carried PM-2.5 emissions from Ground Zero to our collection site.

For the 9/26, 10/4, 10/6, 10/13, and 10/21 sample sets, bar graphs depicting organic carbon (OC), elemental carbon (EC), inorganic concentrations and the unidentified mass fraction of the PM-2.5 are also shown in Plate 1. The integrated bulk chemical constituent concentrations and the PM-2.5 mass obtained from the nephelometer data for these aerosol collection periods are also given in Table II. Recall on 9/26 and 10/4 that a 12 h sample is collected. Hence, the PM-2.5 mass concentration in the bar graph of Plate 1 for these dates is different than the 24-h mean value given in Table II. The PM-2.5 mass concentration range for the sampling periods is 4.0-50.0 μg/m^3. As expected, the lowest PM-2.5 mass occurs on 10/6 due to a precipitation event and a northerly wind direction unfavorable to WTC emissions traversing the collection site. The limited particulate mass associated with the 10/6 sample precludes a more detailed inorganic speciation. Fortunately, this sample comprises the largest carbon fraction by mass, 86% (possibly attributable to the wet deposition of inorganic salts). Note that the OC values are converted to organic compound mass by multiplying by 1.4, a factor that estimates the average molecular weight per carbon weight (*15*). With the exception of 10/6, the species distribution by mass is fairly constant. Though, the OC/EC ratio (2.3-5.1) increases with time, possibly indicating;

1. an increase in the formation of secondary PM in the atmosphere (*16*),
2. diesel emissions at the site varying with operating load (EC is common to diesel exhaust),
3. carbon emissions varying daily from a complex source array,
4. gradual quenching of the smoldering fire at the site, and/or
5. gas-phase adsorption of semivolatile organics by the quartz filters due to changing atmospheric or source conditions.

Generally, much of the fine aerosol mass could be categorized as OC, EC, or inorganic matter. The largest fraction of unidentified mass is 17% observed on both 10/4 and 10/20. Sulfate ion (SO_4^{2-})and organic matter are abundant, together accounting for as much as 54% of the PM-2.5 mass. The sulfate ion concentration is approximated from the XRF-determined S value and confirmed by IC over similar sampling periods.The IC data on similar but staggered daily periods verify a large sulfate component on 10/6-10/7, while meteorological data show the WTC site emissions did not directly impact the sampling site during this period.

In PM-2.5, the SO_4^{2-} is usually in acidic form (H_2SO_4, NH_4HSO_4, and $(NH_4)_2SO_4$) and is emitted from the combustion of fossil fuels that contain sulfur (*17*). The diesel trucks and generators used for the debris removal operations likely contributed to the SO_4^{2-} at our monitoring site over the interval identified. Calcium sulfate ($CaSO_4$) dihydrate, a slightly water-soluble salt compound found in wallboard and added to cement to control hydration rate, may also contribute to the SO_4^{2-} in the PM-2.5 from the WTC site. However, calcium (Ca) is detected at low levels, and calcium hydroxide ($CaOH_2$), and calcium carbonate ($CaCO_3$) in cement or the reduction of larger diameter Ca-containing crustal particles to PM-2.5 due to abrasion may account for the Ca observed. We hesitate to conclude that the Ca is combustion-related because combustion-related fine aerosol emissions are commonly Ca-depleted (*18*).

Table II. Concentrations of Chemical Species in the Fine Particulates Measured in lower Manhattan.

Species	*09/26-09/27*	*10/04-10/05*	*10/06-10/07*	*10/12-10/13*	*10/20-10/21*	*Method*
[a]PM-2.5	24.5	90.2	-	-	-	Nephelometer
[b]PM-2.5	35.4	48.4	4.0	30.0	49.6	Nephelometer
EC	2.91	5.78	0.796	2.9	2.57	OC/EC
$OC_{1.4}$	7.76	13.3	2.6	11.6	13.2	OC/EC
Na	0.2	0.2	NA	0.2	0.1	XRF
NH_4^+	NA	NA	NA	2	4	IC
K	0.7	0.4	NA	0.4	0.4	XRF
Ca	0.2	0.2	NA	0.2	0.9	XRF
Cl	5.9	1.7	NA	1.3	2.4	XRF
NO_3^{2-}	NA	NA	NA	3	4	IC
SO_4^{2-}	8.5	9.9	NA	4.8	10.8	XRF
Fe	0.3	1.0	NA	0.8	1.8	XRF
Zn	0.9	0.6	NA	0.5	0.5	XRF
Br	0.6	0.2	NA	0.1	0.2	XRF
Pb	0.9	0.3	NA	0.1	0.2	XRF

NOTE: Units are $\mu g/m^3$. NA is not available or not analyzed.

NOTE: [a]PM-2.5 is given as the 24-hr daily mean. [b]PM-2.5 is given as the mean over the sampling period. Concentrations of NH_4^+ and NO_3^{2-} for 10/12 – 10/13 were estimated from values obtaind 10/11-10/12. Concentrations of NH_4^+ and NO_3^{2-} for 10/20 – 10/21 were estimated from values obtained on 10/21-10/22.

The XRF data confirm the presence of chlorine (Cl) in the atmosphere. Natural sea salt and/or polychlorinated furans and dioxins formed during incomplete combustion of fuels in the presence of polyvinyl chloride polymer (used in plastics) may account for the occurrence of Cl here. Lead (Pb), zinc (Zn), and iron (Fe) metals and bromine (Br) are also detected in the fine aerosols assigned to WTC emissions. These elements are associated with several industrial materials likely used to construct and maintain the WTC office buildings. For example, oxides and alloys of Pb are used in the manufacture of cable sheathing, pigments, plastics, paints, solder, and glass. Zinc is commonly used as an anti-corrosive agent in coatings, in roof cladding, and as a polymer stabilizer. The Br may have come from fire retardant treatment of materials in the WTC and released with Pb and Zn as a fine aerosol during the ongoing fires. Iron is terrestrially abundant, a major constituent of steel, and may be from the cutting of the WTC steel framework, which was twisted but largely intact after the collapse. Lead, Zn, and Fe are also observed in the settled dust, but Pb and Zn are in smaller proportion (the Fe mass level is not provided) (*8*). Though enriched in the airborne PM-2.5, the relative distribution of Zn and Pb in the dust and aerosol samples is somewhat comparable and may suggest the detachment and resuspension of fine settled dust particles (due possibly to wind erosion at the debris pile surface-air interface or agitation of debris by the motion of heavy-duty vehicles).

Organic Acids

In the period following the destruction of the World Trade Center, 32 organic acids were collected in the lower Manhattan atmosphere using the HiC IOGAPS. Classified as *n*-alkanoic (17), alkanedioic (6), alkeneoic (1), aromatic (6), and resin (1) acids, they are identified and quantified as their methyl esters by GC/MS. The blank-corrected concentrations for all acids and levoglucosan are presented distributed by phase for 9/26 and 10/4 in Table III and in total for 10/6, 10/12, and 10/20 in Table IV. Below, data for each homologous acid series are presented chronologically and discussed within the context of the related investigations.

09/26/2001 – 10/4/2001

Results from the 9/26 and 10/4 sample set are given in Figure 1 panels A (9/26) and B (10/4). Each panel indicates the mass of analyte a ssigned to a specific substrate in the HiC IOGAPS sampling array. As described, we allocate the gas- and particle-phases to the denuder set and quartz-quartz XAD-4 filter series, respectively. For the majority of the acid compounds measured, the

Table III. Concentrations of Organic Acids and Levoglucosan in the Gas and Particle Phases of Samples collected on 9/26 and 10/4/2001 at 290 Boadway.

Species	9/26-g	9/2-p	10/4-g	10/4-p
octanoic	12	1	21	12
nonanoic	8	5	22	7
decanoic	7	4	10	14
undecanoic	5		6	2
dodecanoic	6	5	14	6
tridecanoic	5		7	0
tetradecanoic	8		15	1
pentadecanoic	4		10	1
hexadecanoic	18	5	22	19
heptadecanoic	5	1	3	2
octadecanoic	1	12	4	16
nonadecanoic	1		1	
eicosanoic	3		1	
heneicosanoic	1			
docosanoic	1			
tricosanoic				
tetracosanoic	1			
butanedioic	10		43	16
pentanedioic	11	2	21	6
hexanedioic	6	5	12	7
heptanedioic				
octanedioic			4	
nonanedioic			18	4
9-octadeceneoic acid	5	3	8	
1,2-benzenedicarboxylic	83	3	55	14
1,3-benzenedicarboxylic	6		9	31
1,4-benzenedicarboxylic	6		6	44
1,2-benzenedicarboxylic,4-methyl	9	1	10	4
1,2,4-benzenetricarboxylic		2	2	1
1,2,4,5-benzenetetracarboxylic				
6,8,11,13-abietatetraen-18-oic		5	7	
pinonic acid	17		33	
levoglucosan	5	83	4	106

NOTE: Units are ng/m^3. Gas phase is denoted by (g), particle phase is (p).

Table IV. Total Concentrations of Organic Acids and Levoglucosan in Fine Aerosols Collected on 10/6, 10/12 and 10/20/2001 at 290 Broadway Compared to Results Obtained in California (CA).

Species	10/6	10/12	10/20	CA
octanoic	10	28	14	1-5[a]
nonanoic	12	44	28	3-10
decanoic	6	14	14	1-3
undecanoic	3	15	13	3-6
dodecanoic	3	23	17	4-7
tridecanoic	2	13	14	3-5
tetradecanoic	1	19	14	14-23
pentadecanoic	0	8	8	4-6
hexadecanoic	15	72	40	118-141
heptadecanoic	1	8	6	3-5
octadecanoic	22	50	53	41-59
nonadecanoic		1		0.8-1
eicosanoic		4	3	3-6
heneicosanoic			1	1-2
docosanoic		2	5	6-10
tricosanoic			2	2-3
tetracosanoic			4	9-17
butanedioic		47	51	51-84
pentanedioic	11	18	26	28-39
hexanedioic	6	18	22	14-24
heptanedioic		3	7	
octanedioic	1	5	6	3-4
nonanedioic	5	17	22	23-45
9-octadeceneoic acid		3		0-34[a]
1,2-benzenedicarboxylic	20	21	99	54-61
1,3-benzenedicarboxylic			12	2-3
1,4-benzenedicarboxylic			41	1-3
1,2-benzenedicarboxylic,4-methyl			28	15-30
1,2,4-benzenetricarboxylic			17	0.5-0.8
1,2,4,5-benzenetetracarboxylic			3	0.4-0.8
6,8,11,13-abietatetraen-18-oic		10	6	
pinonic acid	23	85	45	
levoglucosan	52	94	247	

NOTE: Units are ng/m^3.

SOURCE: Data obtained in California (particle phase) is from Reference *4*. California data indicated by ([a]) is from Reference *23*.

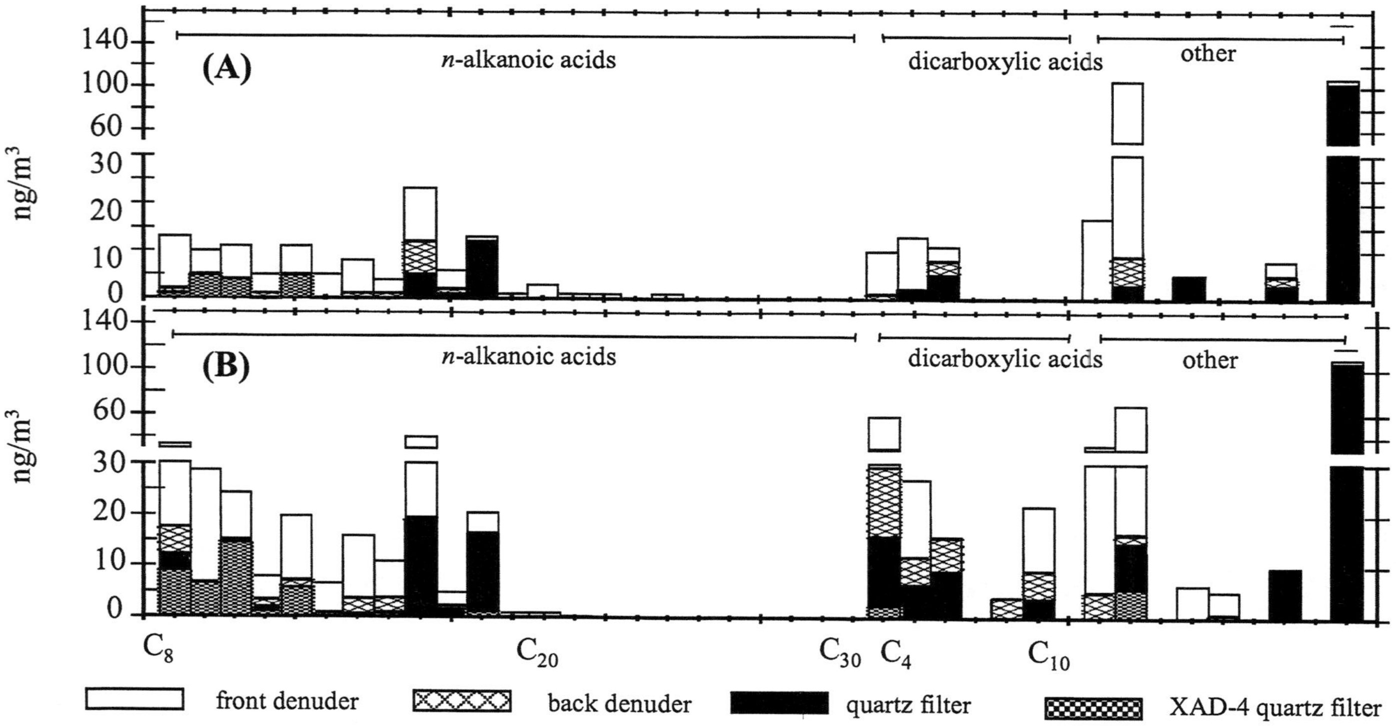

Figure 1. Concentrations of organic acids and levoglucosan by substrate for 9/26 (A) and 10/4 (B). Compounds listed as "other" are in the order shown; pinonic acid, phthalic acid, abietic acid, abetatetra-18-oic acid, 9-hexadecenoic acid, 8-11-octadecadienoic acid, 9-octadecenoic acid, 9, 12, 15-octadecatrienoic acid, and levoglucosan.

material is distributed among the substrates, attesting to the semivolatile nature of the organic acids. Phase distributions, as measured by the HiC IOGAPS, vary by compound, as expected. The total acid concentrations vary temporally. The sum of analytes collected on all substrates for 9/26 is 314 ng/m^3 and for 10/4 is 714 ng/m^3, reflecting respective semi-direct and direct influences from WTC site emissions consistent with the wind direction. Figure 1 also shows that on the different days the compound phase distribution on the substrates is somewhat variable. Sampling bias, temperature, humidity, and particle interface and surface chemistries are among the factors that can effect the gas-particle partitioning. Local climate data indicate a higher average dew point (12.6), relative humidity (36-84), and temperature (26 °C) for 10/4.

To better observe the ability of the collection system to measure the partitioning behavior, we plot an estimate of the organic species' gas vapor pressure versus the experimentally determined gas-particle partitioning coefficient. This plot is defined as;

$$K_p(m^3/\mu g) = [F(ng/m^3)/TSP(\mu g/m^3)]/\ A(ng/m^3) \qquad (1)$$

where F is the mass of species on the filters, A is the mass of species in the gas-phase, and TSP is the mass of total suspended particulate matter. Figure 2 illustrates the results of this data treatment for 9/26 and 10/4. Gas vapor pressures of the *n*-alkanoic and dialkanoic acids are approximated using the methodology and data of Makar (*19*), whose formula presumes vapor pressure is a linear function of carbon number within a given molecular class. The subcooled liquid vapor pressure ($p_L{}^o$) value is also adjusted to reflect the different ambient temperatures (17 and 26 °C) at the time of sample collection. Pankow et al. define the regression line, theorizing for

$$\log K_p = m_r \log p_L{}^o + b_r \qquad (2)$$

that $m_r = -1$ and b_r is compound class dependent (*20*).

For 9/26 and 10/4, the results are repeatable; K_p is consistent from 17-26 °C. However, comparisons are limited to the nine compounds detected on both days. For the n-alkanoic acids with higher vapor pressures, the experimentally based K_p exceeds the theoretical value. This behavior may be proof of gas-phase penetration of the denuder, for example. All of the acid compounds detected absorb to the XAD-impregnated quartz filters. There is no trace of acids on the last XAD-quartz filter, indicating no material loss. Transmission of particle-phase through the quartz filter is discounted due to evidence provided by past experiments (*21*). For a subset of less volatile n-alkanoic and dioic acids, the calculated values of log K_p are lower than theory predicts, possibly suggesting that not all particles pass through the annular denuder system or that these higher

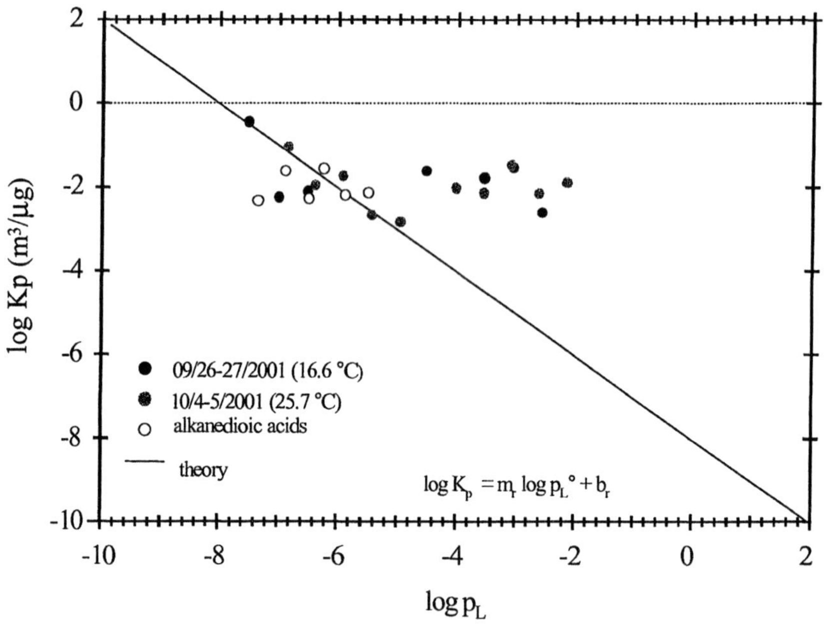

Figure 2. Gas-particle partitioning coefficient as a function of the estimated supercooled liquid vapor pressure(p_L) for the n-*alkanoic and dialkanoic acids.*

molecular weight acids are stripped from the particle surface. These nonideal results seem to support conclusions reached earlier that suggest other physicochemical properties of particles, not just vapor pressure, can effect the phase distributions of organic acids (*21*).

The quantified particle-phase organic acids account for 1% (9/26) and 2% (10/4) of the organic carbon mass, corresponding to 0.2% and 0.6% of the PM-2.5 mass. Atmospheric alkanedioic acids in specific are diminished in the 9/26 PM-2.5 sample (38 ng/m^3). Dissolution of the water-soluble diacids in cloud/fog water droplets likely preceded their wet deposition, as NYC climate data show that rain, fog, and haze conditions continued from 9/20 until 9/25/01. Pinonic acid, a product of the chemical reaction between phytogenic α-pinene and ozone, is detected over the entire sampling period, indicating that emissions from a distant pine forest impacts the NYC atmosphere at this time of year. The existence of this natural molecule in the NYC airshed just weeks following the disaster may point to a gradual clearing of the atmosphere near the measurement site due to wind advection and possibly to normal atmospheric processing.

10/6/2001

Starting on 10/6/01, the NYC aerosols are collected with a quartz filter upstream of three XAD-impregnated quartz filters (*Scheme 2*). Data given in Table IV and Figure 3 confirm that the total acid concentrations on 10/06/01 are mostly lower due to the rain episode and prevailing northerly wind direction that carries the WTC emissions away from our sample collection site. Thus, for the 10/6 collection period we designate the acid aerosol concentrations in the NYC air as background. For studies of urban aerosols of southern California, both Rogge et al. (*22*) and Fraser et al. (*23*) use a remote offshore site (San Nicolas island) to assess background concentrations of organic acids in air parcels entering their urban test region on an annual and episodic (2-day photochemical

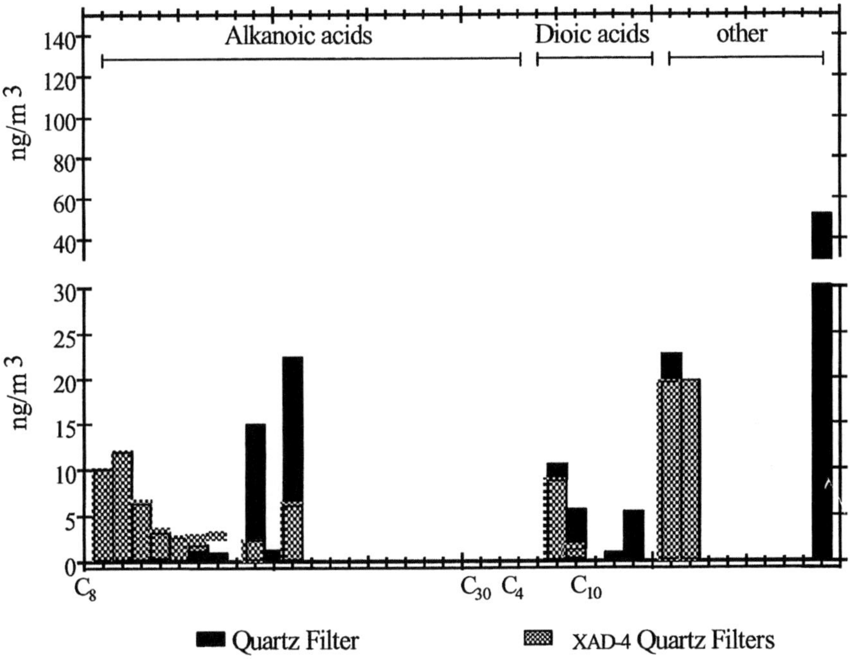

Figure 3. Concentrations of organic acids and levoglucosan by substrate for 10/6. Compounds listed as "other" are in the order shown; pinonic acid, phthalic acid, abietic acid, abetatetra-18-oic acid, 9-hexadecenoic acid, 8-11-octadecadienoic acid, 9-octadecenoic acid, 9, 12, 15-octadecatrienoic acid, and levoglucosan.

smog event) basis. Though differences in topography, landscape, climate, prevailing meteorology, sampling periods, conditions, and equipment preclude a perfectly robust comparison among these studies, we note when considering only quartz filter (particle-phase) data that the acid levels in the NYC air on 10/6 are lower than those reported for either California study (*22, 23*) in all (data for more than 35 acids were compared) but six cases (C_{17}-C_{18} *n*-alkanoic, adipic, suberic, azelaic and phthalic acids). The remarkably low level of organic acids in the NYC air on 10/6 makes this day suitable for approximating baseline pollution levels.

Despite these lower concentrations of organic acids in the atmosphere on 10/6, the PM-2.5 appears comparatively organic acid-enriched (1.1%), whereas the acids/OC ratio (1.7%) for this day is mostly unchanged. It seems clear that the airborne PM-2.5 comprises water-soluble salts, which are removed to a greater extent than the OC mass. We suspect that some of the lower molecular weight acid mass is inaccessible to water because it is interspersed with condensed PM that is hydrophobic. Either that or an entirely different set of PM emission sources with unique chemical properties and spatial influences affects the collection site on this day.

10/12/01 and 10/20/01.

Analagous to the 10/6 data, HiC IOGAPS sampling *Scheme 2* is implemented on 10/12 and 10/20 for 24 h periods. Total acid compound concentrations are given for these days in Table IV. Figure 4 shows the distribution of the acids over the substrate array. With the exceptions of the n-alkanoic acids ($\geqslant C_{16}$) and alkanedioic acids ($\geqslant C_4$), most of the organic acids deposit on either of the first two XAD-impregnated quartz filters. No transmission of the acids to the filter farthest downstream is observed with this sampling scheme. Again, on 10/12 and 10/20, southerly wind flow transmits the WTC emissions to our sample collection site at 290 Broadway and, as expected, relatively high concentrations of acids are observed at the monitoring site on these days (680 ng/m^3 on 10/12 and 857 ng/m^3 on 10/20). In fact, the proportion of PM-2.5 and OC mass ascribed to the acids quantified on 10/12 (0.8, 1.9%) and 10/20 (0.6, 2.2%) is similar to that of 10/4 (0.6, 2.0%), which, as stated earlier, is also a period when air parcels from the WTC area reached the site at 290 Broadway.

Levoglucosan

The anhydro sugar levoglucosan is classified as an organic molecular marker of pyrolized cellulose due to its persistence in air over long transport distances (*24,25*). Over the temporal span of this work, we detected levoglucosan in the

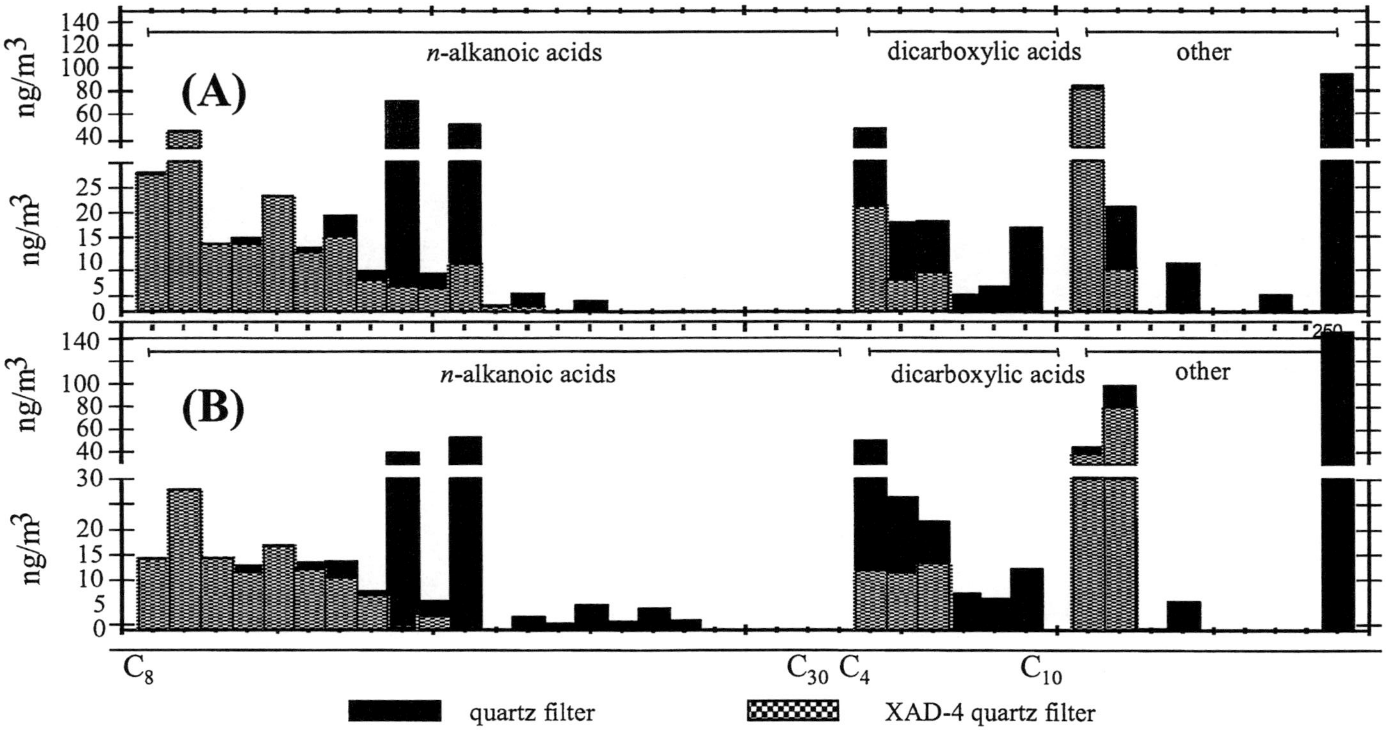

Figure 4 . Concentrations of individual organic acids and levoglucosan by substrate for 10/12 (A) and 10/20 (B). Compounds listed as "other" are in the order shown; pinonic acid, phthalic acid, abietic acid, abetatetra-18-oic acid, 9-hexadecenoic acid,8-11-octadecadienoic acid, 9-octadecenoic acid, 9, 12, 15-octadecatrienoic acid, and levoglucosan.

particle-phase in abundance (Figures 1, 3, 4). In general, levoglucosan is consistently enriched in the fine particulates over the 9/24/01-10/20/01 interval, making up approximately 2% of the OC mass (0.2-1%). Though long-range transport of levoglucosan emissions from wildland fires into the WTC air shed cannot be entirely ruled out, we expect the occurrence of levoglucosan is more likely due to the consumption of wood fuel (and paper) mass during the WTC fires and partly indicates the extent to which these fires perpetuate. The particulate sample collected on 10/6 comprises the largest fraction of levoglucosan (1%). Apparently, the aerosol component inherent to water-soluble levoglucosan (*26*) is unaffected by this day's rain event, which otherwise seems to cause the wet deposition of some chemical fraction of the PM-2.5.

Study Comparisons and Analysis of Data Trends

Aliphatic Monocarboxylic Acids.

There is considerable evidence of organic acid emissions from the WTC site during the removal operations. The prevalence of C_{16} and C_{18} in the *n*-alkanoic acid group can indicate a strong primary anthropogenic emissions component despite the ubiquity of causes of monocarboxylic acids in the atmosphere. A seasonal dependence on component concentrations in this aliphatic acid group in a California-based air shed are demonstrated by Rogge et al. (*22*), whose work further reveals these seasonal patterns varying by carbon number. More specifically, the total concentration of the C_9-C_{19} group exceeds that of C_{20}-C_{30} group in the summer, while C_{20}-C_{30} acid levels rise as the fall season approaches, reaching a maximum in the winter months. Our *n*-alkanoic acid data for NYC corroborate this trend. Note the scarcity of the C_{20}-C_{25} group until 10/20 (Figures 1, 3 and 4). Further, the average *n*-alkanoic acid levels in the fine particulates for the 5-day aggregate (9/26-10/20) are lower than both the low-end range values for an annual 4-site (downtown and west LA, Pasadena, and Rubidoux) ambient particulate study conducted in 1982 (*4*) and a two-day photochemical smog episode in southern California in September 1993 (*23*). From this comparison, we show that the *n*-alkanoic acid concentrations in PM-2.5 collected at 290 Broadway on the days most affected by WTC emissions (9/26, 10/4, and 10/20/2001) are generally below those levels normal to a highly polluted urban atmosphere. In our evaluations, we constrain the fine particulate definition to include the quartz filter substituents only because these media are common to the studies available for comparison and theoretically should classify absolute solid-phase matter. Of course, the aforementioned limitations known to affect comparison across these studies also apply here and indirectly point to the lack of similar measurements of airborne organic acids in the metropolitan NYC area. In summary, we remark that our total acid concentration values are in many cases within range of those quantified on the quartz filter by Schauer et al. (*4*) and Fraser et al. (*23*). However, to reiterate, the precise link between our total acid concentrations and their quartz filter data is uncertain.

Due to their relatively long atmospheric residence times, deposition (wet or dry) is the key atmospheric loss mechanism for the aliphatic monocarboxylic acids (*27*). As discussed, we suspect wet deposition due to climate occurred, affecting the atmospheric acid concentrations on 9/26 and 10/6. Other acid removal mechanisms are plausible in this environment and may serve to explain the particularly low *n*-alkanoic acid levels observed at this site. The alkaline nature of settled cement dust, which was verified by Lioy et al. (*8*) and then reconfirmed in our laboratory, due to $Ca(OH)_2$ and $CaCO_3$ may have neutralized the organic acids via an acid-base reaction pathway. Tests examining evidence of alkaline components in the ambient air are inconclusive. However, prior to becoming airborne, the emissions from subsurface fires may have first permeated and reacted with the microporous, alkaline cement particles in the settled debris. Also, recent efforts with marine aerosols show hygroscopic sea salt (NaCl) particles can react with gaseous OH at the air-particle interface to form basic NaOH, which could also counteract the survival of organic acids in the NYC aerosols we sampled (*28*).

Aliphatic Dicarboxylic Acids

Aliphatic dicarboxylic acids originate from biogenic and anthropogenic (incomplete combustion) sources in the ambient environment. They are also secondary photochemical reaction products, thereby interfering with their use in source apportionment studies that use a chemical mass balance framework. Succinic acid (C_4) is the most abundant alkanedioic acid measured here and is detected in urban air sheds over a global scale (*27*), where its concentration is typically second only to that of oxalic acid. Particularly evident on 10/4, 10/12, and 10/20 is an inverse, progressively nonlinear correlation between carbon number and the total (gas- and particle-phase) dialkanoic acid concentration, with only the C_9 constituent opposing the trend. In general, the abundance of lower mass molecular entities in emissions from incomplete combustion is expected. Notably, this identical trend is identified by Fraser et al. (*23*). Rogge et al. (*22*) also observed this trend and a seasonal affect on the aliphatic dicarboxylic acid concentrations, which are lowest in the months of September and October. Using the same rationale underlying the interpretation of aliphatic monocarboxylic acid comparisons, the aliphatic dicarboxylic acid concentrations in PM-2.5 (defined by quartz filter only) quantified in the NYC sky in the weeks following the 9/11 disaster are frequently below those measured in the California air space over a decade (spanning 1982-1993) (*4,23*).

Aromatic Polycarboxylic Acids

Aromatic polycarboxylic acids are found in emissions from a variety of sources (*22*). Phthalate esters are used as plasticizers in plastics production and are a potential source of these acids, and this may explain their presence in the quartz and XAD-impregnated quartz filter field blanks. Aromatic acids do appear in the NYC air over the 9/26-10/20 period, but a lapse in their detection from 10/6 to 10/12 is observed. It is unclear why this lapse is broadened to include the 10/12 sample. Otherwise, many of the data trends and their possible causes applied to the straight chain acids can also be used to interpret the aromatic dicarboxylic acid observations. We do note in the aromatic acid group that correlating their concentrations among studies is dependent on the degree of substitution with alkyl or carboxyl constituents.

Source Contributions Analysis

Due to the variety of production routes and sources in the troposphere and the complicated preparatory steps preceding their analysis, carboxylic acids are at times disregarded in source apportionment models. However, some resolution of source contributions to the troposphere can be realized by studying distribution patterns within homologous compound series (*29*). Specifically, a carbon preference index (CPI) for the *n*-alkanoic acids, defined as the ratio of Σconcentrations of odd carbon number homologs to the Σconcentrations of even number homologs, can be considered. Our data set returns closely distributed, low CPI values of 2.3–3.2, which imply a pyrogenic burden coincident with a lack of biogenic material. Borrowing *n*-alkane data from earlier work at this site (*7*), we also note that the *n*-alkanoic acid/*n*-alkane ratios near the Ground Zero site (≤1) are consistent with diesel emissions (*30*). This result is expected given the large number of diesel generators and diesel vehicles used during the recovery and removal phase operations.

Conclusions

An evaluation of the bulk chemical properties of airborne PM-2.5 collected roughly 500 m from Ground Zero over a 30 day period (9/26-10/20/2001) is achieved. As much as 97% of PM-2.5 material could be classified, with the largest weight fraction assigned to SO_4^{2-} and organics. Analysis by XRF further identifies several elemental constituents in airborne PM-2.5, which may be tentatively ascribed to cement additives and the burning of synthetic plastics and coatings materials.

For the first time, an assessment of the polar organic compounds in NYC air measured proximal to Ground Zero following 9/11 is offered. Evidence showing that these polar compounds originate at Ground Zero is considerable. Emissions from the WTC site appear to include individual acid components from several classes (*n*-alkanoic, dialkanoic, alkeneoic, and aromatic acids) and levoglucosan. Emissions signatures of these chemical compounds are relevant to mobile (diesel and vehicle exhaust) and biogenic sources as well as burning cellulose.

Though a topic of some debate, the HiC IOGAPS indicates the organic acid compounds undergo gas-, particle-phase partitioning and thus are semivolatile in nature. However, the relationship between compound vapor pressure and equilibrium partitioning proved non-ideal, suggesting that other physicochemical factors affect this system. Frequently, organic acid concentrations measured (on quartz filters) at 290 Broadway are below those reported for a closely examined urban California air shed, which is considered polluted (*4, 22, 23*). Of course, comprehension of the human health effects of Ground Zero emissions in NYC air by using just organic acid data is impractical, as they are chemically unrepresentative of the entire aerosol mixture in the atmosphere. Additionally, the biotoxicity of airborne organic acids is generally considered unresolved. Though natural reaction products are detected that point to the transport of aerosols into the NYC atmosphere, analysis of distributions within homologous compound series validates a robust pyrogenic influence.

References

1. Dockery, D. W.; Pope, A. C.; Xu, X.; Spengler, J. D.; Ware, J. H.; Fay, M. E.; Ferris, B. G., Jr.; Speizer, F. E. *N. Engl. J. Med.* **1993**, *329*, 1753-1759.
2. Swift, D. L. *Inhal. Toxicol.* **1995**, *7*, 125-130.
3. Tsuda, A.; Rogers, R. A.; Hydon, P. E.; Butler, J. P. *Proc. Natl. Acad. Sci. U.S.A.* **2002**, *99*, 10173-10178.
4. Schauer, J. J.; Rogge, W. F.; Hildemann, L. M.; Mazurek, M. A.; Cass, G. R. *Atmos. Environ.* **1996**, *30*, 3837-3855.
5. Fraser, M. P.; Kleeman, M. J.; Schauer, J. J.; Cass, G. R. *Environ. Sci. Technol.* **2000**, *34*, 1302-1312.
6. Swartz, E.; Stockburger, L.; Gundel, L. *Environ. Sci. Technol.* **2003**, *37*, 597.
7. Swartz, E.; Stockburger, L.; Vallero, D. A. *Environ. Sci. Technol.* **2003**, *37*, 3537-3546.

8. Lioy, P. J.; Weisel, C. P.; Millette, J. R.; Eisenreich, S.; Vallero, D. A.; Offenberg, J. H.; Brian, B.; Turpin, B. J.; Zhong, M.; Cohen, M. D.; Prophete, C.; Yang, I.; Stiles, R.; Chee, G.; Johnson, W.; Porcja, R.; Alimokhtari, S.; Hale, R. C.; Weschler, C.; Chen, L. C. *Environ. Health Perspect.* **2002**, *110*, 703-714.
9. Dalton, L. W. *Chemical and Engineering News*. October 20, 2003, p. 26.
10. Hays, M. D.; Geron, C. D.; Linna, K. J.; Smith, N. D.; Schauer, J. J. *Environ. Sci. Technol.* **2002**, *36*, 2281-2295.
11. In "Test Methods for Evaluating Solid Waste (SW-846) Physical/Chemical Methods," Office of Solid Waste; U. S. Environmental Protection Agency.
12. Rogge, W. F.; Hildemann, L. M.; Mazurek, M. A.; Cass, G. R. *Environ. Sci. Technol.* **1991**, *25*, 1112-1125.
13. In *NIOSH Method 5040 in NIOSH Manual of Analytical Methods (NMAM)*; 4th, 2nd supplement ed.; Cassinelli, M. E., O'Connor, P. F., Eds., 1998; Vol. Supplement to DHHS (NIOSH) Publication No. 94-113.
14. Birch, M. E.; Cary, R. A. *Aerosol Sci. Technol.* **1996**, *25*, 221-241.
15. Turpin, B. J.; Lim, H.-J. *Aerosol Sci. Technol.* **2001**, *35*, 602-610.
16. Chen, S.-J.; Hsieh, L.-T.; Tsai, C.-C.; Fang, G.-C. *Chemosphere* **2003**, *53*, 29-41.
17. Hazi, Y.; Heikkinen, M. S. A.; Cohen, B. S. *Atmos. Environ.* **2003**, *37*, 5403.
18. Schauer, J. J. Source Contributions to Atmospheric Organic Compound Concentrations: Emissions Measurements and Model Predictions. Ph.D. thesis, California Institute of Technology, Pasadena, CA, 1998.
19. Makar, P. A. *Atmos. Environ.* **2001**, *35*, 961-974.
20. Pankow, J. F.; Storey, J. M. E.; Yamsaski, H. *Environ. Sci. Technol.* **1993**, *27*, 2220-2226.
21. Limbeck, A.; Puxbaum, H.; Otter, L.; Scholes, M. C. *Atmos. Environ.* **2001**, *35*, 1853-1862.
22. Rogge, W. F.; Mazurek, M. A.; Hildemann, L. M.; Cass, G. R.; Simoneit, B. R. T. *Atmos. Environ.* **1993**, *27A*, 1309-1330.
23. Fraser, M. P.; Cass, G. R.; Simoneit, B. R. T. *Environ. Sci. Technol.* **2003**, *37*, 446-453.
24. Fraser, M. P.; Lakshmanan, K. *Environ. Sci. Technol.* **2000**, *34*, 4560-4564.
25. Simoneit, B. R. T.; Schauer, J. J.; Nolte, C. G.; Oros, D. R.; Elias, V. O.; Fraser, M. P.; Rogge, W. F.; Cass, G. R. *Atmos. Environ.* **1999**, *33*, 173-182.
26. Abel, C.; Pio, C.; Carla, S. *Atmos. Environ.* **2003**, *37*, 1775-1783.
27. Chebbi, A.; Carlier, P. *Atmos. Environ.* **1996**, *30*, 4233-4249.
28. Laskin, A.; Gaspar, D. J.; Wang, W.; Hunt, S. W.; Cowin, J. P.; Colson, S. D.; Finlayson-Pitts, B. J. *Science* **2003**, *301*, 340-344.
29. Simoneit, B. R. T. *J. Atmos. Chem.* **1989**, *8*, 251-275.
30. Schauer, J. J.; Kleeman, M. J.; Cass, G. R.; Simoneit, B. R. T. *Environ. Sci. Technol.* **1999**, *33*, 1578-1587.

World Trade Center Exposure Assessments

Chapter 11

Evaluation of Potential Human Exposures to Airborne Particulate Matter Following the Collapse of the World Trade Center Towers

Joseph P. Pinto[1], Lester D. Grant[1], Alan F. Vette[2], and Alan H. Huber[3]

[1]National Center for Environmental Assessment, U.S. Environmental Protection Agency, B–243–01, Research Triangle Park, NC 27711
[2]National Exposure Research Laboratory, U.S. Environmental Protection Agency, Research Triangle Park, NC 27711
[3]Atmospheric Oceanic and Atmospheric Administration. On assignment to the U.S. Environmental Protection Agency, National Exposure Research Laboratory, Research Triangle Park, NC 27711

The World Trade Center attack on September 11, 2001 resulted in dispersal of numerous potentially toxic materials in the dust and smoke cloud that enveloped lower Manhattan and extended over other New York City areas. This chapter illustrates approaches used, under challenging circumstances and time constraints, to evaluate human exposures of the general population to WTC-derived airborne particulate matter and potential associated human health impacts, based on integrating information derived from (a) analyses of composition and toxicity of deposited dust; (b) data from ambient air monitoring at Ground Zero, and at sites in lower Manhattan and elsewhere in the metropolitan area; (c) atmospheric dispersion modeling of the WTC plume movement/dispersal; (d) comparison of concentrations against peak urban pollutant levels and against health benchmark values judged to be indicative of risk for adverse effects due to short-term and/or prolonged particulate exposures.

Introduction

The collapse of the World Trade Center (WTC) towers resulted in vast quantities of structural materials and building contents being crushed and pulverized into airborne particles in a wide range of sizes, including both fine and coarse particles. These particles, along with particles and gases emitted from burning jet fuel, aircraft parts, and building debris, formed the immense dust and smoke cloud that rapidly spread across the New York City (NYC) area and dispersed over hundreds of square miles. Thus, particulate matter (PM) air monitoring was essential to help characterize human exposure and potentially associated health effects of the initial dust and smoke cloud, particles produced by the ensuing fires, and reentrained particles stirred up into the air during recovery activities and transport of debris away from the WTC site.

This assessment of the potential exposures to PM produced by the collapse of the WTC towers and the subsequent fires and recovery and rescue operations is among several efforts undertaken by the EPA's Office of Research and Development (ORD) to evaluate the overall impacts of this disaster. A broader inhalation risk assessment, which includes several WTC contaminants of concern including asbestos, polychlorinated biphenyls (PCBs), dioxins, and metals, in addition to PM, was conducted by the EPA's National Center for Environmental Assessment (*1*). The PM data in this assessment was largely obtained through efforts of ORD's National Exposure Research Laboratory, which established key monitoring sites around Ground Zero to characterize PM and its components. These efforts have been described elsewhere (*2, 3*).

Experimental Methods

Monitoring of Ambient Particulate Matter

At the time of the September 11, 2001 WTC attack, the only monitoring site in Manhattan measuring airborne PM-2.5 or PM-10 concentrations was in operation at the Canal Street Post Office located about a half mile north of the WTC. However, there were numerous New York state-operated PM sampling sites monitoring ambient air PM concentrations at various other locations throughout the five boroughs of NYC, with most measuring PM-2.5 levels by Tapered Element Oscillating Microbalance (TEOM) monitors.

Following September 11, substantial efforts were made to augment existing PM monitoring capabilities by adding PM sampling sites immediately around the WTC Ground Zero work zone and in the surrounding community in lower Manhattan. Saturation samplers were set up to determine concentrations and elemental composition of PM-2.5 at three surface sites triangulating the WTC Ground Zero work zone perimeter, as shown in Plate 1 (Sites A, K, C), starting on September 21 (*2, 3*). Site A was located at Barclay and West Broadway just north of Ground Zero, Site C at Liberty and Trinity to the southeast of Ground Zero, and Site K at Albany and West Streets to the southwest of Ground Zero. Only the battery-powered saturation samplers could be operated within the WTC perimeter, because power outages precluded operation of conventional samplers, such as EPA Federal Reference Method (FRM) samplers in use at preexisting New York State-operated PM monitoring sites. Although the saturation sampler has been shown to achieve accuracy within 6 to 10% compared to PM-2.5 FRM measurements at concentrations typically found in ambient air (*4*), uncertainty in measured PM-2.5 concentrations increases as ambient concentrations increase, especially above 100 $\mu g/m^3$ (24-hour average). Indeed, even the FRM is subject to increasing uncertainties at high PM-2.5 levels. In addition to the surface sites, PM sampling was initiated using a modified dichotomous versatile air pollutant sampler (VAPS) on September 22 at the 16th floor of the EPA Region II facilities in the Federal Building at 290 Broadway, about six or seven blocks northeast of the WTC (*5, 6*). The VAPS collected PM-2.5 and PM-(10-2.5) samples. These samples were collected on an almost daily basis for sampling periods of about 22 h/day until January 31, 2002. Along with the VAPS sampler, high time-resolution measurements of light scattering by particles, which is related approximately to the concentration of fine PM, and light extinction, which is related to the concentrations of black carbon and to complex organic compounds such as polyaromatic hydrocarbons (PAHs), were also made. By October 2, the U.S. EPA Region II and EPA Office of Air Quality Planning and Standards (OAQPS), along with New York State Department of Environmental Conservation (NYSDEC) personnel, set up PM monitors at the Borough of Manhattan Community College (BMCC) just north of Chambers Street at Park Row, and at the Coast Guard Station (CGS) in Battery Park.

Concentrations of PM-10 differ from those of PM-2.5 in that PM-10 includes particles whose aerodynamic diameters range from 2.5 μm up to 10 μm [PM-(10-2.5)] in addition to those smaller than 2.5 μm (PM-2.5). Measurements of PM-10 concentrations were made either as part of routine monitoring networks or in response to the WTC disaster. Long-term measurements of

Plate 1. Particulate matter monitoring sites established in response to the WTC attack.

PM-10 were obtained at the Canal Street Post Office, IS 52 in the Bronx and at PS 274 in Brooklyn. Additional PM-10 monitors were set up at the BMCC, the CGS in Battery Park, and on Wall Street. Two collocated PM-10 monitors were also set up on Albany Street-Battery Park City.

Numerical Modeling

To aid in assessing the potential effects of WTC-related air pollutants on air quality around NYC, numerical modeling of the dispersal of WTC Ground Zero-generated plumes was conducted based on prevailing meteorological conditions. The modeling was conducted in two phases: (1) modeling of the release of continuous puffs of material throughout every hour that develop into plumes, which provide regional-scale hour-by-hour plots roughly delineating the spread and direction of the WTC plume, and (2) high resolution local-scale computational fluid dynamic modeling, which was expected to provide improved, detailed estimation of the dispersal of emissions from Ground Zero in the street canyons of lower Manhattan south of Canal Street (*7*). A full description of the plume modeling is provided elsewhere (*8, 9*). It is important to note that the results of the regional-scale plume modeling primarily allow the estimation of the likely direction and width of the WTC plume at particular times, but they do not alone enable one to conclude if or where the plume may have touched down and resulted in ground-level increases in pollutant levels. They alone also do not allow one to conclude what the PM concentrations would have been, only what the estimated dilution would have been relative to WTC Ground Zero concentrations. The results are not meant to describe the transport of pollutants in the street canyons of lower Manhattan. During periods when emissions from Ground Zero were low, any increases in concentrations within the plume would also be low. Other inputs (e.g., evaluation of ground-level PM measurements at various sites and/or photos of the WTC plume) are also needed to aid in characterizing likely occurrences of plumes affecting the surface concentrations or possible human exposures.

Results and Discussion

The collapse of the WTC buildings and the associated fires resulted in the initial dispersion of large quantities of various-size particles in the massive dust and smoke cloud that enveloped lower Manhattan. It is clear from Plates 2 and 3 that the densest portion of the dust and smoke cloud initially spread in all directions and impacted most of lower Manhattan, especially below Chambers Street. As determined from analysis of the settled dusts reported in Chapter 2,

Plate 2. Spread of dense dust and smoke cloud over lower Manhattan, drifting to the east-southeast immediately after the September 11, 2001 collapse of the World Trade Center buildings. (NYPD file photo).
(See page 19 of color inserts.)

Plate 3. Spread of dense dust and smoke cloud over lower Manhattan, drifting to the east-southeast immediately after the September 11, 2001 collapse of World Trade Center buildings. (NYPD file photo).
(See page 20 of color inserts.)

most of the mass in the settled dust was due to particles > 53 μm diameter (~51 to 64% of total mass) and 10 to 53 μm (~35 to 45% of total mass), followed by lesser percentages for 2.5 to 10 μm (0.3 to 0.4% of total mass), and < 2.5 μm (~0.9 to 1.3% of total mass) particles (*10*). These results for the size distribution should be viewed with caution. Resuspension of settled dust does not yield the original atmospheric size distribution because of coagulation of smaller particles to produce larger particles and of smaller particles with larger particles, as described in Chapter 1. Given the tendency of large coarse particles to settle out of the atmosphere closer to their emission source(s) than smaller fine particles, it is likely that higher percentages of small coarse particles (> 2.5 but < 10 μm) and fine particles (< 2.5 μm) were more widely dispersed in the plume of dust and smoke that spread primarily to the southeast over Brooklyn and to the south over New York Harbor during the first 18 to 24 hours after the collapse of the WTC buildings on September 11.

PM-2.5 Concentrations

Plate 4 shows the results of plume dispersion modeling for September 11. The shading in Plate 4 represents the most likely areas of plume dispersion and the numbers shown are hourly PM-2.5 concentrations measured at preexisting state-operated PM monitoring stations. The dark area closest to Ground Zero is an estimated dilution of the source emission to 100 - 500, and the dark area at the outer rim of the plume is an estimated dilution to < one millionth of the pollutant concentration at the WTC source. The predominant wind direction was to the south-southeast and this direction continued well into the next day. As seen in Plate 4, the predominant direction of the modeled WTC plume flow is to the southeast, based on wind directions and speeds indicated by black arrows in this plate. This result is consistent with photo images, such as Plate 5, a one-meter resolution satellite image of Manhattan, New York collected at 11:43 AM EDT on September 12, 2001 by Space Imaging's IKONOS satellite. The image shows an area of white dust and smoke at the location where the 1,350-foot towers of the WTC once stood. Note the very concentrated vertical convection of dust and smoke particles upwards and the flow in a well-defined plume towards the south-southeast.

Although no direct measurements of PM concentrations are available for nearby areas in lower Manhattan during the collapse of the WTC, some rough estimates can nevertheless be made of what concentrations may have been reached. The dust cloud was optically dense, as can be seen from the airborne images (Plates 2 and 3). Under such conditions, sunlight does not reach the surface, and visibilities are greatly reduced, perhaps to meters. Conditions such

Plate 4. Modeled WTC plume dispersion for12:00 PM September 11, 2001. The numbers are hourly PM-2.5 concentrations ($\mu g/m^3$).

Plate 5. Satellite photograph of the WTC plume lofting from Ground Zero at 11:43 AM on September 12, 2001. (Reproduced from Space Imaging) (11).
(See page 21 of color inserts.)

as these are found in dust storms and dense fogs. Total suspended PM concentrations in such conditions can approach hundreds of mg/m^3. In the absence of direct measurements during air pollution emergencies, visibility can be used to crudely estimate particle loadings for conditions in which visibility is not totally obscured (*12*). There are a number of simple formulas that relate visibility to the concentration of PM-2.5, such as (*13*):

$$0.5\ (\text{km mg/m}^3) = V\ (\text{km}) * C\ (\text{mg/m}^3) \quad (1)$$

where V is the visibility range and C is the concentration of PM-2.5. During the collapse of the WTC towers, visibilities were reduced to less than 100 m (about one city block) on many streets. The application of the above formula indicates that PM-2.5 concentrations could have been about 5 mg/m^3 (5,000 μg/m^3) on these streets. In the areas closer to Ground Zero where visibilities were reduced to meters, PM-2.5 concentrations would have been proportionately higher. It should be stressed that such formulas ignore the effects of larger particles, which are less effective (per unit mass) in limiting visibility. However, the collapse of the WTC towers produced a huge volume of larger, coarse particles (*10*). Both highly effective PM-2.5 particles and much less-effective coarser particles contributed to visibility reductions and the concentrations given above represent lower limits on the abundance of total PM. It should also be noted that the above estimate of visibility is based on the loss of contrast between light and dark objects. In the streets in which sunlight was blocked by PM, total PM concentrations could have been much higher than given above.

Thus, individuals engulfed in the initial dust and smoke cloud could have been exposed for several hours to concentrations of both fine and coarse inhalable particles in the range of milligrams per cubic meter (> 1,000 μg/m^3) and perhaps to much higher concentrations of total suspended PM. However, it does not appear that people outside the lower Manhattan area (except possibly very briefly for those on Governor's Island or in Brooklyn Heights) experienced such extreme PM exposures.

This estimation is based on hourly PM-2.5 levels observed at several NYSDEC monitoring sites, shown in Figure 1, during September 11 to 13. Of most note, as seen in Figure 1, PM-2.5 concentrations measured at NYC sites generally remained under 25 μg/m^3 during most of September 11 and 12. However, hourly PM-2.5 concentrations were in the 50 to 100 μg/m^3 range for a few hours on September 12 and 13 at PS 64, which is located about 2.5 km northeast of Ground Zero, and at PS 199, which is located in Queens several miles east-northeast of Ground Zero. These PM-2.5 increases most likely reflected east-northeast dispersal of not only windblown fine PM from settled dust but also probably of newly formed fine PM generated by the intense fires at

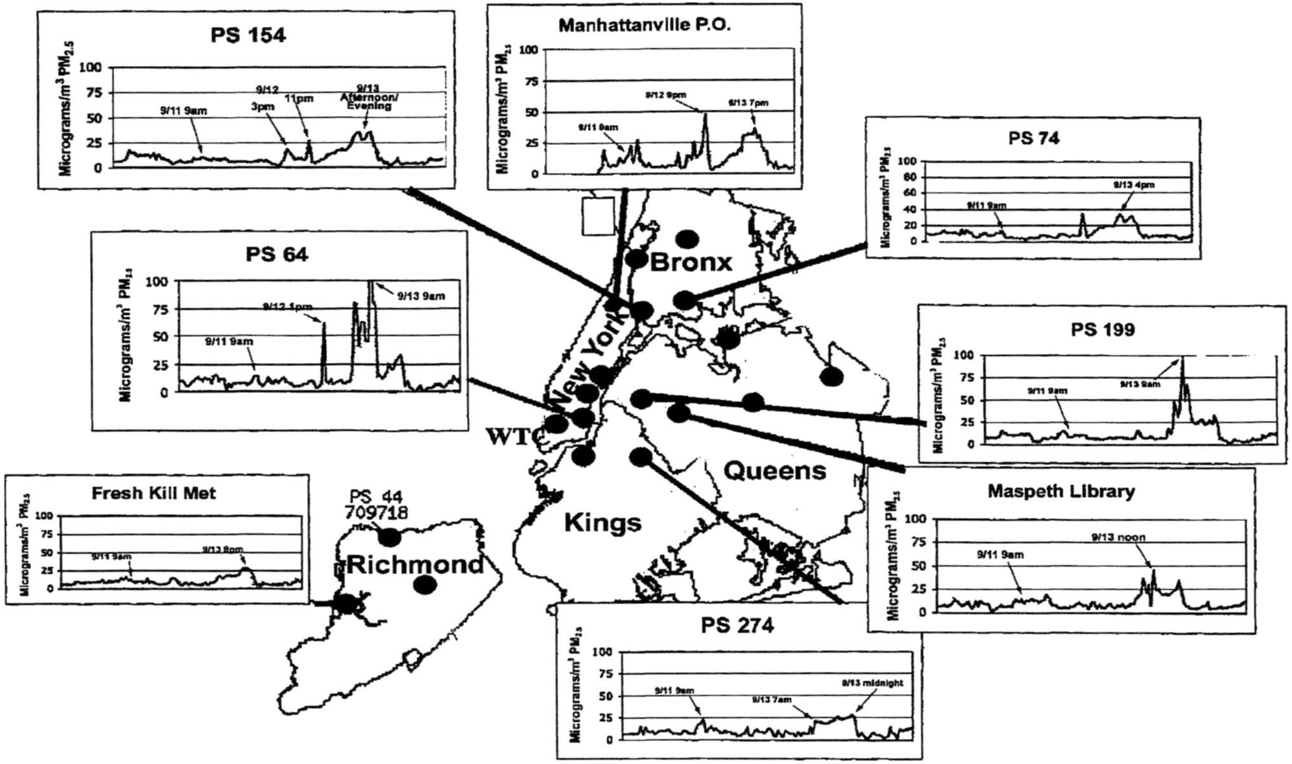

Figure 1. Increased hourly PM-2.5 concentrations measured on September 12 and/or 13 at several NYSDEC monitoring sites.

WTC Ground Zero, which burned at temperatures > 1000 °F. This would be consistent with dispersal to the east-northeast of the WTC plume, as indicated by the modeled plume for September 13 at 9:00 AM shown in Plate 6. Note the increases in measured hourly PM-2.5 concentrations at PS 64 (166 $\mu g/m^3$) and PS 199 (100 $\mu g/m^3$), consistent with the east-northeast direction of the modeled plume dispersion. The increased hourly measured PM-2.5 levels at PS 64 and PS 199 indicate that the WTC plume likely briefly fumigated the surface for a few hours at those and/or intervening locations during the morning of September 13. Changes in wind direction later in the day on September 13 resulted in rotation of the WTC plume back to a flow predominantly to the south-southwest (mainly over New York Harbor) through September 14 and 15, as indicated by the modeled winds shown in Plates 7 and the satellite photograph in Plate 8. Note the very low PM-2.5 hourly values (almost all < 6 $\mu g/m^3$) at NYSDEC monitoring sites throughout the NYC area following the rain associated with a frontal passage on September 14 and, also, likely reflecting in part the decreased vehicular traffic in the aftermath of September 11.

During the next several days (after September 14), ORD plume dispersion modeling indicates that the plume rotated in such a manner as to result in transport in varying directions, including sometimes to the north-northwest (over Manhattan and northern New Jersey), but there was little indication of the plume fumigating the surface, based on surface PM measurements. During the rest of September and on into October, 24-hour PM-2.5 concentrations at NYSDEC monitoring sites did not exceed the daily PM-2.5 National Ambient Air Quality Standard (NAAQS) of 65 $\mu g/m^3$. In fact, daily PM-2.5 values from most preexisting fixed sites throughout the NYC area did not show marked elevations in comparison to historical PM-2.5 levels for NYC areas. The occurrence through mid- to late October (at different locations and times) of a few 24-hour PM-2.5 values approaching the 24-hour Air Quality Index (AQI) limit of concern (LOC) for highly susceptible individuals of 40 $\mu g/m^3$ were not notably out of line with the past frequency of such excursions in NYC.

Starting September 21, the EPA/ORD WTC perimeter monitoring sites at Sites A, C, and K (see Plate 1), within 100 to 200 m of Ground Zero, allowed tracking of WTC-related ambient PM emissions in the immediate WTC vicinity. The Daily PM-2.5 concentrations around the Ground Zero perimeter and at 290 Broadway, six blocks northeast of Ground Zero, is shown in Plate 9A. Results from the ORD parimeter stations yielded widely varying PM concentrations across the WTC perimeter sites from day to day, with high 24-hour PM-2.5 concentrations seen at one or another perimeter site downwind from Ground Zero on given days. As shown in Plate 9, exceedances of the 24-hour PM-2.5 NAAQS level of 65 $\mu g/m^3$ occurred at some Ground Zero perimeter sites during late September and into October, but on only a few occasions thereafter. A general downward trend in daily PM concentrations, as well as a decreasing

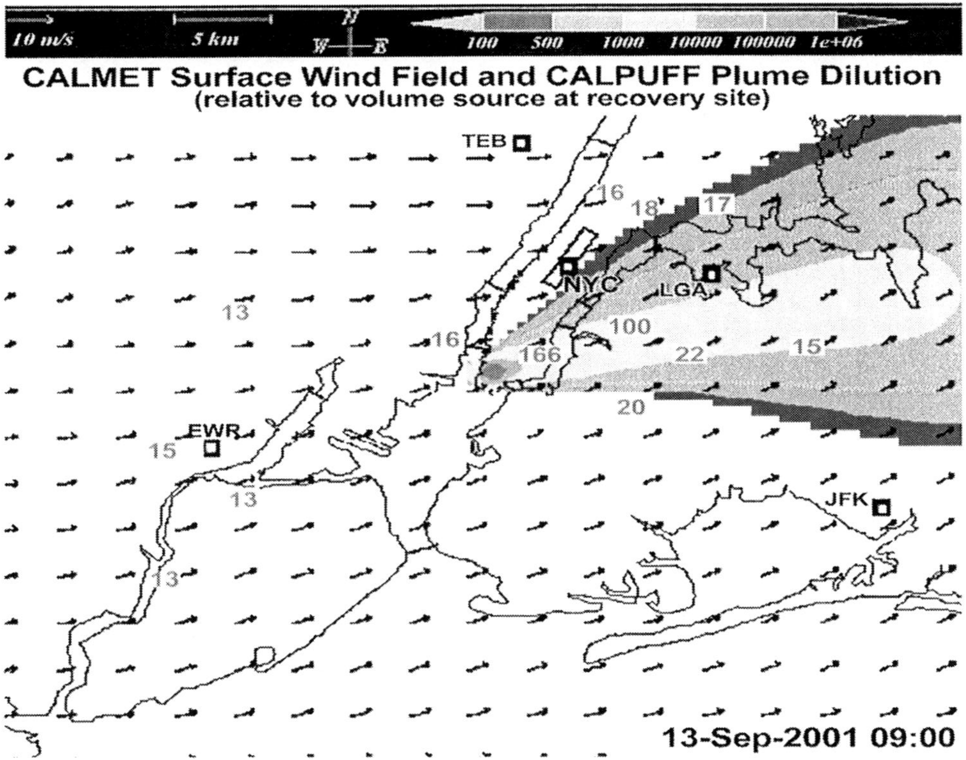

Plate 6. Modeled WTC plume dispersion for 9:00 AM September 13, 2001. The numbers are hourly PM-2.5 concentrations (μg/m³).

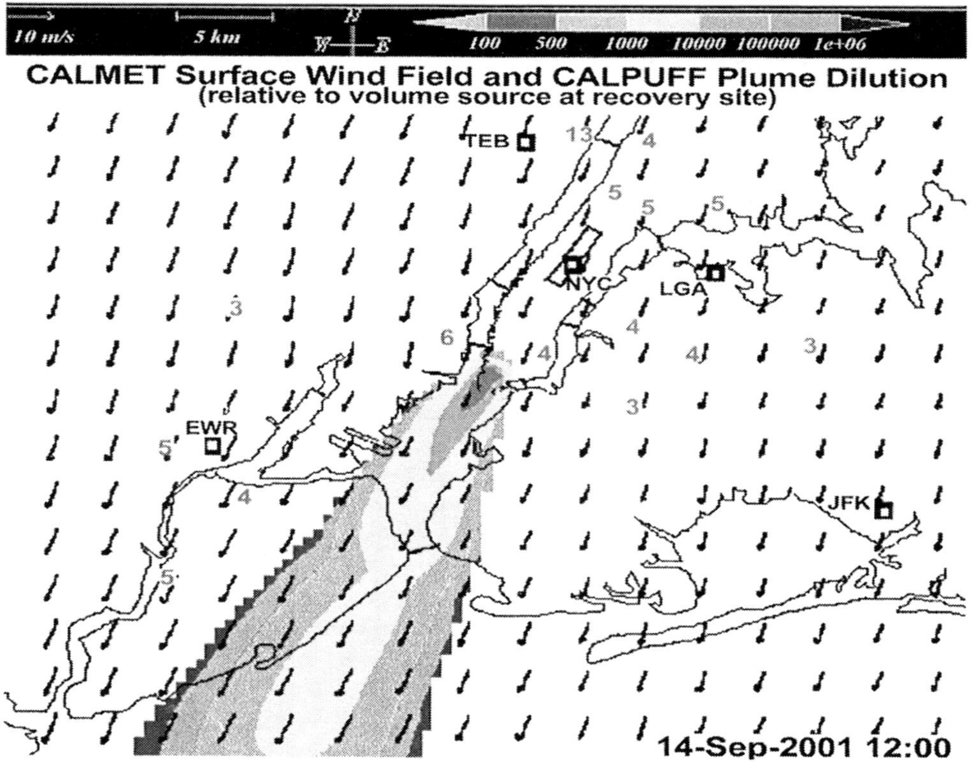

Plate 7. Modeled WTC plume dispersion for 12:00 noon September 14, 2001 The numbers are hourly PM-2.5 concentrations (μg/m³).

Plate 8. Satellite photograph of WTC plume lofting from Ground Zero at 11:54 AM on September 15, 2001, and dispersing to the south-southwest out over the New York Harbor (Reproduced from Space Imaging) (11)
(See page 24 of color inserts.)

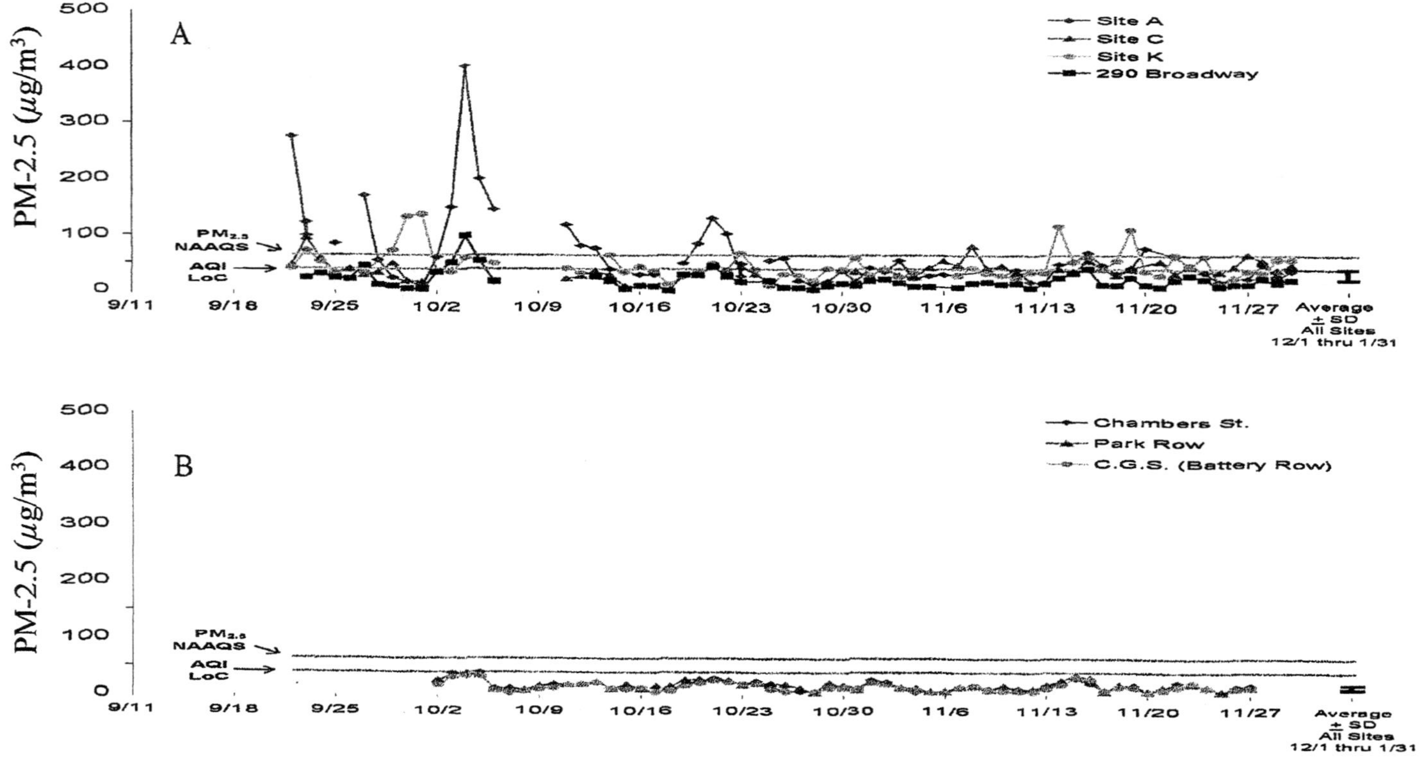

Plate 9. Daily PM-2.5 concentrations (A) at sites A, C, and K around Ground Zero perimeter and at 290 Broadway six blocks northeast of Ground Zero and (B) at several extended monitoring network sites in lower Manhattan within three to 10 blocks of WTC Ground Zero. (See page 25 of color inserts.)

range of 24-hour variations, were seen for PM-2.5 concentrations at these WTC perimeter sites from early October 2001 onward. The range of 24-hour values among the ORD sites generally remained below the AQI LOC of 40 $\mu g/m^3$ during December 2001 and January 2002 (as depicted by the black bar in the lower right of Plate 9A). In contrast to the results seen for the ORD/WTC perimeter sites, distinctly lower PM-2.5 concentrations were observed at the 290 Broadway site, about six blocks northeast of Ground Zero. The 24-hour PM-2.5 concentrations recorded there by VAPS sampling exceeded 65 $\mu g/m^3$ only once, on October 4.

The 24-hour AQI LOC was approached or exceeded at the 290 Broadway site on only a few occasions (e.g., September 27; October 3–5; October 20; November 15–16). These 24-hour values often reflect high hourly values occurring overnight mainly during early morning before 7:00 or 8:00 AM. The overall pattern of results from ORD perimeter monitoring stations near the WTC, coupled with distinctly lower PM-2.5 concentrations monitored by ORD on the 16th floor at 290 Broadway (about six blocks northeast of the WTC) suggest occasional short-term increments in fine PM values at WTC Ground Zero and, at times, along the WTC fire plume path, with the impacted areas shifting with prevailing winds. For example, when low wind speeds and a shallow mixing layer height associated with a high-pressure system settled off the southeastern coast of the United States during October 3–5, generally increased region-wide PM-2.5 concentrations were observed across much of northern New Jersey and NYC (see Chapter 8). As a result, PM concentrations were generally elevated in the New York metropolitan area. During such stagnant weather conditions, plumes tend to remain intact and are more identifiable for longer distances than during other weather conditions (14). In particular, 24-hour PM-2.5 at WTC Site A (on the northern perimeter of Ground Zero) reached 400 $\mu g/m^3$ on October 3–4, but 24-hour PM-2.5 values dropped off to 90 $\mu g/m^3$ at the 290 Broadway site several blocks northeast of WTC and to 53 $\mu g/m^3$ at PS 64 about 1.5 km further to the northeast of Ground Zero and only reached 60 $\mu g/m^3$ at Site K (on the southwest perimeter of Ground Zero). This was consistent with prevailing winds to the northeast and the modeled plume dispersion depicted in Plate 10 for October 4. Note the general regional elevation of PM-2.5 levels for monitoring sites scattered across both northern New Jersey and NYC areas, indicated by the numbers in Plate 10. This occurs even outside modeled areas of likely greatest plume intensity.

Daily average PM-2.5 data obtained at additional sites in lower Manhattan (Chambers St., Park Row, and the U.S. Coast Guard Station at Battery Park) are shown in Plate 9B. These sites are located from three to ten blocks to the north, east, and south of the WTC (see Plate 1). It can readily be seen in Plate 9 that concentrations of PM-2.5 were much lower at these sites than at the WTC perimeter, indicating a very rapid decline with distance from Ground Zero. The

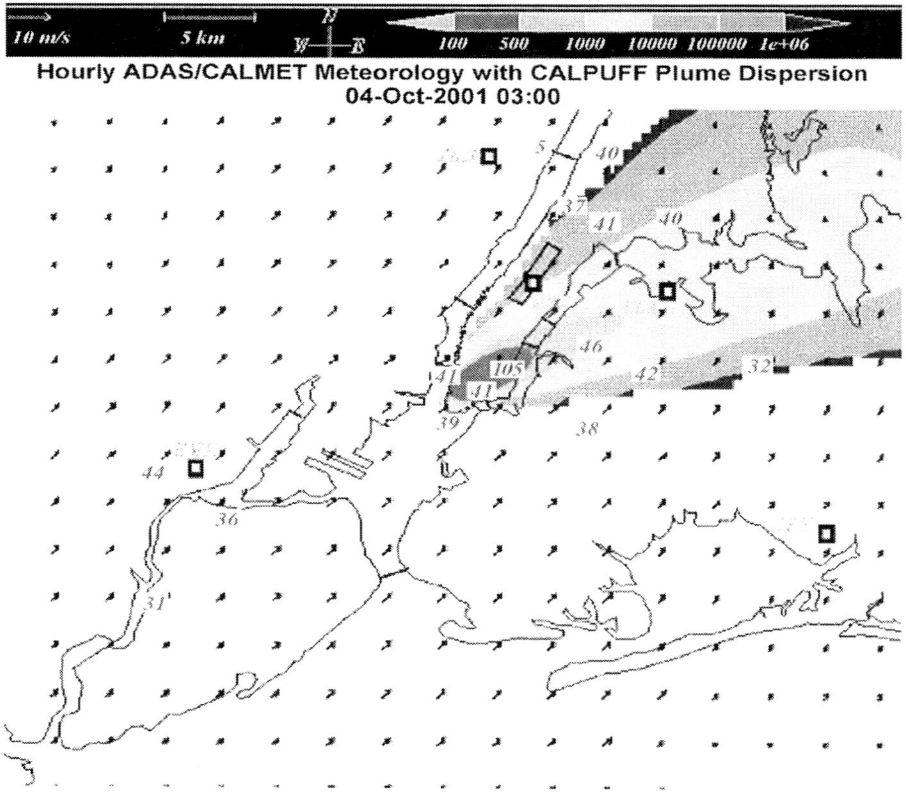

Plate 10. Modeled WTC plume dispersion for 3:00 to 4:00 AM October 4, 2001. The numbers are hourly PM-2.5 concentrations ($\mu g/m^3$).

PM-2.5 concentrations at these three sites can also be seen to rise and fall together. Correlation coefficients between pairs of these sites are all > 0.9 and the concentrations are all very similar, suggesting that these sites were responding mainly to variations in urban and regional background sources rather than to WTC emissions. None of the daily PM-2.5 values exceeded the PM-2.5 24-hour NAAQS of 65 $\mu g/m^3$. Only a few daily PM-2.5 values for the NYSDEC lower Manhattan sites to the north, east, and south of the WTC, even approached the 40 $\mu g/m^3$ AQI LOC value. Many of the 24-hour PM-2.5 levels for such sites were below 20 $\mu g/m^3$. For example, Figure 2 shows the 24-hour average concentrations for PM-2.5 at the NYSDEC monitoring site PS 64 from September 11 to October 2, 2001 compared to historic levels of PM-2.5 concentrations seen at PS 64 during the two previous years (February 23, 2000 to September 1, 2001.

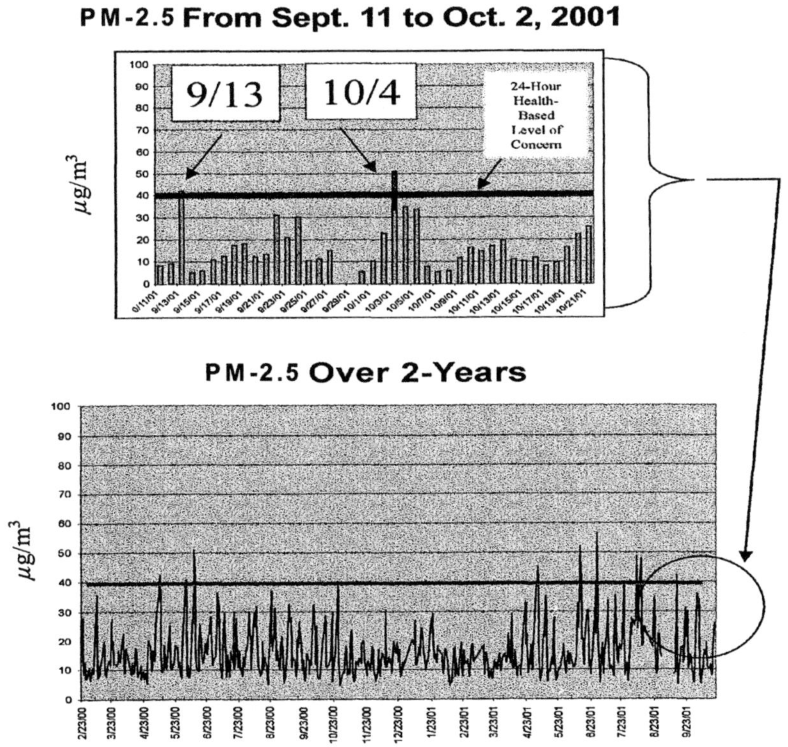

Figure 2. Daily PM-2.5 concentrations recorded at PS 64 from September 11 to October 27, 2001 (upper) compared to 24-hour PM-2.5 recorded at the same site from February 23, 2000 to September 1, 2001(lower).

PM-10 Concentrations

The time series of daily average PM-10 concentrations are shown in Plate 11 and the monthly mean PM-10 concentrations are given in Table I. The subscripts in Table I refer to the number of measurements taken in that month. The entries without number of measurements indicated have hourly measurements so there were over 700 measurements per month taken at these sites.

As can be seen from Plate 11 and Table I, PM-10 concentrations obtained at sites in lower Manhattan tended to be higher during the autumn of 2001 than during the following winter and spring. Table I shows that 24-hour average PM-10 concentrations were higher at sites in lower Manhattan than at the sites in the other boroughs, with the difference generally narrowing with time. However, the difference did not disappear entirely. The PM concentrations are expected to be somewhat higher in lower Manhattan, because of trapping of emissions from motor vehicles and resuspended dust in street canyons. The highest daily average

Table I. Monthly Mean PM-10 Concentrations

Site	*Sep*	*Oct*	*Nov*	*Dec*	*Jan*	*Feb*	*Mar*	*Apr*	*May*	*Jun*
290BW	16_{5}	42_{25}	30_{70}	22_{25}	23_{30}	20_{18}	23_{21}			
CSPO		39_{25}	34_{25}	25_{25}	27_{10}	25_{10}	25_{10}	25_{10}	26	26
CS		42_{31}	35_{28}	32_{27}	36_{15}	34_{27}	34_{27}	35_{29}	31_{29}	
CGS		36_{30}	32_{29}	28_{21}						
PS274		28_{28}	25_{28}	22_{28}	24_{8}	15_{25}	20_{31}	35_{29}	25_{29}	
PR		61_{28}	38_{29}	32_{41}	34_{25}	22_{24}	25_{27}	29_{30}	29_{28}	
ASBP		67	31	26	25	25	29	27	27	30
WS			27	26	25	25	26	28	30	
IS52		22	25	19	20	19	21	24	26	
AS14				30_{9}	29_{18}	22_{5}	33_{26}	29_{29}	28_{29}	

NOTE: Monitoring sites are: 290 Broadway (290BW), Canal Street Post Office (CSPO), Chambers Street (CS), Coast Guard Station at Battery Park (CGS), PS 274, Park Row (PR), Albany Street Battery Park (ASBP), Wall Street (WS), IS 52, and Albany Street, Site 14 (AS14).

NOTE: Units are $\mu g/m^3$. Subscripts are the number of measurements. Sites without subscripts are hourly with >700 measurements. September sampling dates for 290BW were 9/25–9/30. October sampling dates for ASBP were 10/30–10/31 and for IS52 were 10/5–10/9. November sampling dates for WS were 11/18–11/30.

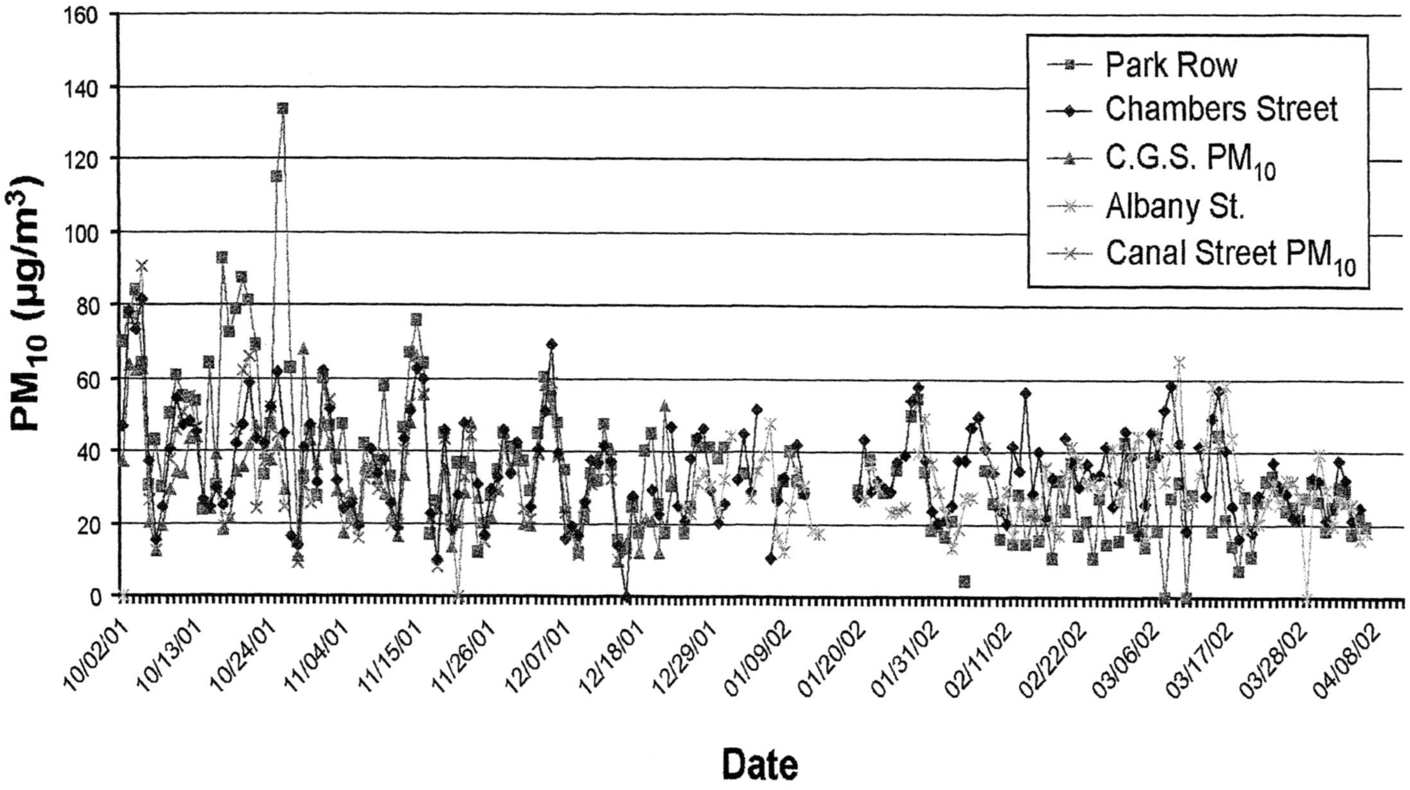

Plate 11. Daily average PM-10 concentrations measured at sites in lower Manhattan (as indicated in Figure 1). (See page 27 of color inserts.)

PM-10 concentrations were observed on October 26, 2001 at Park Row (~135 $\mu g/m^3$), a level a bit below the 24-hour NAAQS for PM-10 of 150 $\mu g/m^3$. Except for two days in October (the 25th and 26th) at the CGS, all 24-hour PM-10 concentrations during the monitoring period shown in Table I were below 100 $\mu g/m^3$. Maxima in 24-hour PM-10 concentrations measured at BMCC and Park Row (~80 $\mu g/m^3$) and the Canal Street Post Office (~90 $\mu g/m^3$) occurred from October 3 through 5 with southwesterly winds. During the same period, 24-hour PM-10 concentrations measured at PS 274 were about 60 $\mu g/m^3$.

The average ratio of PM-2.5 to PM-10 concentrations observed at sites that had collocated PM-2.5 and PM-10 monitors in lower Manhattan was about 0.5 from October–December 2001. The ratio of PM-2.5 to PM-10 measured at 290 Broadway increased steadily from about 0.5 to 0.7 from October 2001–March 2002. The mean ratio of PM-2.5 to PM-10 concentrations in the Northeast is about 0.7, based on available monitoring data from state and local PM monitoring networks. The PM derived from combustion or as the result of photochemical reactions involving gaseous precursors tends to be found mainly in the PM-2.5 size range, whereas resuspended dust tend to be found as larger particles (see Chapter 1). Thus, monitoring sites that exhibit a high ratio of PM-2.5 to PM-10 are likely dominated by combustion products and this ratio would be lowered to the extent that these sites are affected by resuspended dust.

Measurements of Particle Composition

Analyses of bulk samples of dust produced by the collapse of the WTC towers were performed by Lioy et al. (*10*). Two bulk dust samples were collected on September 16 and another on September 17. The samples were collected at weather-protected sites located less than 1 km to the east of the WTC. The particle samples were separated according to size by aerodynamic and gravimetric methods. As noted earlier, results of the aerodynamically separated samples indicate that only a very small fraction (about 1%) of PM was in the PM-2.5 size range and less than 0.5% was in the PM-(10–2.5) size range (see Chapter 2, Table I). The overwhelming fraction of the mass of PM found in settled dust particles was larger than 10 μm. However, it should be noted that it is difficult or impossible to reproduce the original, airborne PM size distribution based on resuspension of settled dust, because the smallest particles have a strong tendency to stick to larger particles. Larger particles also settle out of the atmosphere at a faster rate than smaller particles. As a result, the contribution of smaller particles will likely be underestimated. Most of the mass consisted of pulverized building and construction materials (e.g., cement and glass fibers) and office materials (e.g., cellulose). High concentrations of inorganic constituents including silica, calcium, and sulfate components of building material and metals, such as lead and zinc, were found. Also, total polycyclic aromatic hydrocarbons (PAHs), which are products of incomplete combustion, constituted more than 0.1% of the total mass of dust. The fraction of adsorbed PAHs in each size fraction is expected to be roughly related to the relative surface area in each size fraction. According to this criterion, smaller particles would have contained proportionately greater concentrations of PAHs than indicated by their relative mass. Pleil et al. (*15*) recently reported measurements of PAH concentrations in the PM-2.5 filter samples that were collected as part of the EPA/ORD effort (*2, 3*). The PAH concentrations were estimated from 3 to 200 days after September 11 based on modeling trends in the filter data set. They predicted individual PAH concentrations to range from 1.3 ng/m^3 for benz(a)anthracene and dibenz(a,h)anthracene to 15 ng/m^3 for benzo(b)fluoranthene on September 14. Values were certainly higher during the prior few days. They also noted that these concentrations were among the highest reported for outdoor sources in the United States, but that concentrations of the PAHs rapidly decreased with time. However, it should also be noted that PAH concentrations in the range of several ng/m^3 have been sustained throughout entire winters in a (formerly) highly polluted city, Teplice, in the Czech Republic (*6, 16*).

Lioy et al. (*10*) provided complete descriptions of numerous other specific compounds that were found in dust particles that settled outdoors. They also noted that penetration of substantial quantities of WTC-derived dust into indoor office or residential spaces likely notably increased the potential for indoor exposures (via ingestion or by inhalation of reentrained particles) to high levels of constituent elements and compounds.

More detailed chemical analyses have been performed on aerodynamically size-separated PM-2.5 derived from bulk dust samples collected on September 12 and 13 from several locations within 0.5 miles of Ground Zero (*1, 17*). These analyses showed that calcium sulfate (gypsum) and calcium carbonate (calcite) were major components of the fine fraction, indicating that very finely crushed building materials were still dominant components even in this size range. This is important because fine particles more easily penetrate into offices and residential spaces and thereby contribute to indoor exposures more readily than coarse particles.

Data for EPA/ORD measurements of PM elemental composition of fine particles (PM-2.5) for samples collected starting September 21 are graphically depicted in Figures 3 to 7. These were selected for illustration from a larger set measured by ORD, based on evident elevations of their concentrations over typical background levels at some point during the sampling campaign. The elements are grouped in each figure roughly according to their relative abundance in the subject air samples, those in Figures 3 to 5 being among the most abundant and those in Figures 6 and 7 being distinctly less abundant. Most of the elements shown were highly correlated ($r > 0.85$) with each other at

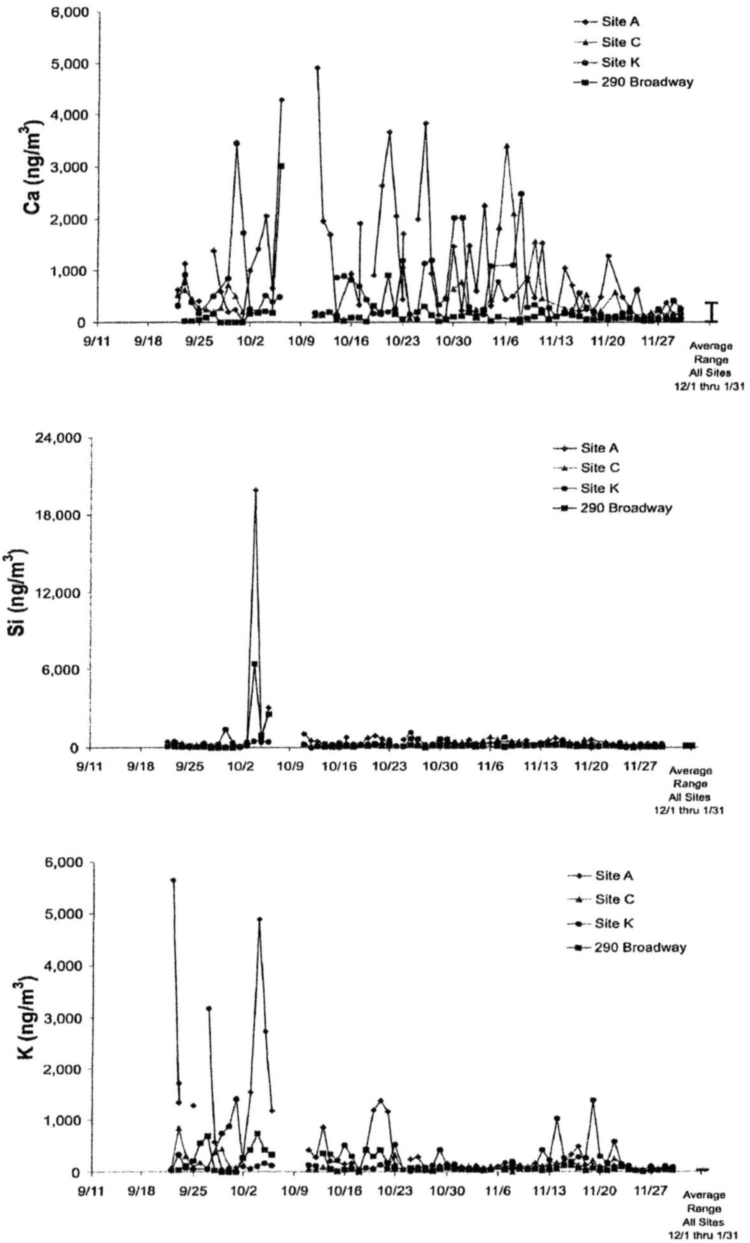

Figure 3. Calcium (Ca), silicon (Si), and potassium (K) concentrations in PM-2.5 at WTC Ground Zero perimeter (Sites A,C,K) and 290 Broadway.

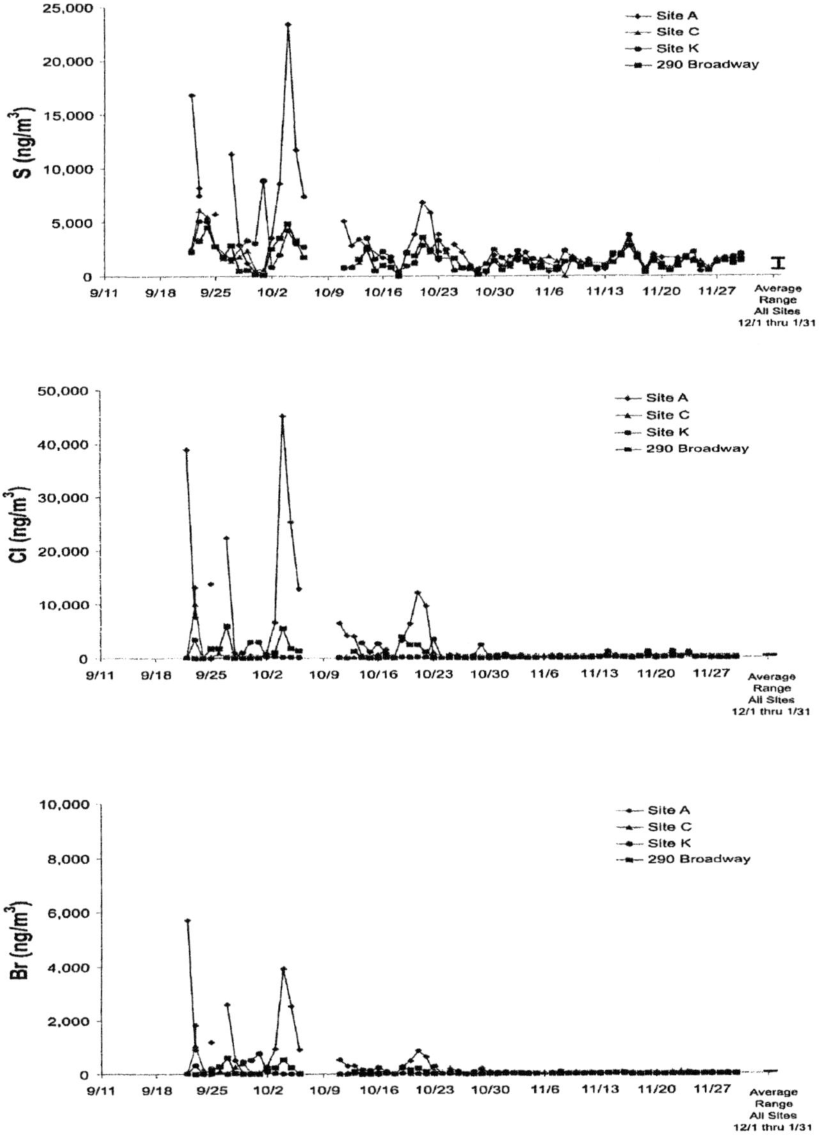

Figure 4. Sulfur (S), chlorine (Cl), and bromine (Br) concentrations in PM-2.5 at WTC Ground Zero perimeter (Sites A,C,K) and 290 Broadway.

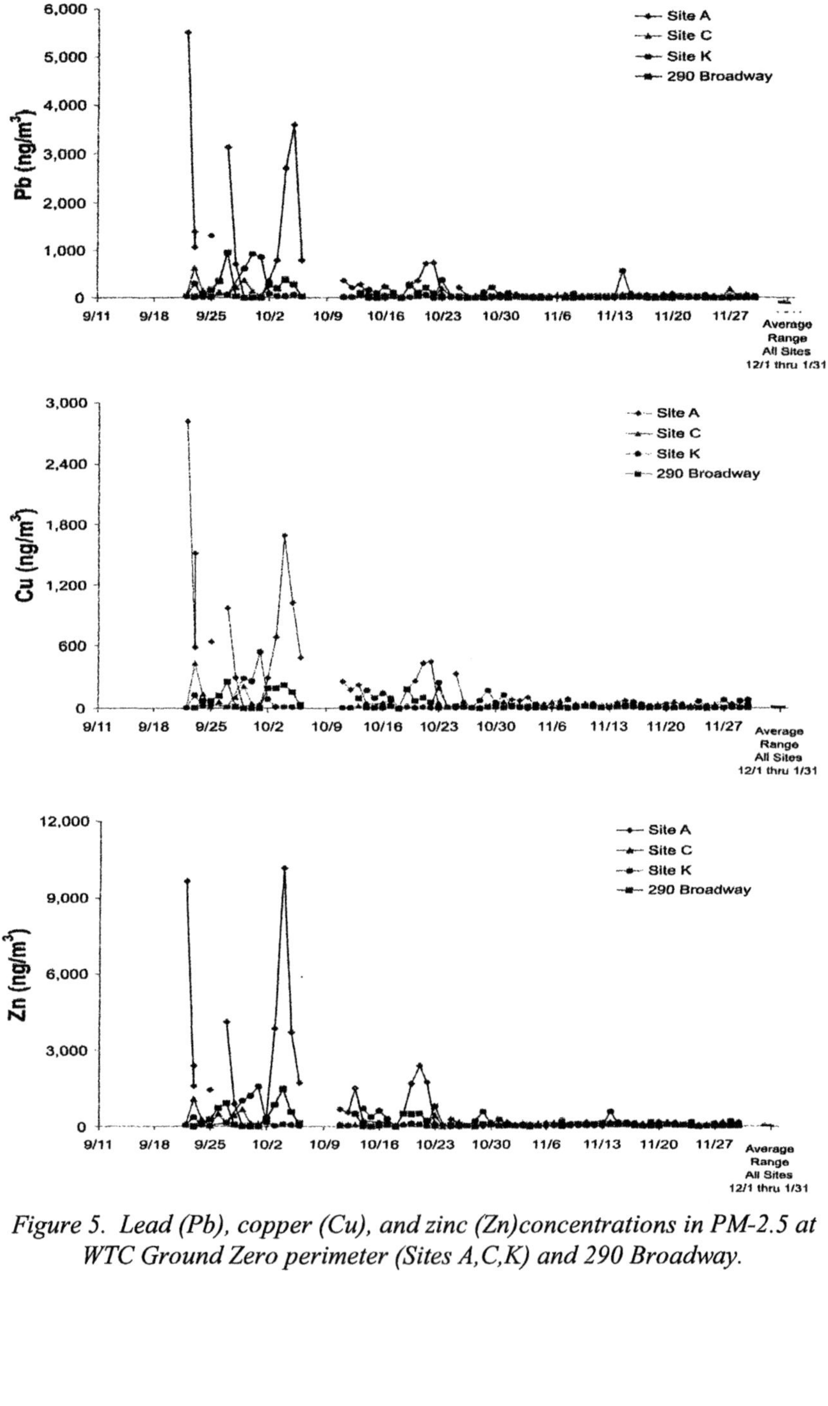

Figure 5. Lead (Pb), copper (Cu), and zinc (Zn)concentrations in PM-2.5 at WTC Ground Zero perimeter (Sites A,C,K) and 290 Broadway.

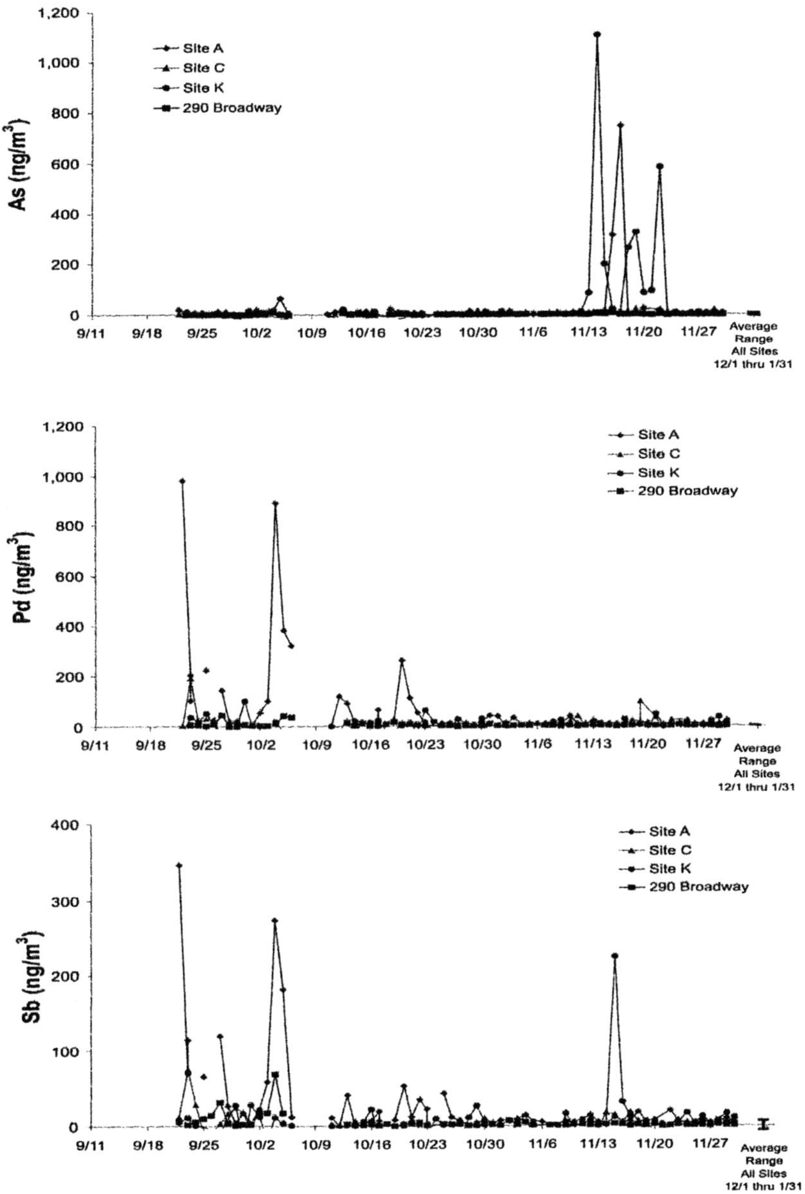

Figure 6. Arsenic (As), palladium (Pd), and antimony (Sb) concentrations in PM-2.5 at WTC Ground Zero perimeter (Sites A,C,K) and 290 Broadway.

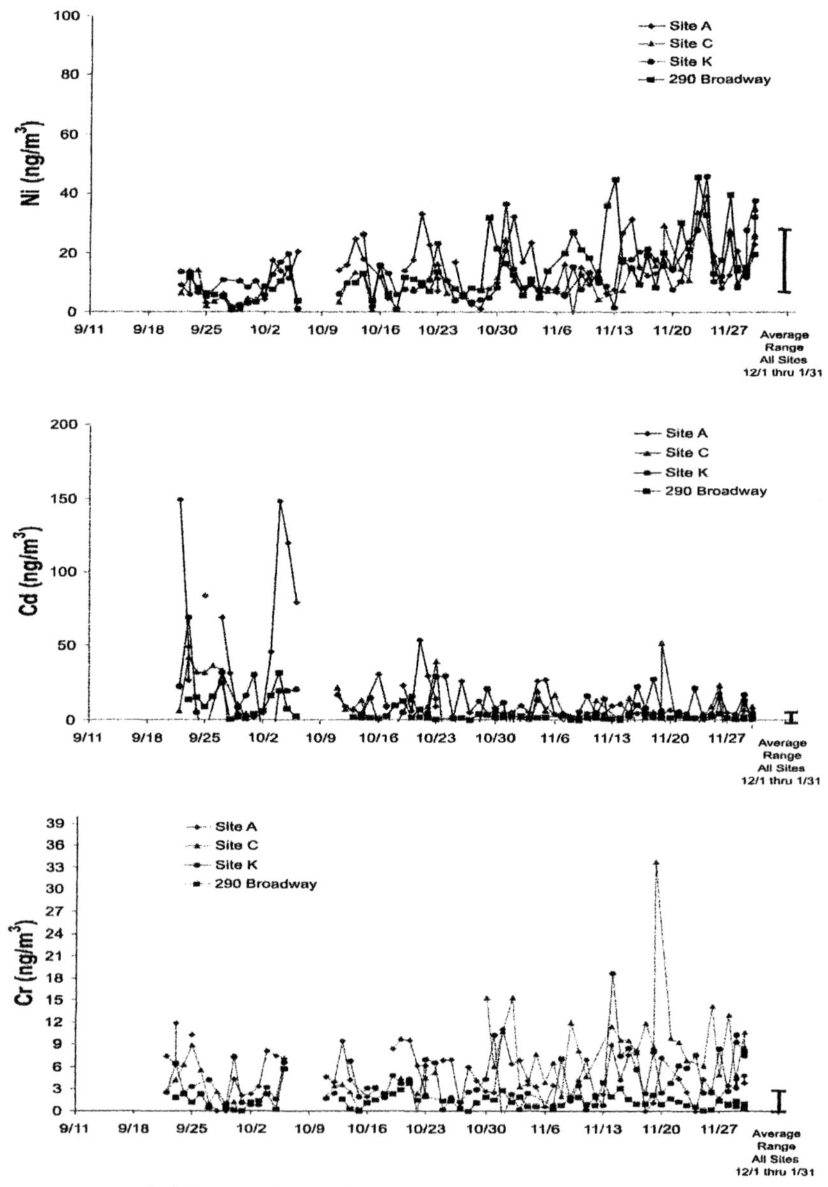

Figure 7. Nickel (Ni), cadmium (Cd), and chromium (Cr) concentrations in PM-2.5 at Ground Zero perimeter (Sites A,C,K) and 290 Broadway.

individual sites, with several being much more highly correlated ($r > 0.95$) with each other and with PM-2.5 throughout the sampling campaign. The very high correlations suggest a common source origin for these elements, that is, the WTC fires. The composition of the emissions from Ground Zero combustion sources changed with time as evidenced by initial peaks in several elements (e.g., calcium, potassium, sulfur, chlorine, bromine, lead, copper, and zinc) during late September and early October (Figures 3 to 5), followed by later peaks in the concentration of chromium, arsenic, and antimony during mid- or late November (Figures 6 and 7).

Consistent with Lioy et al.'s finding of highly enriched calcium in both fine and coarse fractions of settled dust near the WTC, markedly increased levels of calcium continued to be seen in airborne fine particles (Figure 3) at ORD's Ground Zero perimeter sites off and on throughout September and October and into much of November, decreasing to low levels by late November. Elevated levels of calcium in the fine fraction are indicative of highly pulverized building materials (e.g., wallboard) from the WTC site. However, except for a few occasions (e.g., on October 5), airborne fine PM calcium levels were not markedly elevated above typical urban levels at the EPA 290 Broadway site several blocks northeast of Ground Zero, whereas calcium in the PM-(10–2.5) coarse fraction (not graphically depicted here) did show rather frequent elevations at the 290 Broadway site but decreased to typical urban levels, as shown in Figure 3, by the end of November.

Fine PM silicon elevations were only evident briefly during October 3–5 at perimeter Site A and at 290 Broadway, in contrast with coarse silicon elevations (generally in the 1000 to 3000 ng/m^3 range) seen at 290 Broadway on a number of days well into late November. The coarse fraction calcium and silicon enrichments most likely reflect (a) windblown reentrainment of calcium and silicon-contaminated dust remaining on rooftops or window ledges, e.g., after hazardous material cleanup of WTC-derived dust in lower Manhattan during the 2 weeks after September 11, and (b) calcium and silicon particle reentrainment into ambient air associated with WTC recovery operations and transport of debris away from Ground Zero, or a combination of these factors.

Data for concentrations of elemental carbon and total organic compounds in the aerosol phase based on analyses of paired filter samples are given elsewhere (*18*). Aethalometer results and the results of analyses of bulk dust composition below suggest that the WTC emissions contained substantial quantities of carbon produced by incomplete combustion. Surprisingly low total carbon levels (1.5 to 8.5%) were found in PM-2.5 samples aerodynamically size-separated from the bulk dust samples collected on September 12 and 13 (*17*). These results indicate that crushed building materials were the dominant sources of fine PM immediately after the collapse of the towers, whereas combustion from ongoing fires and increased emissions from diesel-powered heavy equipment and generators was a relatively more important source of PM-2.5 in later emissions from the WTC disaster site (*18*). Potassium enrichments were also especially notable for measurements made at Site A into early October consistent with combustion of organic materials such as wooden furniture, paper, etc. However, much lower potassium concentrations occurred at 290 Broadway, and air levels of potassium at all ORD sites returned to very low background levels by late November.

Elevations of sulfur, chlorine, and bromine (Figure 4) were clearly evident at the Ground Zero perimeter sites and sometimes at 290 Broadway during late September and gradually decreased into October, again consistent with the notable enrichments seen by Lioy et al. in WTC dust particles. The sulfur was likely in oxidized form, some perhaps having been converted from primary emissions of SO_2 into secondary sulfate particles, consistent again with both reports of elevated sulfate levels in settled WTC dust particles (*10, 17*) or airborne particles (including very fine fraction) collected at the Varick Street site on October 3 (*19*). Also consistent with the findings of chlorine and bromine enrichments in WTC settled dust are ORD measurements of unusually elevated chlorine and bromine at Ground Zero perimeter sites (see Chapter 9). The WTC sources of these halides are not clear, but chlorine from burning plastics and paper is not unlikely. The specific enrichment of chlorine in fine particles (versus more typical sodium chloride present as coarse particles) with no notable sodium enrichment rule out attribution of the chlorine levels simply to airborne sea salt influxes into the WTC fire site.

As seen in Figures 3 to 7, lead and certain other metals (copper, zinc, antimony, palladium, and cadmium) were notably elevated on some days in late September and into early October at Site A, as compared with concentrations at Sites C and K, but the concentrations of these metals had generally decreased to background levels by mid-October. The late September/early October elevations at Site A on the WTC north perimeter indicate that the WTC fires were likely a common source of emissions of these metals, because the winds were mainly from the southwest at this time. Meaningful average concentrations for particulate mercury could not be calculated because its concentrations were usually below the detection limit of about 12 ng/m^3. Highest concentrations (70 ng/m^3) were measured on or about October 3. Mercury in gaseous form was not measured. The detection of elevated levels of arsenic and antimony in mid-November at Site K on the southwest perimeter of Ground Zero (but not at Sites A or C or at 290 Broadway to the north, northeast, or southeast of the WTC) indicates both a different source of emissions and the shift to winds coming mainly from the north/northeast. The chromium elevations seen around November 20, mainly at Site C, suggest possibly yet another later shift in the composition of Ground Zero sources of WTC-generated airborne particle emissions. Hence, the composition of emissions from the WTC site appears to have varied over time.

In contrast to the above patterns of element levels that indicate that they may have originated from the WTC fires, the gradually increasing concentrations of nickel up to a range sustained during December and January (after the WTC fires were out) seem to argue against any notable airborne fine-particle nickel emissions from the WTC fires subsequent to the collapse of the WTC buildings on September 11. Still, enrichments in samples of settled dust from sites east of the WTC (*10*) may be indicative of nickel having been among the metals present in high concentrations of airborne particles in the initial dust and smoke that enveloped lower Manhattan on September 11. This raises the possibility of (a) any remaining nickel-containing dust being reentrained into outdoor air during later rescue/recovery operations and/or (b) continued elevations of nickel concentrations in WTC-derived indoor dust and reentrained indoor air particles. However, the general pattern in nickel concentrations suggests that regional sources, perhaps related to the combustion of fuels to meet seasonal heating demands, may be responsible for the increase of nickel with time.

The measurements of PM-2.5 and its components discussed above were obtained over sampling periods of close to 24 hours. Additional data must be examined to determine more precisely the duration and nature of enhanced concentrations of PM constituents and to help understand possible public health impacts of such excursions. For example, of much interest are PM elemental composition data for the period October 3–5, when NYC was under the influence of a high-pressure system that had settled off the southeastern coast of the United States. Concentrations of sulfur were close to six times higher at the downwind WTC site (Site A) than at the upwind site (Site K) on October 4. Concentrations of a number of other elements were also much higher at Site A compared to Site K, as indicated by the large enrichment factors (noted in parentheses) for the following elements: silicon (41), chlorine (500), potassium (47), calcium (5), bromine (350), copper (130), zinc (110), palladium (> 100), cadmium (> 100), antimony (> 100), and lead (66).

The fractional contribution of major elements determined by X-ray fluorescence spectroscopy to PM-2.5 on October 3 is shown in Figure 8. These values were calculated based on the difference between those measured at Site A (downwind of Ground Zero) and Site K (upwind of Ground Zero). Overall, the above results are most consistent with brief, episodic increases from WTC Ground Zero during early morning hours on October 3 and 4 leading to elevated concentrations of PM and its constituent elements at the WTC perimeter sites and one or another sites in lower Manhattan located downwind of the WTC on those days. These enhanced concentrations were superimposed on generally higher concentrations of PM and its constituents found upwind of the WTC on those days. The higher concentrations found at the upwind sites were associated

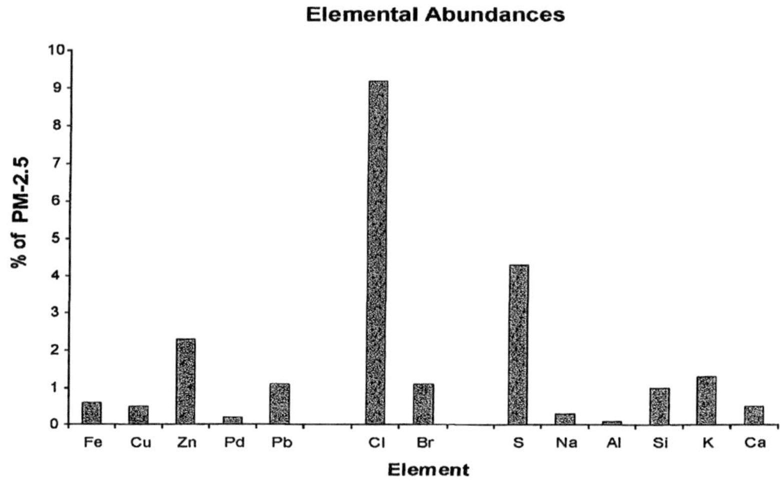

Figure 8. Elemental composition of emissions from the WTC site on October 3 determined by X-ray fluorescence spectroscopy.

with the high-pressure system mentioned above. In any case, the results suggest strongly that the WTC site was the dominant source of these elements on October 3–4 at sites in close proximity to the WTC (see Chapter 8).

Measurements of the size distribution of PM were made on the roof of the Federal Building at 201 Varick Street,approximately 2.0 km north-northeast of Ground Zero starting on October 2, as described in Chapters 8 and 9 (19). These data indicate a sharp increase in concentrations of PM-2.5, mainly in the size range between 0.34 to 0.56 μm on the morning of October 3, as shown in Figure 9. This very brief excursion only lasted a few hours. Concentrations of sulfur and silicon at this time were notably elevated at Varick Street as compared to days immediately before and after. Also, data obtained by nephelometer and aetholometer at 290 Broadway indicated substantial elevations of PM and light-absorbing components (primarily elemental carbon) within the space of a few hours early in the morning of October 4. Hourly PM-2.5 levels were also elevated at PS 64, some in excess of 100 $\mu g/m^3$, during or shortly after the same hours on October 4, suggesting brief surface fumigation by the WTC plume to the north-northeast of Ground Zero.

It has been suggested that the high concentrations of sulfur observed at the Federal building on Varick Street were related to transport from power plants either in the Northeast or in the Ohio Valley. This is possible because steady, light southwesterly winds associated with the high-pressure system were capable

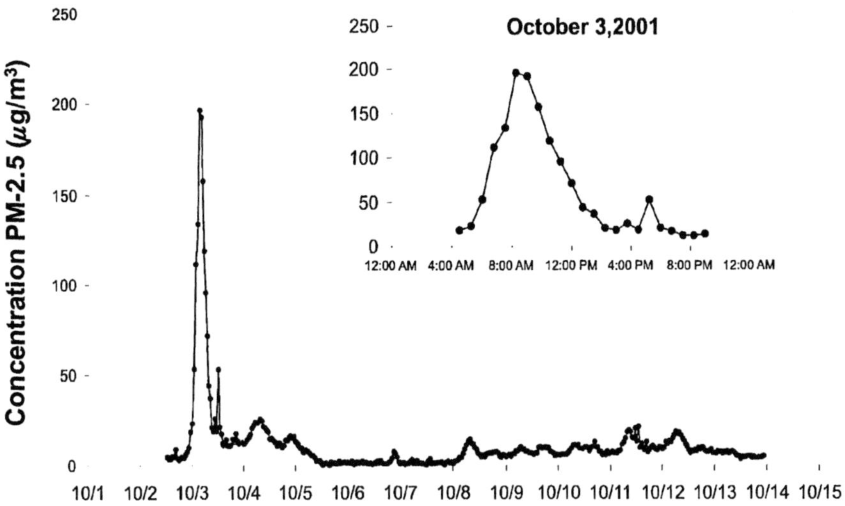

Figure 9. PM-2.5 concentrations measured at 201 Varick Street. (Reproduced from reference 19.)

of transporting pollutants long distances to NYC. However, the extremely high abundance of sulfur on the Ground Zero perimeter and the ratio of sulfur observed at Site A compared with that observed at Site K suggests that Ground Zero was a large source of sulfur on October 3 and could have also contributedto sulfur readings at the Varick Street rooftop site.

The peak elevations of airborne fine PM silicon limited to October 3–4 at Site A and at 290 Broadway are notable, suggesting high temperature volatilization of silicon from glass and/or cement by intense WTC fires on those dates and transport of the WTC plume in a north-to-northeasterly direction. This is consistent both with the plume trajectory plotted in Plate 10 and the marked increases in silicon levels reported by Cahill et al. (*19*) at the Varick Street site, to the north-northeast of the WTC, including unusual measured elevations of silicon in very fine (0.09 to 0.50 μm) and, possibly ultrafine (< 0.01 μm) PM size ranges. Typically silicon is mainly associated with coarse fraction particles (> 2.5 μm). There were also marked increases in concentrations of various metals on October 3–4, reinforcing the conclusion that the Varick Street data for October 3 reflected emissions from intense Ground Zero fires. The very brief increase in PM-2.5 values and high levels of silicon at the Varick Street site on October 3 apparently did not occur again at that site. Interestingly, Lioy et al. (*10*) reported lead and other substances as being unusually congealed with silicon in particles from the WTC settled dust, likely due to the vaporization of silicon, lead, and/or other metals by intense heat, followed by their condensation and coagulation into particles with unusual composition.

The above high-resolution measurements suggest that WTC emissions in late September and early October varied greatly at times over 24-hour periods and that some of the more notable emissions probably occurred in discrete events that resulted in air pollutant elevations that lasted only a few hours. Such events were likely related to activities of rescue and recovery operations at the WTC, such as the removal of large pieces of debris perhaps resulting in increased oxygen flow and brief flareups of fires within the WTC rubble pile.

When NYC weather is dominated by high-pressure conditions in the cooler months of the year, especially at nighttime, there would be a greater tendency for emissions to form steady, coherent plumes, which can "snake their way" through the street canyons of lower Manhattan. When winds are highly variable and there is thermally enhanced vertical mixing, emissions have a greater tendency to disperse in all directions and to dilute at a faster rate. The results obtained at 201 Varick Street, 290 Broadway, and at East Third Street and Avenue B (PS 64) suggest that such plumes could have maintained their integrity for several km. Further high temporal resolution analyses, including more detailed local-scale WTC plume plots, will be needed to better understand the specific lower Manhattan areas impacted by short-term WTC emission events and the implications of these sporadic events for potential human exposures and health impacts.

Evaluation of Human PM Exposures and Potential Health Impacts

Exposures to airborne PM are of concern for human health, because they can be associated with a wide variety of adverse human health effects. Health impacts may include pulmonary effects (such as lung inflammation and exacerbation of asthma) and cardiovascular effects (including exacerbation of preexisting chronic heart disease). As with exposures to other environmental contaminants, potential health impacts depend on PM concentrations and durations of exposure, as well as the size of the particles inhaled and many other factors (including the age and health status of exposed individuals). Depending on age and health status, some groups (such as infants and children, the elderly, or individuals with preexisting cardiovascular or respiratory diseases) may be considered sensitive or susceptible to exposures of PM.

Probably the most useful first step in evaluating the potential impacts of pollutants derived from events such as those of September 11, 2001 is the comparison of monitored concentrations against existing air quality standards and guidelines, e.g., the NAAQS and the associated AQI. Particulate matter is one of six common, widespread air pollutants for which the EPA has set

NAAQS (the others are ozone, carbon monoxide, nitrogen oxides, sulfur dioxide, and lead). These air quality standards, set under the Clean Air Act, are designed to protect public health. In 1987, the EPA set the PM-10 standard at 150 $\mu g/m^3$ as a 24-hour average and 50 $\mu g/m^3$ as an annual average (averaged over 3 years) to protect against health risks associated with inhalable particles that can deposit in lower (thoracic) portions of the human respiratory tract. These health-related PM-10 particles include both fine particles (< 2.5 μm aerodynamic diameter) and a subset of coarse particles (2.5–10 μm aerodynamic diameter). After reviewing the scientific bases for the PM NAAQS in 1996, the EPA concluded that fine and coarse components of PM-10 particles should be treated as separate classes of pollutants. Thus, to decrease health risks associated with fine particle exposures, the EPA moved in 1997 to set an annual PM-2.5 NAAQS at 15 $\mu g/m^3$, annual average (averaged over 3 years) to protect against both short- and long-term exposures and a supplemental 24-hour average PM-2.5 NAAQS at 65 $\mu g/m^3$ to protect against unusually high peak levels. The PM-10 NAAQS were retained to address risks related to coarse particles. As more scientific evidence becomes available, consideration may be given to setting PM standards for shorter averaging periods (i.e., < 24 h).

The EPA established an AQI LOC for daily PM-2.5 ambient concentrations at 40 $\mu g/m^3$ to provide real-time, day-to-day information to state and local health officials and the public. The AQI is meant to provide reference points for judging levels of potential health concern and to guide actions by citizens or government officials to protect the health of the public, including susceptible groups. Thus, in order to minimize risk of potential health effects among highly susceptible individuals (e.g., persons > 65 years old or individuals with preexisting chronic cardiovascular or respiratory disease), actions should be taken to reduce or avoid exposures of such persons to 24-hour PM-2.5 concentrations above 40 $\mu g/m^3$.

Because no direct measurements were obtained for airborne particle concentrations present in the dense dust and smoke cloud that enveloped lower Manhattan for up to about 4 hours after the collapse of the WTC buildings on September 11, estimates of likely exposures to airborne PM for individuals caught in the initial dust and smoke cloud can only be deduced from indirect evidence and are subject to great uncertainty. Nevertheless, several tentative conclusions appear to be warranted on the basis of available inputs thus far.

First, it is likely that many persons caught outdoors (and even some indoors) in the initial dust and smoke cloud were exposed for several hours to extremely high levels of airborne particles. This exposure probably included inhalation of PM concentrations in the milligrams per cubic meter range, well in excess of 1 to 2 mg/m^3 (1000 to 2000 $\mu g/m^3$), for *both* fine (PM < 2.5 μm diameter) and coarse (PM > 2.5 μm) inhalable particles, perhaps extending into the range of hundreds of mg/m^3 within close proximity to the collapse of the towers, as noted earlier. Such a conclusion is also supported by analyses of bulk dust samples conducted by Lioy et al. (*10*).

The coarse inhalable particles likely included substantial quantities of particles capable of reaching lower respiratory tract (thoracic) regions of the lung. Individuals who inhaled such high concentrations of WTC dust particles, even for a few hours, would logically be expected to be at potential risk for immediate acute respiratory and other symptoms and/or, possibly, more chronic health impacts (i.e., lung disease) that potentially could occur years later.

Persons exposed to the very high PM levels in the initial dust cloud and who continued to work within the perimeter of Ground Zero or returned to work there without wearing adequate protective respiratory gear might be at especially increased risk for potentially acute or chronic health effects, depending on the extent of any ensuing exposures to high PM levels on or immediately around the Ground Zero rubble pile. The latter could include additional exposures to coarse PM constituents (e.g., calcium or silicon) present in reentrained dust particles from the initial WTC building collapse and/or exposures to newly formed fine particle constituents (e.g., metals), as well as to organic constituents (e.g., PAHs) emitted from the WTC fires.

Evaluation of potentially acute and chronic health impacts associated with the above types of PM exposures should be further facilitated by disease registry efforts and retrospective epidemiologic analyses of physician/emergency department visits and hospital admission records being sponsored by the Centers for Disease Control and Prevention (CDC), the Agency for Toxic Substances and Disease Registry (ATSDR), the National Institute of Environmental Health Sciences (NIEHS), and other federal, state, and NYC agencies and that are now under way. Recent reports (*20, 21*) indicate that large percentages of firemen caught in the initial WTC dust cloud and others who worked at Ground Zero during the first 2 to 7 days post September 11 experienced respiratory (e.g., "WTC cough" or bronchial hyperactivity) or other symptoms that continued to persist for some individuals several months after cessation of exposures at WTC Ground Zero.

During the week following September 11, the plume from the initial high-intensity WTC fires appears to have been largely convected upwards and dispersed mainly to the south-southeast or south-southwest without much evident ground level contact, except perhaps for a few hours on the mornings of September 12 and 13, when it flowed to the east-southeast of the WTC. This resulted in briefly increased hourly PM-2.5 levels at sites in lower Manhattan (166 $\mu g/m^3$ at PS 64 on September 13) and in Queens (100 $\mu g/m^3$ at PS 199 on September 13). Probably few people were exposed around the PS 64 site, given the restrictions in effect on motor vehicular or pedestrian traffic below 14th Street until September 14, but some may have been briefly exposed in the vicinity of the PS 199 site and/or at locations between the two sites. Although it is doubtful that the brief, several-hour PM-2.5 excursion on the morning of September 13 resulted in harmful PM exposures, retrospective examination of physician and emergency department visits and/or hospital records in the affected areas may help to verify this.

After September 21, EPA/ORD monitoring indicated initially high levels of WTC-derived airborne particles (especially at certain Ground Zero perimeter sites) during late September and early October, but occurrences of PM excursions decreased over time through late October and into November. The rate of decrease in concentrations was not uniform throughout the monitoring period; rather, there were episodes of high PM levels spaced between periods of much lower concentrations. The frequency of the episodes was highest during the first month following the collapse of the WTC buildings and then declined afterwards. For example, notable PM emission episodes occurred during the first month of sampling, as shown in Plate 9. The PM-2.5 concentrations varied over a wide range (sometimes exceeding the relevant AQI 40 $\mu g/m^3$ action level) during late September and early October at Site A. The concentrations of a number of elements measured at this site also showed large day-to-day variability, as shown in Figures 3 to 7. On a number of days in late September and early October, concentrations of several elements were many times higher than the more typical urban levels recorded during December and January, after the WTC fires had largely or entirely burned out.

On the basis of overall air quality results summarized above, it appears that 24-hour PM-10 and PM-2.5 values throughout most all of the NYC metropolitan area generally remained at or returned rather quickly to historical background levels and WTC PM emissions posed no increased health risks beyond those due to usual PM levels for most areas of NYC. On the other hand, high PM-2.5 concentrations recorded on the perimeter of Ground Zero during late September and early October may imply increased health risks for the most highly exposed individuals (i.e., persons who spent extended periods of time within the WTC Ground Zero work zone without wearing protective respirators). Specifically, acute exposures to irritating materials present in either the PM-2.5 or the coarse particle components of PM-10, especially during any high hourly peak excursions, may have contributed to acute or continuing respiratory symptoms reported by some workers and/or residents in lower Manhattan areas in the immediate WTC vicinity. It is much less likely that any markedly increased health risks were posed by ambient air PM exposures elsewhere in the lower Manhattan neighborhoods surrounding the WTC, although more thorough analysis and modeling of potential PM exposures and correlation with health records is needed to evaluate this issue more fully.

It may be useful to place the above potential airborne PM exposures in perspective by comparing them to (a) exposures that occurred during some past notable PM air pollution episodes and (b) more recent historical data recorded for NYC areas. A number of past severe air pollution episodes involved extended periods of exposure of urban populations to high concentrations of airborne PM and associated air pollutants such as sulfur dioxide (*22–24*). In contrast to the relatively brief periods (< 8 hours) of inhalation exposures on September 11 to concentrations in the range of milligrams to hundreds of mg/m^3 of WTC-derived airborne coarse and fine PM, a number of past air pollution episodes in U.S. cities (e.g., Donora, PA in 1948; NYC in 1953 and 1962/63) and abroad (e.g., Neuse Valley, Belgium, 1930; London, UK in 1952, 1957, 1963) involved exposures to very high levels of PM that lasted for at least several days. Probably the most famous such episode occurred in London in December 1952, when millions of Londoners were exposed to daily PM levels (measured as British Smoke), which included high percentages of fine particles of 1000 to 4000 μg/m^3 (sometimes reaching hourly peaks of 6000 μg/m^3 or more) on 3 to 5 consecutive days in the presence of 1000 to 4000 μg/m^3 of sulfur dioxide.

In NYC, PM-2.5 values recorded on some days at the Ground Zero perimeter and occasionally elsewhere in lower Manhattan during late September and early October clearly exceeded the more usual background levels of fine PM seen in NYC since implementation of the PM NAAQS in the 1970s began to substantially reduce ambient PM concentrations in U.S. urban areas. For example, some perimeter site 24-hour PM-2.5 measurements of more than 100 μg/m^3 likely exceeded most — but not necessarily all — values recorded at the New York University Medical Center (NYUMC) in an aerosol sampling study conducted in August 1976 (*25*) as per the mean value shown for PM-2.5 in Table II. However, the 24-hour PM-2.5 concentrations (predominantly below 30 to 40 μg/m^3) usually seen during most of the rest of October and into November at the perimeter sites were notably lower than the 1976 values; and the PM-2.5 levels reached by December and January generally compare favorably with PM-2.5 values obtained at a monitoring site at the Bronx Botanical Gardens (BBG) during February–June 2000 (see Table II). This BBG site can be considered to be a relatively "clean" urban background site largely free of the effects of strong local sources. Maximum PM-2.5 values ranged from 35.4 to 43.3 μg/m^3 and average 24-hour PM-2.5 values ranged from 12.5 to 15.6 μg/m^3. Also, PM-10 values for lower Manhattan sites, which were mainly in the range of 50 to 90 μg/m^3 during October, and mostly below 50 μg/m^3 in November, were not markedly different from historical values observed in NYC. For example, the fourth highest and the maximum 24-hour PM-10 values reported in the EPA AIRS database for the five NYC boroughs during 1996–2001 ranged from 40 to 89 μg/m^3 and from 51 to 121 μg/m^3, respectively.

During September–October 2001, concentrations of many elements, including heavy metals, at the Ground Zero perimeter sites were at times much greater than those observed at several sites in the northeast in February-June

2000. As part of a pilot study for EPA's PM-2.5 speciation network, concentrations of PM-2.5 and a number of key elements were measured from February–June, 2000, in NYC (at BBG), Boston, and Philadelphia, at sites that were likely characteristic of urban backgrounds. As shown in Table II, the average concentrations of PM-2.5 and the individual elements measured at these three sites varied relatively little in contrast to those measured near the WTC site. The highest measurements of PM-2.5 and heavy metals at the WTC

Table II. Maximum and Average PM-2.5 Concentrations and Elemental Composition Measured in September and October, 2001 at the WTC Perimeter Sites (A, C, K) Compared to other Urban Sites.

	Max	*Site*	*Date*	*A*	*C*	*K*	*NYU*	*BRX*	*BST*	*PHL*
PM	400	A	10/4	85	34	50	81.7	12.5	10.7	14.7
Na	870	A	10/4	273	157	169	570	72	178	63
Mg	490	K	11/13	101	67	79	103	4.8	16	7.7
Al	670	A	10/20	198	74	113	187	9.2	25	18
Si	20000	A	10/4	943	224	333		75	92	118
S	23000	A	10/3	4796	1808	2524	6820	1200	933	1500
Cl	45000	A	10/4	7247	540	845	119	9.8	68	7.7
K	5600	A	9/22	988	147	260	194	38	38	60
Ca	4900	A	10/11	1304	345	749		38	50	57
Cr	34	C	11/19	5	4	3	28	0.3	0.5	1.1
Fe	9400	K	12/12	1745	904	975	400	91	76	103
Co	24	A	10/3	4	Bdl	Bdl	3.2	0.4	0.4	
Ni	50	K	12/11	11	9	11	18	12	2.8	4.4
Cu	2800	A	9/22	435	59	92		2.8	2.2	4.5
Zn	10000	A	10/4	1526	164	307	224	21	9.7	16
As	1100	K	11/14	9	Bdl	Bdl	3	1.1	0.9	1
Br	5700	A	9/20	800	82	124	133	2.5	2.5	3.4
Pd	900	A	10/4	132	2	7				
Cd	150	A	9/22	33	9	9	7.7	1.7	1.7	1.8
Sb	350	A	9/22	50	2	1	11	3.6	2.4	3.5
Pb	5500	A	9/22	791	83	167	1170	4.2	3.4	5.6

NOTE: PM is PM-2.5. PM units are $\mu g/m^3$, all other units are ng/m^3. Bdl is below detection limits.

NOTE: NYU is New York Medical Center Summer Aerosol Study, 1976. BRX is Bronx, NY February–June, 2000. BST is Boston, MA February–June, 2000. PHL is Philadelphia, PA, February–June, 2000.

SOURCE: NYU is from Reference 25. BRX, BST, and PHL are from Reference 26.

parimeter sites are also shown in Table II for comparison. During the first weeks after September 11, PM-2.5 levels and concentrations of several elements often were many times higher than those obtained at the other northeastern sites. As also seen in Table II, concentrations recorded at the NYUMC in 1976 (*25, 27*) were distinctly higher than those obtained at the other northeastern sites approximately 25 years later. These data were collected before pollution control measures, such as the removal of lead from gasoline, were implemented. These control measures were introduced because of concern about health effects associated with exposures to high concentrations of pollutants such as lead.

Measurements of the composition of coarse particles (10–2.5 μm) are available only from the site at 290 Broadway. The composition of PM-2.5 and PM-(10–2.5) particles at this site is shown in Table III. As expected, the

Table III. Average Mass (PM) and Elemental Composition of < 2.5 μm (F) and 10-2.5 μm (C) Particulate Matter Measured from September – October and January, 2001 at 290 Broadway,

Species	*Sept-Oct(F)*	*Jan (F)*	*Sept-Oct (C)*	*Jan (C)*
PM	23	16.5	16	7
Na	111	364	73	265
Mg	30	243	28	111
Al	72	589	27	281
Si	369	1438	69	711
S	1806	987	1106	142
Cl	983	670	74	353
K	193	142	49	57
Ca	221	2154	58	472
Cr	1	2.6	1	1
Fe	429	643	221	339
Co	Bdl	Bdl	Bdl	Bdl
Ni	9	5	29	10
Cu	65	26	7	5
Zn	258	81	55	17
As	Bdl	Bdl	Bdl	Bdl
Br	100	1	5	1
Pd	Bdl	Bdl	Bdl	Bdl
Cd	Bdl	Bdl	Bdl	Bdl
Sb	Bdl	Bdl	Bdl	Bdl
Pb	118	23	9	5

NOTE: PM units are $\mu g/m^3$, all other units are ng/m^3. Bdl is below detection limits.

composition of particles in the size range from 2.5 to 10 μm is dominated by common crustal elements such as aluminum, silicon, iron, and calcium. Metals are found mainly in the size range $\leq$ 2.5 μm. The concentrations of all components are markedly lower than those found at the surface monitoring sites shown in Table II. Because the samples at 290 Broadway were collected on the 16th floor, they were likely to contain a higher fraction of regional background particles than those collected at the surface, and so their composition may differ a bit from those shown in Table II. The concentrations in both size fractions shown in Table III decrease from the September–October period to January. However, there are shifts in the ratios of elements. The ratio of calcium to silicon in the coarse mode particles is about 2:1 during October and decreases to about 0.5:1 during January. Usually the ratio of calcium to silicon is about one-half that in airborne samples, reinforcing the finding that the coarse fraction contained a higher fraction of alkaline material than normal during the September–October 2001 time frame.

During December and January, the concentrations of most of the elements measured by ORD decreased to levels similar to those measured at the three northeastern sites, suggesting that the WTC was no longer a significant source of these elements. However, it should be noted that even the markedly elevated element concentrations over typical background values noted mainly in September and October did not exceed applicable Permissible Exposure Limits (PEL) set by the U.S. Department of Labor, Occupational Safety and Health Administration (OSHA) for an 8-hour time-weighted average or other more broadly applicable health benchmark values, suggesting a generally low health risk for those in lower Manhattan around the WTC. Workers in the piles clearly were exposed to much higher concentrations because of their close proximity to the source.

Alkalinity of the dust from the WTC disaster may have been a possible health concern for exposed individuals. Some reported symptoms (eye, nose, and throat irritation; nose bleeds; cough) may have been due to exposure to unusually elevated quantities of certain crustal materials derived from pulverized concrete, wallboard, and other WTC structural components present in airborne particles and settled dust in neighborhoods near the WTC. Aqueous solutions of WTC settled dust showed an initially high alkalinity (pH 10.0), which decreased for outdoor samples taken after rainfall (*10, 17*), as reported by the U.S. Geological Survey (USGS) (see Chapter 12). Because much of the outdoor settled dust was removed by hazardous material cleanup procedures or was washed away by rainfall during late September, outdoor exposures to highly basic PM components would seem to be of much less potential health concern beyond late September when restricted zone shrinkages allowed more people and traffic in neighborhoods immediately surrounding the WTC work zone. However, the USGS noted higher alkalinity for dusts sampled indoors, raising the possibility of greater risk of acute irritation symptoms being associated with indoor exposures to WTC dusts than with outdoor dusts leached by rainfall.

Lioy et al. (*10*) more broadly highlighted indoor exposures to WTC-derived dust PM as posing potential increased health risks. Individuals visiting, residing, or working in buildings not adequately cleaned before reoccupation could have been subjected to repeated, long-duration exposure to many of the components from the original WTC collapse found in settled dust to the east of the WTC. It has been noted that long, narrow glass fibers in the WTC-derived dust had various potentially toxic materials attached to them and could contribute to acute short-term irritative effects and possibly to more chronic health risks (*10*). On the other hand, asbestos concentrations were generally low in comparison to occupational standards (see Chapter 3).

Also of potential concern would be any extended indoor air exposures to finely pulverized building materials (e.g., calcium, silicon, iron, and sulfate), to PM of either fine or coarse size containing marked elevations of certain metals, or to fine PM containing usual combinations of silicon coagulated with metals or other toxic materials. Indoor dust loadings of lead have been identified as posing potential chronic health risks (*10*) and the possible contributions of certain other metals (e.g., nickel, chromium) found in settled dusts or airborne PM to irritative symptoms also need further evaluation. The issue of potentially greater toxicity being associated with unusually increased quantities of very fine or ultrafine particles present in airborne PM also needs to be evaluated further.

Some newly available findings from laboratory toxicity studies of WTC-derived dusts may offer insights into potential health responses associated with exposures on September 11 to the initial WTC building collapse dust cloud and later exposures to WTC-particles deposited indoors. The ORD National Health and Enviromental Effects Research Laboratory (NHEERL) scientists have analyzed chemical and toxicological properties of PM-2.5 derived from the collapsed WTC buildings (*1, 17, 28*). Deposited dust samples were collected from sites within a half-mile of Ground Zero on September 11 and 12 and size-separated to collect the PM-2.5 fraction. Gavett et al. (*28*) evaluated responses of young adult female mice to bolus doses of WTC-derived PM-2.5 dusts administered by intratracheal instillation directly into the lungs and to 5-hour inhalation exposures to WTC PM-2.5. Only PM-2.5 was assessed because of the limited availability of samples and because the PM-2.5 fraction was most relevant for health effects studies. However, it should be noted that the composition of the resuspended PM-2.5 may not be entirely representative of the PM-2.5 that was originally airborne. On the basis of the overall pattern of results obtained, Gavett et al. observed both a small degree of pulmonary inflammation in response to WTC dust, which was distinctly less compared with exposure to residual oil fly ash (ROFA), and some notable increases in airway hyperresponsiveness to methacholine, a nonspecific bronchoconstricting agent, greater than that observed in mice exposed to ROFA or other PM samples. These effects may be interpreted as being consistent with reports of airway hyperresponsiveness and irritant responses in people exposed to high concentrations of WTC-derived dusts. Whereas the mild pulmonary inflammation in mice diminished from 1 to 3 days after exposure, the airway hyperresponsiveness appeared to persist longer.

These results suggest possible limited, short-term lung inflammation effects from exposures to high concentrations of WTC dust (as may have occurred mainly on September 11) or possible long-term airway hyperresponsiveness that might portend more prolonged sensitivities and irritative symptoms for persons experiencing extended high-level exposures to WTC-derived dusts indoors. The human exposure equivalent of the doses that led to these responses in mice was estimated (*28*), and an 8-hour exposure of a concentration of 425 $\mu g/m^3$ at moderate activity level was calculated using a multiple-path particle deposition model (*29*). The dose of WTC PM-2.5 per tracheobronchial (TB) surface area in mice was assumed to cause the same effects in humans. Because inflammation was observed mainly in the airways of mice, and airway hyperresponsiveness is mainly due to dysfunction of airway smooth muscle, this dose metric is probably most relevant for assessing health risk in humans. The dose per TB surface area of mice was known because precise doses were administered directly to the airways, and the equivalent dose per TB surface area in humans was used to back-calculate the air concentration of WTC PM-2.5 using the assumptions and procedures outlined previously (*28*). In brief, the mouse/human dose per TB surface area (mg/m^2) was multiplied by the human TB surface area (m^2) to obtain the total human TB dose (mg) equivalent to the mouse TB dose. This dose was divided by the deposition fraction in the human TB region to obtain the total inhaled dose (*29*). This total inhaled dose was divided by the amount of air breathed over an 8-hour work shift at a moderate degree of ventilation to obtain the concentration of WTC PM-2.5 necessary to produce a human TB dose equivalent to the mouse TB dose.

Individuals who are especially sensitive to inhalation of dusts, such as asthmatics, might experience these effects at lower concentrations over more prolonged periods of exposure. As mentioned previously, it is likely that persons caught in the initial dust and smoke cloud could have been exposed to PM-2.5 concentrations in excess of 1000 $\mu g/m^3$. However, a dose equivalent to a human exposure of 130 $\mu g/m^3$ over 8 hours caused no significant effects in the mice, suggesting that most healthy people would not be expected to respond to this moderately high exposure level with any adverse respiratory responses (*1, 28*). No interspecies adjustment was carried out in these calculations, because the degree of responsiveness to PM in mouse versus human airways has not been established. These studies did not address the effects of the coarse fraction of

WTC PM on responses in mice, which may be very important, considering the upper airways responses and ocular irritant effects that have been reported.

Conclusions

People in the immediate vicinity of Ground Zero were exposed to extremely high PM concentrations in the dust and smoke cloud produced by the collapse of the WTC towers on September 11, 2001. Notable PM-2.5 elevations continued to occur in the immediate vicinity of Ground Zero during late September to early October, with concentrations at WTC perimeter sites on some days exceeding the 24-hour PM-2.5 NAAQS.

Such high PM-2.5 concentrations were not observed at other lower Manhattan monitoring sites within 3–10 blocks (approximately 0.3–1.0 km) of Ground Zero after additional monitoring was initiated on October 2. Daily PM-2.5 concentrations approached or exceeded the AQI LOC of 40 $\mu g/m^3$ at one or two sites (e.g., 290 Broadway or PS 64) lying along the path of the plume emitted by the WTC at Ground Zero in addition to elevations seen at WTC perimeter sites on only a few occasions. The frequency of such excursions was generally in line with the historical frequency of PM-2.5 values approaching or exceeding 40 $\mu g/m^3$ either before the September 11 WTC attack or since the WTC fires ended. These excursions most likely occurred when there was a well-defined plume during periods of light winds and minimal vertical mixing in the atmosphere near the surface (i.e., within the planetary boundary layer).

No notable elevations in PM-2.5 concentrations were seen at NYSDEC lower Manhattan sites located 3 to 10 blocks to the north, east, and south from the WTC, with no PM-2.5 values exceeding either the PM-2.5 daily NAAQS or the AQI LOC from the start-up of PM monitoring on October 1 onward.

Average PM-10 concentrations at sites near Ground Zero were higher during October and most likely during September than at later times. Concentrations never exceeded the PM-10 NAAQS at the PM-10 monitoring sites in lower Manhattan. However, this does not preclude the possibility that the PM-10 NAAQS was exceeded either earlier at these sites, before monitoring was initiated or at other locations in lower Manhattan.

The ratio of PM-2.5 to PM-10 concentrations was lower than typically found at other sites in the northeastern United States. As the larger particles were more likely to be more alkaline than the smaller particles, this finding supports the supposition that PM-10 particles were more alkaline than usual.

Acknowledgments

We wish to acknowledge the support and data provided by Mr. Kenneth Fradkin and other staff of U.S. EPA's Region II Office. We also wish to acknowledge material provided by Dr. Steven Gavett of the U.S. EPA National Health and Ecosystem Exposure Research Laboratory based on his work on the toxic properties of WTC-derived material. The research contributed by Alan Huber and presented here was performed under the Memorandum of Understanding between the U.S. Environmental Protection Agency (EPA) and the U.S. Department of Commerce's National Oceanic and Atmospheric Administration (NOAA) under agreement number DW13921548.

Disclaimer

The views expressed in this chapter are those of the authors and do not necessarily reflect the views or policies of the U.S. Environmental Protection Agency or the National Oceanic and Atmospheric Administration.

References

1. *Exposure and Human Health Evaluation from the World Trade Center Disaster* (External Review Draft); EPA 600/P-2/002A; U.S. Environmental Protection Agency, Office of Research and Development, U.S. Government Printing Office: Washington, DC, 2002. [http:cfpub2.epa.gov/ncea/] accessed December 2004.
2. Vette, A.; Seila, R.; Swartz, E.; Pleil, J.; Webb, L.; Landis, M.; Huber, A.; Vallero, D. *Environ. Manager* **2004,** *February*, 23–26.
3. Vette, A.; Gavett, S.; Perry, S.; Heist, D.; Huber, A.; Lorber, M.; Lioy, P.; Georgopoulos, P.; Rao, S. T.; Petersen, W.; Hicks, B.; Irwin, J.; Foley, G. *Environ. Manager* **2004,** *February*, 14–22.
4. Hill, J. S.; Patel, P. D.; Turner, J. R. In *Proceedings of the 92nd Annual Meeting of the Air & Waste Management Association,* Washington University School of Engineering & Applied Science Air Quality Laboratory [WUAQL report no. 0399-01, 1–17]: St. Louis, MO, 1999. [http://www.airmetrics.com/products/studies/1.html] 28 September, 2004.
5. Stevens, R.; Pinto, J.; Mamane, Y.; Ondov, J.; Abdulraheem, M.; Al-Majed, N.; Sadek, M.; Cofer, W.; Ellenson, W.; Kellogg, R. *Water Sci. Technol.* **1993,** *27,* 223–233.

6. Pinto, J. P.; Stevens, R. K.; Willis, R. D.; Kellogg, R.; Mamane, Y.; Novak, J.; Santroch, J.; Benes, I.; Lenicek, J.; Bures, V. *Environ. Sci. Technol.* **1998,** *32,* 843–854.
7. Huber, A.; Georgopoulos, P.; Gilliam, R.; Stenchikov, G.; Wang, S.; Kelly, B.; Feingersh, H. *Environ. Manager* **2004,** *February,* 35–40.
8. Gilliam, R. C.; Childs, P. P.; Huber, A. H.; Raman, S. *Pure Appl. Geophys.* **2004,** in press.
9. Gilliam, R. C.; Huber, A. H.; Raman, S.; Neogi, D. *Pure Appl. Geophys.* **2004,** in press.
10. Lioy, P. J.; Weisel, C. P.; Millette, J. R.; Eisenreich, S.; Vallero, D.; Offenberg, J.; Buckley, B.; Turpin, B.; Zhong, M.; Cohen, M. D.; Prophete, C.; Yang, I.; Stiles, R.; Chee, G.; Johnson, W.; Porcja, R.; Alimokhtari, S.; Hale, R. C.; Weschler, C.; Chen, L. C. *Environ. Health Perspect.* **2002,** *110,* 703–714.
11. "Space Imaging: One Year Viewed from Space" [http://www.spaceimaging.com/gallery/9-11] Space Imaging.
12. *World Health Organization. Health Guidelines for Vegetation Fire Events: Guideline Document;* Schwela, D.; Goldammer, J. G.; Morawska, L.; Simpson, O., Eds.; Institute of Environmental Epidemiology, Ministry of the Environment: Singapore, 1999.
13. Stevens, R. K.; Dzubay, T. G.; Lewis, C. W.; Shaw, R. W., Jr. *Atmos. Environ.* **1984,** *18,* 261–272.
14. Stull, R. *Meteorology for Scientists and Engineers: a Technical Companion Book with Ahrens' Meteorology Today,* 2nd ed.; Brooks/Cole: Pacific Grove, CA, 2000.
15. Pleil, J. D.; Vette, A. F.; Johnson, B. A.; Rappaport, S. M. *Proc. Natl. Acad. Sci., U. S. A.* **2004,** *101,* 11685–11688.
16. Stevens, R. K.; Pinto, J. P.; Shoaf, C. R.; Hartlage, T. A.; Metcalfe, J.; Preuss, P.; Willis, R. D.; Mamane, Y.; Novak, J.; Santroch, J.; Benes, I.; Lenicek, J.; Subert, P.; Bures, V. *Czech Air Quality Monitoring and Receptor Modeling Study;* EPA/600/R-97/084; U.S. Environmental Protection Agency: Research Triangle Park, NC, 1997.
17. McGee, J. K.; Chen, L. C.; Cohen, M. D.; Chee, G. R.; Prophete, C. M.; Haykal-Coates, N.; Wasson, S. J.; Conner, T. L.; Costa, D. L.; Gavett, S. H. *Environ. Health Perspect.* **2003,** *111,* 972–980.
18. Olson, D. A.; Norris, G. A.; Landis, M. S.; Vette, A. F. *Environ. Sci. Technol.* **2004,** *38,* 4465–4473.
19. Cahill, T. A.; Cliff, S. S.; Perry, K. D.; Jimenez-Cruz, M.; Bench, G.; Grant, P.; Ueda, D.; Shackelford, J. F.; Dunlap, M.; Kelly, P. B.; Riddle, S.; Selco, J.; Leifer, R. Z. *Aerosol Sci. Technol.* **2004,** *38,* 165–183.
20. Prezant, D. J.; Weiden, M.; Banauch, G. I.; McGuinness, G.; Rom, W. N.; Aldrich, T. K.; Kelly, K. J. *New Engl. J. Med.* **2002,** *347,* 806–815.

21. Landrigan, P. J.; Lioy, P. J.; Thurston, G.; Berkowitz, G.; Chen, L. C.; Chillrud, S. N.; Gavett, S. H.; Georgopoulos, P. G.; Geyh, A. S.; Levin, S.; Perera, F.; Rappaport, S. M.; Small, C.; *Environ. Health Perspect.* **2004,** *112*, 731–739.
22. *Air Quality Criteria for Particulate Matter and Sulfur Oxides*; EPA-600/8-82-029aF-cF. 3v.; U.S. Environmental Protection Agency, Office of Health and Environmental Assessment, Environmental Criteria and Assessment Office; Research Triangle Park, NC, 1982. Available from: NTIS, Springfield, VA; PB84-156777.
23. *Second Addendum to Air Quality Criteria for Particulate Matter and Sulfur Oxides (1982): Assessment of Newly Available Health Effects Information;* EPA-600/8-86-020F; U.S. Environmental Protection Agency, Office of Health and Environmental Assessment, Environmental Criteria and Assessment Office: Research Triangle Park, NC, 1986. Available from: NTIS, Springfield, VA; PB87-176574.
24. Elsom, D. M. *Atmospheric Pollution: a Global Problem,* 2nd ed.; Blackwell: Oxford, U.K., 1992.
25. Lippmann, M., Kleinman, M. T. Bernstein, D. M., Wolff; G. T., and Leaderer, B. P., In *The New York Summer Aerosol Study,* Kneip, T. J. and Lippmann, M. Eds. 1976, New York Academy of Sciences, N.Y., 1979; pp. 29-43.
26. Coutant, B.; Zhang, X.; Pivetz, T. *Summary Statistics and Data Displays for the Speciation Minitrends Study: Final Report;* contract no. 68-D-98-030. U.S. Environmental Protection Agency, Office of Air Quality Planning and Standards: Research Triangle Park, NC, 2001.
27. Bernstein, D. M.; Rahn, K. A. In *New York Summer Aerosol Study, 1976;* Kneip, T. J.; Lippmann, M., Eds.; New York Academy of Sciences: N.Y., 1979; pp. 87–97.
28. Gavett, S. H.; Haykal-Coates, N.; Highfill, J. W.; Ledbetter, A. D.; Chen, L. C.; Cohen, M. D.; Harkema, J. R.; Wagner, J. G.; Costa, D. L. *Environ. Health Perspect.* **2003,** *111*, 981–991.
29. Freijer, J. I.; Cassee, F. R.; Subramaniam, R.; Asgharian, B.; Miller, F. J.; Van Bree, L.; Rombout, P. J. *A. Multiple Path Particle Deposition Model (MPPDep version 1.11): a Model for Human and Rat Airway Particle Deposition;* RIVM publication 650010019; RIVM: Bilthoven, The Netherlands, 1999.

Chapter 12

Inorganic Chemical Composition and Chemical Reactivity of Settled Dust Generated by the World Trade Center Building Collapse

Geoffrey S. Plumlee, Philip L. Hageman, Paul J. Lamothe, Thomas L. Ziegler, Gregory P. Meeker, Peter M. Theodorakos, Isabelle K. Brownfield, Monique G. Adams, Gregg A. Swayze, Todd M. Hoefen, Joseph E. Taggart, Roger N. Clark, Stephen E. Wilson, and Stephen J. Sutley

Denver Microbeam Laboratory, U.S. Geological Survey, MS 964, Box 25046 Denver Federal Center, Denver, CO 80225

Samples of dust deposited around lower Manhattan by the September 11, 2001, World Trade Center (WTC) collapse have inorganic chemical compositions that result in part from the variable chemical contributions of concrete, gypsum wallboard, glass fibers, window glass, and other materials contained in the buildings. The dust deposits were also modified chemically by variable interactions with rain water or water used in street washing and fire fighting. Chemical leach tests using deionized water as the extraction fluid show the dust samples can be quite alkaline, due primarily to reactions with calcium hydroxide in concrete particles. Calcium and sulfate are the most soluble components in the dust, but many other elements are also readily leached, including metals such as Al, Sb, Mo Cr, Cu, and Zn. Indoor dust samples produce leachates with higher pH, alkalinity, and dissolved solids than outdoor dust samples, suggesting most outdoor dust had reacted with water and atmospheric carbon dioxide prior to sample collection. Leach tests using simulated lung fluids as the extracting fluid suggest that the dust might also be quite reactive in fluids lining the respiratory tract, resulting in dissolution of some particles and possible precipitation of new phases such as phosphates, carbonates, and silicates. Results of these chemical characterization studies can be used by health scientists as they continue to track and interpret health effects resulting from the short-term exposure to the initial dust cloud and the longer-term exposure to dusts resuspended during cleanup.

Introduction

The September 11, 2001, attack on and collapse of the World Trade Center (WTC) towers produced a massive dust cloud that engulfed much of lower Manhattan, leaving behind extensive deposits of settled dust and debris. This chapter summarizes inorganic chemical characterization studies of a subset of settled dust samples from 33 localities (31 outdoor, 2 indoor) around lower Manhattan, two samples of material coating steel girders at the WTC (thought to be primarily composed of spray-on girder insulation), and two samples of WTC concrete. All but one of the samples studied were collected by a U.S. Geological Survey (USGS) field crew on September 17 and 18, 2001 (*1, 2*). The sample collection procedures are detailed further in Chapter 3. One sample studied was collected the following week from a 30th floor apartment southwest of the WTC by the apartment's owner.

This work was carried out at the request of the U.S. Environmental Protection Agency (EPA) and U.S. Public Health Service, as part of an interdisciplinary environmental characterization of the WTC area after the attacks. Preliminary results and interpretations were released via the Internet to emergency response teams on September 27, 2001. After additional work to fill key data gaps, and further detailed peer review, the study results were released publically on November 27, 2001 (*2*). Since 2001, limited further studies have been conducted on the samples to refine earlier interpretations. Chapters 3-5 summarize the sample locations, sample collection methods, laboratory spectral reflectance characterization studies, and materials characterization studies of the deposited dust samples (*1*), as well as remote sensing results (*3*).

Experimental Methods

Inorganic Chemical Analyses

Twenty-two of the 37 samples collected were analyzed for bulk chemical composition. Representative subsamples of each were obtained using a standard cone and quartering approach and then were ground and homogenized prior to chemical analysis. Subsample splits were analyzed for their total major elemental composition by wavelength-dispersive X-ray fluorescence (WD-XRF), total major and trace element composition by inductively coupled-plasma mass spectrometry (ICP-MS) after a 4-acid (including HF) dissolution procedure, total carbon and sulfur by combustion, and carbonate carbon by coulometric titration. Due to limitations in sample volume and other factors, not all samples were analyzed by all methods. Duplicate splits of 6 samples were analyzed by ICP-MS to assess compositional variability within individual samples. Standard reference soil materials were also analyzed with each batch of samples, and duplicates were run to assess the analytical variability. Further details of the analytical methodologies used for the bulk chemical analyses are presented elsewhere (*2,4*).

Chemical Leach Tests

Most samples were subjected to leach tests using deionized water (DIW) as the extracting fluid. The leach procedure was developed as a screening method to quickly assess potential metal release from mine wastes during runoff from rain storms or snowmelt (*5*), and is a modification of the US EPA Method 1312 (*6*). As applied to the WTC materials, this test can be used to help understand potential chemical reactions that occurred when the dusts came into contact with water (such as rainfall or the water used in fire fighting or street washing). It may also provide an indication of metals that are potentially bioaccessible should the dusts be inhaled, ingested, or discharged into ecosystems.

Dust samples were leached at a 1:20 ratio (2.5 g dust / 50 mL) with DIW (pH ~5.5). Following shaking for 5 minutes, the leachate solution was allowed to settle for 5 minutes. The leachate solution was then filtered using a plastic syringe and a 0.45 µm pore size nitrocellulose membrane filter. Sample pH and specific conductance were measured immediately after filtration. Sub-samples of the filtered leachate were collected and preserved by acidifcation with nitric acid for further analysis. All leachate samples were analyzed for major cations and trace metals by ICP-MS, and a subset for anions by ion chromatography (IC), following the protocols outlined elsewhere (*4*). Leachate solutions for a subset of the dust samples were also analyzed for dissolved mercury using mercury-specific sample filtration and preservation techniques (*7*), followed by analysis using a Lachat® QuikChem Mercury Analyzer with a fluorescence detector (*4*).

Since the release of our preliminary findings in November 2001, we also analyzed a subset of leachates from both indoor and outdoor dust samples for total alkalinity with an automatic titrator using fixed titration increments to an endpoint pH of 4.5 (*4*). Total hydroxide, carbonate, and bicarbonate alkalinity were calculated by both the inflection point and fixed endpoint methods (*8*) with essentially equivalent results. For comparison, we subjected a sample of commercial slag wool (similar to that found in the WTC dust) and a variety of common materials (e.g. cement powder, milk of magnesia, liquid drain cleaner, household bleach, solid lye drain cleaner) to the same leach tests using the same weight ratios as we used for the WTC dust samples, and measured the pH and alkalinity of these leachates.

Chemical speciation calculations run on the program Geochemist's Workbench were used to aid in the interpretation of the leach test results (*9*). For example, speciation calculations were used to assess charge balance of the analyses and to estimate the saturation states of the fluids with respect to the minerals found in the dusts and other minerals that might precipitate through interactions of the dusts with the leachate fluids.

Size-Fractionated Samples

Two outdoor dust samples were size-separated using steel sieves into various size fractions, including >800 µm, 425-800 µm, 150-425 µm, 75-150 µm, 38-75 µm, 20-38 µm, and <20 µm, as described in Chapter 5. The dust samples were not separated into fractions finer than 20 µm due to limitations in the amount of the bulk samples available for analysis. These size fractions were subjected to total inorganic chemical analysis and a DIW leachate test using the same methods as described above for the bulk samples. These tests provide information on the variations in chemical composition and chemical reactivities of the dust samples as a function of particle size.

In Vitro Bioaccessibility Leach Tests

Bioaccessibility leach tests performed on a subset of the WTC settled dust samples provided insights into how the dusts might react chemically with electrolytes and some of the many organic species that are present in the fluids lining the respiratory tract. To date, tests have been conducted on one indoor and one outdoor sample comparing DIW and simulated lung fluids (SLF) as the extraction fluids (*10*). One part dust sample was added to 20 parts deionized water or SLF by weight, and mixed at 37°C for 24 hours, with the leachate filtered to 0.45 µm. The pH of the leachate was measured and the leachate analyzed for cations and trace metals by ICP-MS. The composition of the SLF used in the extraction was: pH 7.4; Na—150.7 mM; Ca—0.197 mM; NH_4—10 mM; Cl—126.4 mM; SO_4—0.5 mM; HCO_3—27 mM; HPO_4—1.2 mM; Glycine—5.99 mM; Citrate—0.2 mM *(11)*. A variety of trace metals are present in the SLF blank and were contributed as trace constituents of the various chemical reagents used to make up the fluids.

Results

Chemical Compositions of Bulk Settled Dust and Girder Coating Samples

Analytical results for the 22 bulk outdoor dust samples, 2 indoor bulk dust samples, and 2 samples from the WTC steel girder coating are summarized in Table I and Figure 1. These results were compared to those from two other WTC studies (*12, 13*) and for an eastern U.S. soil (*14*) and two standard reference materials developed by the National Institute of Standards and Technology (NIST) (SRM 1648, an urban dust from St. Louis, and SRM 1649, urban atmospheric particulates from Washington DC, both collected in the mid 1970's) (*15, 16*). The NIST samples were sieved to <125 µm before analysis. Gaps in the element profile in Figure 1 for a given sample type indicate that the element was below detection limits or was not measured.

The results for the outdoor dust samples, given in full in Appendix I, indicate that the settled dust deposits can be quite variable in composition from locality to locality, with the concentrations of some elements varying as much as an order of magnitude between samples. Silicon and Ca are the dominant elemental components of the settled dusts, with lesser S, Mg, Al, Fe, organic C, carbonate C, Na, and K. Volatile compounds (water bonded in minerals, water adsorbed onto materials, and organic materials such as papers, plastics, etc.) are measured as loss on ignition (LoI) during XRF analysis, and can approach nearly 20% by weight. However, the exact types and amounts of these volatile components in the samples were not determined. A wide variety of other elements are present at lesser concentrations in the dusts. Of these, Ti, Sr, and Zn are typically the most abundant (several hundred to several thousand ppm), with Mn, Ba, Pb, Cu, and Cr occurring in somewhat lower concentrations (hundreds of ppm). Nickel, Sb, V, Mo, Li, and Rb concentrations are typically several tens of ppm.

The two indoor dust samples have element concentration profiles that generally mimic those of the outdoor dust samples (see Figure 1). However, concentrations of many elements in the indoor dust samples, especially Si, Fe, K, P, Ba, Pb, Cr, Ni, V, Co, and As, generally fall at the lower end of the outdoor dust concentration ranges. In contrast, Ca and S concentrations tend to fall at the upper end of the outdoor dust sample ranges.

Analysis of duplicate splits of several outdoor dust samples (see Appendix I) suggest that individual samples can be quite variable in composition, particularly with respect to trace metals. Because the analytical results for the standards analyzed with the WTC samples routinely showed good agreement with accepted values, it can be concluded that these differences reflect heterogeneities in the original sample. This is consistent with observations made in microscopic studies of the dust samples described in Chapter 5 in which many of the trace metals were seen to occur in discrete particulates that likely are not evenly distributed throughout a given sample.

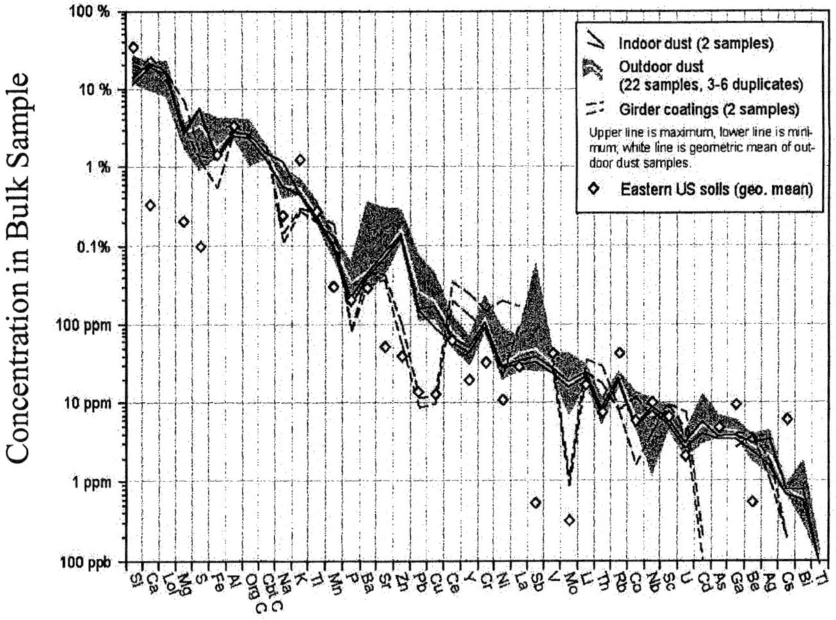

Figure 1. Concentrations of elements in the bulk WTC dust samples and girder coatings compared to the mean bulk concentrations in an eastern U.S. soil (14). LoI is loss on ignition; Org C is organic carbon; Cbt C is carbonate carbon.

Table I. Elemental Composition for Outdoor and Indoor Dust Samples and WTC Steel Girder Coatings Compared to an Eastern U.S. Soil and NIST Urban Dust and Particulate Standard Reference Materials.

	Outdoor				*Indoor*		*Girder*		*Soil*	*NIST-SRM*	
Element	*mean*	*sd*	*min*	*max*	*min*	*max*	*min*	*max*		*1648*	*1649*
Si	15.0	3.9	11.4	26.3	11.7	14.2	15.0	15.5	34.0	12.5	—
Ca	18.8	2.9	9.6	22.0	19.4	21.3	20.7	26.0	0.3	—	—
LoI	16.6	3.5	8.0	22.8	15.7	16.9	13.0	15.8	—	—	—
Mg	2.6	0.4	1.7	3.5	2.6	2.9	3.2	6.9	0.2	0.8	0.9
S	3.3	1.5	0.9	5.4	5.5	5.8	1.2	1.4	0.1	5.0	3.3
Fe	1.7	0.6	1.3	4.1	1.3	1.4	0.6	1.3	1.4	3.9	3.0
Al	2.8	0.4	2.3	4.1	2.5	2.9	2.9	3.6	3.3	3.4	—
Org C	2.5	0.8	1.0	4.0	2.3	2.7	2.5	2.5	—	—	—
Cbt C	1.5	0.2	1.2	1.9	1.3	1.5	1.9	1.9	—	—	—
Na	0.57	0.15	0.40	0.93	0.58	1.16	0.12	0.16	0.25	0.43	—
K	0.53	0.06	0.46	0.69	0.46	0.46	0.28	0.32	1.20	1.05	—
Ti	0.27	0.04	0.20	0.39	0.23	0.25	0.21	0.28	0.28	0.40	—
Mn	0.11	0.02	0.07	0.15	0.10	0.11	0.14	0.19	0.03	0.08	0.02
P	341	119	218	594	175	218	87	87	200	—	—
Ba	461	683	260	3670	390	438	317	472	290	737	569
Sr	764	513	409	3130	706	823	378	444	53	—	—
Zn	1641	527	1110	2990	13300	14000	57	101	40	4760	1680
Pb	260	160	109	756	153	159	9	12	14	6550	124000
Cu	198	89	117	438	95	176	10	13	13	609	223
Ce	83	19	51	132	62	70	202	356	63	55	52
Y	49	10	30	68	44	53	134	243	20	—	—
Cr	117	27	80	224	94	107	87	153	33	403	211
Ni	29	14	19	88	29	30	23	202	11	82	166

Table I. Continued.

	Outdoor				Indoor		Girder		Soil	NIST-SRM	
Element	*mean*	*sd*	*min*	*max*	*min*	*max*	*min*	*max*		*1648*	*1649*
La	42	10	26	70	31	36	102	175	29	42	33
Sb	48	117	26	596	34	39	1	1	1	45	30
V	32	4.4	25	43	25	29	31	40	43	127	345
Mo	18	8.8	7	42	16	19	1	1	—	—	14
Li	23	3.3	17	30	22	25	25	36	17	—	—
Th	8	1.6	5	11	7	9	18	31	8	7.4	6.6
Rb	21	1.8	18	25	19	21	8	8	43	52	48
Co	6.5	2.0	4.9	13.9	5.0	5.3	1.7	12.3	5.9	18	16.4
Nb	7.0	3.1	1.2	11.0	8.0	9.0	4.4	6.3	10.0	—	—
Sc	6.6	1.1	4.4	9.3	5.4	6.4	9.2	9.8	6.5	7	8.7
U	3.0	0.5	2.0	4.1	2.7	3.2	4.7	7.6	2.1	5.5	2.7
Cd	5.1	2.2	3.0	13.0	4.2	5.8	0.1	0.2	0.1	75	22
As	4.6	0.9	3.6	6.8	3.5	3.8	<2	<2	4.8	115	67
Ga	4.3	0.5	3.8	6.0	3.6	4.0	2.8	4.2	9.3	—	—
Be	2.8	0.5	1.8	3.8	2.5	3.1	4.0	4.2	0.6	—	—
Ag	2.1	1.0	1.2	4.4	1.6	3.5	1.0	1.8	0.1	6	3.5
Cs	0.76	0.06	0.68	0.88	0.72	0.78	0.18	0.22	6.0	3	3
Bi	0.57	0.38	0.25	1.87	0.64	0.82	0.01	0.01	—	—	—
Tl	0.10	0.01	0.09	0.13	0.09	0.09	0.02	0.02	—	—	—

NOTE: Units for Si through Mn are percent dry weight. Units for P through Tl are ppm. LoI is loss on ignition, sd is standard deviation, (—) indicates parameter not measured. Org C is organic carbon, Cbt C is carbonate carbon.

SOURCE: Results for Eastern U.S. soil are reproduced from Reference 14. Results for NIST SRM1648 and 1649 are reproduced from References 15 and 16, respectively.

Perhaps in part because of the sample heterogeneities, systematic spatial variations in dust composition are not readily discernible between the various outdoor sample sites. For example, a sample with one of the highest lead concentrations (sample 2) was the farthest sample from the WTC complex, but was also adjacent to another sample (sample 4) with relatively low lead concentrations. Maps showing spatial variations in concentrations of the elements are presented elsewhere (*2*). Total bulk elemental concentration ranges measured for the dust samples in this study generally agree with those obtained previously on three settled dust samples (*12*) and, with the exception of Mg and Al, those reported by Lioy *et al.* for three settled dust samples (*13*).

The girder coatings are substantially different from the settled dust deposits in concentrations of some major elements and trace elements. They are enriched in Ca, Mg, Mn, Ce, Y, Cr, Ni, La, Li, Th, and U relative to the dust samples, and are depleted in Fe, Na, K, P, Ba, Sr, Zn, Pb, Cu, Sb, Mo, Co, and Cd relative to the dusts (see Figure 1). These differences illustrate that the girder coatings, presumed to be primarily composed of spray-on insulation material, do not contribute substantially to the overall chemical makeup of the dusts.

Chemical Compositions of Size-Fractionated Settled Dust Samples

The chemical analyses of size-fractionated WTC settled dust sample 18 are summarized in Table II and Plate 1 and those for sample 19 are summarized in Table III and Plate 2. Also shown in Plates 1 and 2 for comparison are the WTC outdoor bulk dust composition ranges from Figure 1 (gray). Gaps in the element profile for a given size fraction in Plates 1 and 2 indicate the element is below detection limits or was not measured. The results show that the chemical makeup of these samples does not vary consistently for most elements as a function of particle size. In sample 18, the range of compositions between the different size fractions is generally small, and the element concentration profile is similar to that of the indoor dust samples. There is a somewhat broader concentration range in sample 19 than in sample 18 between the different size fractions for many elements, and the chemical profile is generally more similar to that of the outdoor bulk dust samples. Limitations in sample volumes available precluded a similar analysis of more samples to determine if the size fractionation results for these two samples are representative of the dusts as a whole.

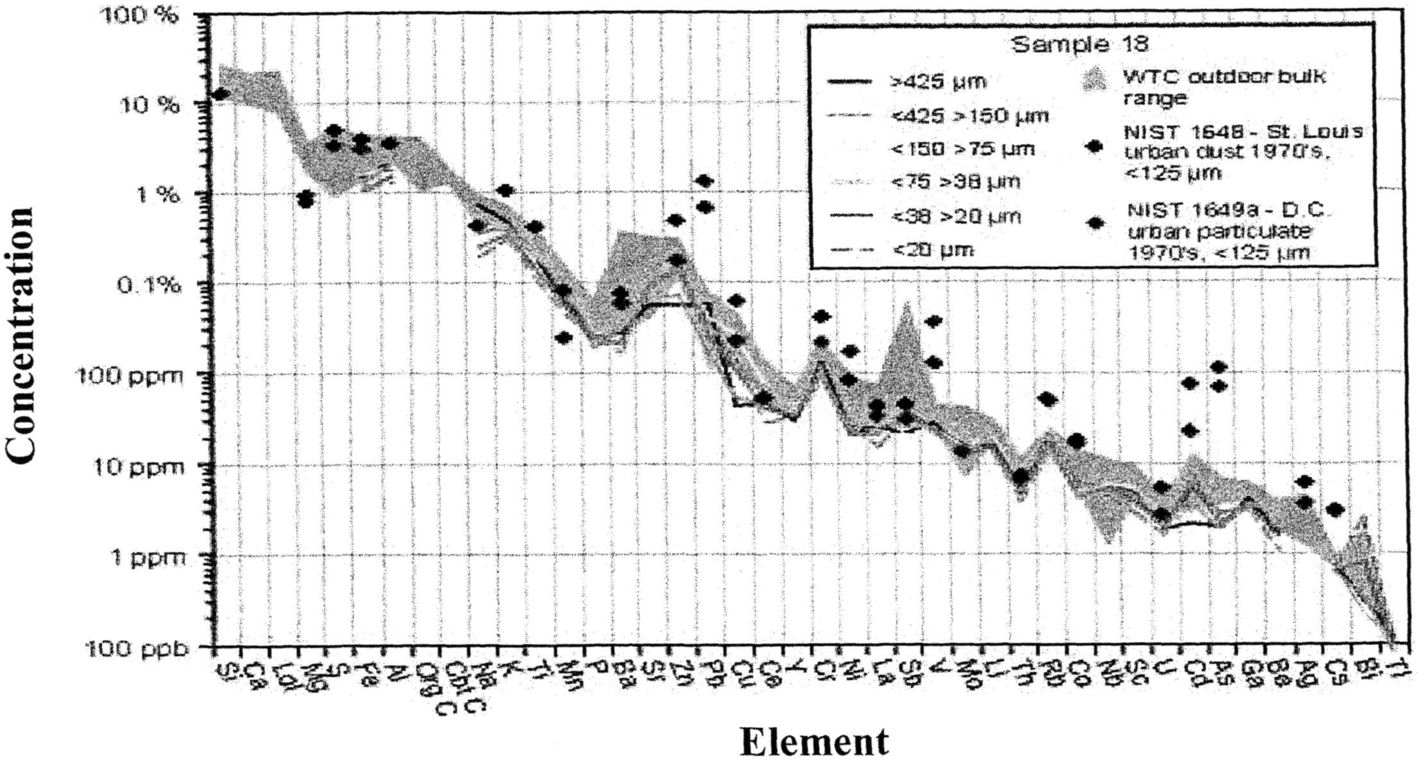

Plate 1. Concentrations of the elements in sieved size fractions of WTC dust sample 18 determined by ICP-MS compared to the WTC outdoor bulk dust composition ranges from Figure 1 (gray), and the concentrations of elements from NIST SRM-1648 and SRM-1649a sieved to <125 µm(15, 16). LoI is loss on ignition; Org C is organic carbon; Cbt C is carbonate carbon.

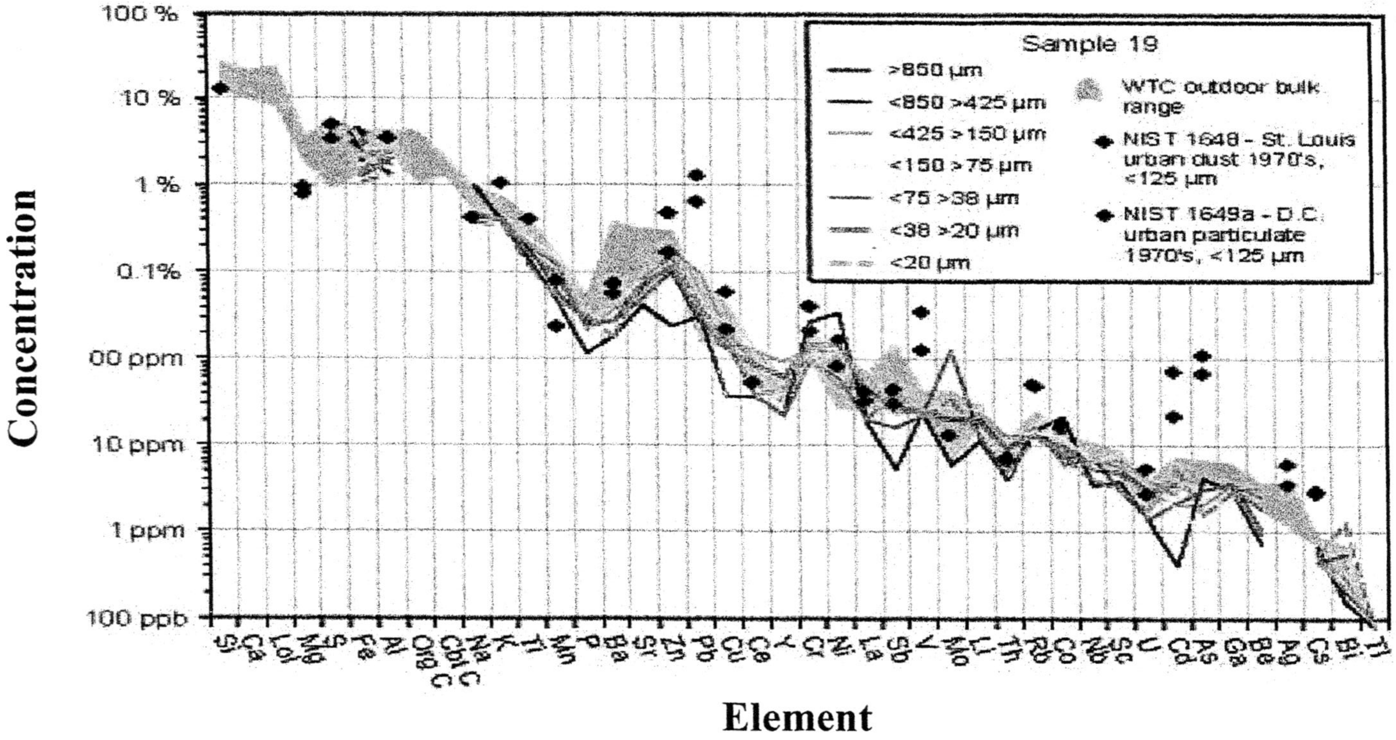

Plate 2. Concentrations of the elements in sieved size fractions of WTC dust sample 19 determined by ICP-MS compared to the WTC outdoor bulk dust composition ranges from Figure 1 (gray), and the concentrations of elements from NIST SRM-1648 and SRM-1649a sieved to <125 µm(15, 16). LoI is loss on ignition; Org C is organic carbon; Cbt C is carbonate carbon.

Table II. Weight Fraction (%) and Elemental Composition of WTC Size Fractionated Settled Dust Sample 18, as Determined by ICP-MS.

	>425μm	*425-150μm*	*150-75μm*	*75-38μm*	*38-20μm*	*<20μm*
Weight	15.3	18.6	22.5	22.6	18.5	2.4
Ca	15.2	17.7	18.9	19.3	20.2	20.1
Mg	1.8	1.7	1.9	2.1	1.4	1.2
Fe	1.5	1.7	1.6	1.3	1.0	1.2
Al	2.1	1.8	2.0	2.2	1.6	1.3
Na	0.73	0.35	0.29	0.37	0.25	0.20
K	0.48	0.45	0.38	0.39	0.35	0.31
Ti	0.20	0.22	0.21	0.19	0.14	0.11
Mn	0.07	0.07	0.09	0.10	0.06	0.04
P	243	186	265	226	236	258
Ba	280	306	305	324	210	170
Sr	553	654	801	892	855	898
Zn	562	693	1460	1360	1410	1630
Pb	580	169	410	266	266	346
Cu	44.0	53.1	262	107	109	143
Ce	47.3	48.1	71.0	71.5	41.0	28.6
Y	29.8	32.9	47.9	50.3	34.2	34.0
Cr	133	121	86	98	105	103
Ni	26.0	22.5	22.2	21.5	21.1	25.7
La	24.9	25.0	36.2	37.0	21.9	15.7
Sb	21.8	37.7	20.4	26.0	33.5	36.7
V	28.7	24.4	24.6	26.1	23.8	22.1
Mo	11.8	16.6	15.5	10.2	13.5	16.2
Li	17.3	17.4	21.2	24.8	15.5	14.1
Th	4.9	5.4	7.5	7.7	5.0	3.8
Rb	20.9	21.8	16.6	16.9	16.1	14.5
Co	6.2	5.9	4.8	4.4	4.3	4.7
Nb	6.0	6.2	6.1	6.9	5.4	4.5
Sc	4.7	4.4	5.5	5.8	3.9	2.8
U	1.9	2.1	2.9	3.0	2.0	1.6
Cd	2.2	5.2	2.7	3.5	6.2	7.7
As	2.0	2.2	2.6	1.9	2.8	2.2
Ga	4.1	3.5	3.5	3.7	3.4	3.1
Be	1.7	1.8	2.8	2.7	1.6	1.1
Ag	< 2	< 2	< 2	< 2	< 2	< 2
Cs	0.76	0.72	0.62	0.64	0.67	0.66
Bi	0.30	0.21	0.20	0.33	0.81	2.56
Tl	0.12	0.11	0.09	0.08	0.09	0.09

NOTE: Units for Ca through Mn are percent dry weight. Units for P through Tl are ppm.

Table III. Weight Fraction (%) and Elemental Composition of WTC Size Fractionated Settled Dust Sample 19, as Determined by ICP-MS.

	>850μm	850-425μm	425-150μm	150-75μm	75-38μm	38-20μm	<20μm
Weight	57.0	9.4	13.2	6.5	5.3	4.0	0.9
Ca	12.3	11.9	16.1	19.2	19.5	19.8	21.5
Mg	6.4	1.8	2.6	3.5	3.6	3.0	1.7
Fe	2.0	2.6	2.5	1.6	1.4	1.3	1.1
Al	1.3	1.8	2.3	2.7	2.5	2.0	1.3
Na	0.99	0.46	0.46	0.47	0.50	0.46	0.4
K	0.37	0.42	0.43	0.41	0.40	0.39	0.3
Ti	0.14	0.18	0.20	0.24	0.21	0.16	0.1
Mn	0.05	0.06	0.12	0.15	0.13	0.09	0.05
P	116	277	236	224	217	248	320
Ba	186	280	374	394	376	274	181
Sr	422	510	599	823	730	730	821
Zn	242	1050	1860	1360	1200	1160	1530
Pb	329	277	1010	402	310	295	407
Cu	38.2	164	231	278	149	138	175
Ce	37.5	40.3	91.6	137.0	130	87.4	39.4
Y	23.7	23.0	61.8	99.1	93.3	66.4	44.4
Cr	279	96.8	103	125	137	150	141
Ni	352	59.6	48.2	109	124	148	103
La	18.7	20.6	45.9	72.0	68.5	44.6	21.5
Sb	5.5	16.6	27.8	25.8	24.0	29.2	42.3
V	25.0	23.4	26.9	26.8	27.4	23.6	21.4
Mo	6.1	130.0	22.0	28.8	14.6	20.7	34.7
Li	11.9	12.7	21.5	26.5	24.7	19.4	14.1
Th	4.2	4.4	9.2	13.5	13.2	9.3	4.8
Rb	15.3	16.6	17.1	15.0	13.6	13.1	13.1
Co	22.3	7.6	6.3	13.0	8.8	10.2	7.8
Nb	3.5	7.4	7.3	6.9	6.5	5.5	4.2
Sc	4.0	3.1	5.9	7.5	7.1	5.1	3.1
U	1.4	1.4	3.2	4.3	4.1	3.0	1.8
Cd	0.4	2.2	2.1	1.9	2.8	3.7	5.9
As	4.3	3.2	2.1	1.0	< 1	< 1	1.6
Ga	3.0	3.8	3.7	3.6	3.6	3.3	3.0
Be	0.8	1.1	2.6	3.2	3.1	1.9	1.0
Ag	< 2	< 2	< 2	< 2	< 2	< 2	< 2
Cs	0.68	0.62	0.58	0.50	0.46	0.47	0.54
Bi	0.16	0.21	0.29	0.23	0.23	0.56	1.36
Tl	0.08	0.10	0.08	< 0.08	< 0.08	< 0.08	0.08

NOTE: Units for Ca through Mn are percent dry weight. Units for P through Tl are ppm.

Comparison of the WTC Settled Dust Compositions to those of Soils and Urban Particulates

Although the materials are not directly comparable in either their origin or potential human exposure mechanisms, duration, and intensity, a comparison of the WTC settled dust compositions to those of soils and urban particulates provides some indication of the relative chemical enrichments or depletions of the dusts compared to natural dust source materials and urban particulates matter.

The WTC bulk settled dust samples tend to be enriched in some elements compared to the soils from the eastern U.S. shown in Figure 1, including Ca, Mg, S, Na, Mn, P, Ba, Sr, Zn, Pb, Cu, Y, Cr, Ni, Sb, Mo, U, and Be (*14*) . When compared to the <125-µm NIST urban dust and particulate standard reference materials (SRM-1648, SRM-1649a) (*15, 16*), the bulk and size-fractionated WTC dust samples are substantially depleted in a variety of heavy metals such as Zn, Pb, Cu, Cr, Ni, V, Co, Cd, and As (Plates 1 and 2). However, since the NIST materials were collected in the mid-1970's prior to the banning of leaded gasoline, and so Pb and possibly some other heavy metals may be elevated compared to present-day urban particulates.

Compositions of WTC Sample Leachates

Results of DIW leach tests are summarized in Tables IV and Figure 2. Gaps in the element profile for a given sample type shown in Figure 2 indicate that the element was below detection limits or not measured. The results show that the settled dust samples contain reactive and soluble components. The leachate solutions developed moderately to very alkaline pH values (8.2–12.4), and high specific conductances (1.31–3.9 milliSiemens/cm) indicative of high dissolved solids (see Figure 2). Although several outdoor samples generated nearly as high pH values as the indoor samples, the majority of the outdoor samples generated substantially lower pH values than the indoor samples and all generated lower conductivities than the indoor samples. Leachate alkalinities show the same trends, with indoor dust samples typically generating substantially higher alkalinities than those outdoor samples which generate lower-pH leachates.

With increasing pH above 10.3, the leachate alkalinity becomes progressively more caustic due to increasing concentrations of hydroxyl ions compared to carbonate ions. Although both indoor samples and several outdoor samples generated leachates with caustic alkalinity, they generated less caustic

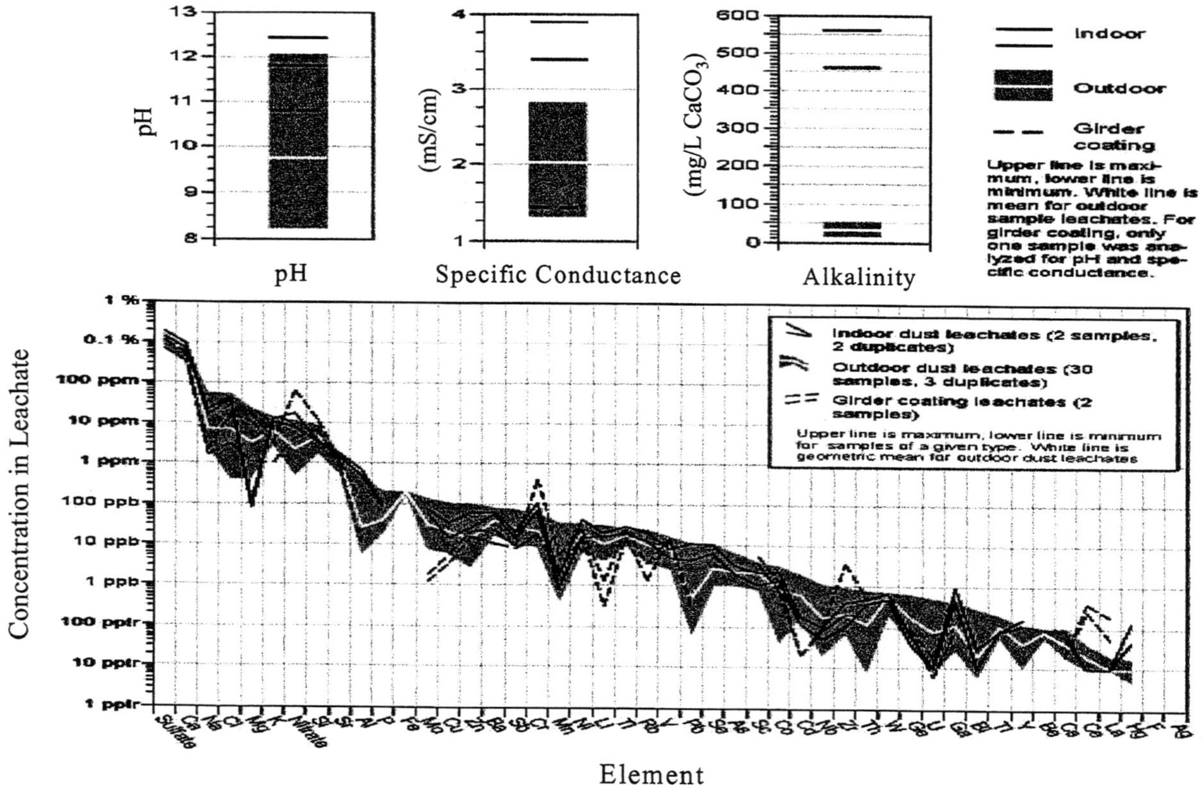

Figure 2. Specific conductance, pH, alkalinity (upper) and chemical compositions (lower) of WTC sample DIW leachates.

Table IV. Chemical Composition of DIW Leachate from Outdoor and Indoor Dust Samples and WTC Steel Girder Coatings.

	Outdoor				*Indoor*		*Girder*	
	min	*max*	*mean*	*sd*	*min*	*max*	*min*	*max*
pH	8.22	12.04	9.73	1.92	11.8	12.4	10.8	10.8
S Cond	1.31	2.79	2.01	0.74	3.4	3.9	1.43	1.43
Alkalinity	20	49	34	15	460	560	—	—
Sulfate	694	1600	1246	457	1320	1840	674	1090
Ca	314	649	530	170	718	913	336	528
Bicarbonate	24	59	41	17	552	671	—	—
Na	2.48	55	7.0	29.1	15.3	22.1	1.5	2.1
Cl	0.4	52	7.1	28.1	40	45	3	16
Mg	0.4	22.4	3.2	12.0	0.08	0.11	1.1	10.3
K	2.84	14.2	6.4	5.8	10.2	12.3	1	3
Nitrate	0.5	11	2.2	5.6	8	17	4.1	62
Si	1.8	8.6	4.6	3.4	3.2	3.5	6.7	11.3
F	—	—	—	—	—	—	—	—
Sr	561	1540	1043	490	1420	1690	758	990
Al	5.3	505	25.9	282	480	702	10.8	121
P	20	200	44.31	97.7	90	100	—	—
Fe	200	200	200	0	—	—	—	—
Mo	7.42	140	31.4	70.6	60.4	73.8	1.2	1.7
Cu	6.2	110	20.3	56.3	15.1	34.2	3.5	5.6
Zn	2.7	104	13.8	55.6	26.2	61.8	15.8	20.1
Ba	17.5	83.7	41.2	33.5	56.7	63.6	10.4	22.8
Sb	9.89	75.6	25.7	34.3	13.6	20.8	7.8	8.7
Cr	9	55.8	21.9	24.2	69.4	110	18	408
Mn	0.4	35.1	2.2	19.5	0.8	1.7	2.1	5.5
Ni	6.9	32.1	19.9	12.6	14.5	42.6	16.6	24.9
Li	4.1	29.7	10.4	13.3	16.3	19.5	0.3	1.3
Ti	10.2	26.3	17.7	8.1	19.3	28.4	15.3	24.9
Rb	3.6	25	9.9	11.0	12.3	20.8	1.4	3.5
V	2.7	16.8	9.2	7.1	4.8	7.8	13.8	14.4
Pb	0.06	11.5	0.5	6.5	4.3	10.9	0.3	0.4
Se	1	10.9	2.8	5.3	6.4	10.5	—	—
As	1	6.8	2.3	3.0	2	3.3	—	—
Sc	0.8	3.9	2.1	1.6	1.1	2.1	3.6	5.5
Co	0.04	3.18	0.8	1.6	1.4	2.2	0.75	1.3
Cd	0.16	1.96	0.6	0.94	0.12	0.18	0.02	0.02
Nb	0.02	1.17	0.2	0.63	0.05	0.1	0.08	0.08
Zr	0.07	0.97	0.2	0.48	0.1	0.4	0.2	3.7
Th	0.008	0.8	0.1	0.43	0.38	0.51	0.18	0.52
W	0.22	0.75	0.6	0.27	0.5	0.71	—	—

Continued on next page.

Table IV. Continued.

	Outdoor				*Indoor*		*Girder*	
	min	*max*	*mean*	*sd*	*min*	*max*	*min*	*max*
Ge	0.04	0.65	0.2	0.32	0.04	0.07	0.1	0.1
U	0.008	0.52	0.1	0.28	0.01	0.01	0.006	0.02
Ga	0.03	0.41	0.1	0.20	0.46	0.97	0.08	0.38
Bi	0.006	0.28	0.02	0.16	0.01	0.03	—	—
Tl	0.06	0.2	0.1	0.07	0.08	0.08	—	—
Y	0.01	0.12	0.04	0.06	0.11	0.16	0.27	0.31
Be	0.06	0.1	0.1	0.02	—	—	—	—
Cs	0.02	0.1	0.04	0.04	0.06	0.09	0.02	0.02
Ce	0.01	0.04	0.02	0.02	0.01	0.01	0.26	0.4
La	0.01	0.02	0.01	0.01	0.01	0.01	0.05	0.18
Hg	0.005	0.018	0.01	0.01	0.05	0.13	—	—
Ag	—	—	—	—	—	—	—	—

NOTE: Units for Alkalinity, sulfate, bicarbonate and Ca through F are ppm. Units for Sr through Cs are ppb. S Cond is specific conductance, sd is standard deviation, (—) indicates parameter not measured or all measurements were below detection limits. Bicarbonate concentrations were calculated based on the measured alkalinity.

alkalinity than equivalent weights of solid and liquid lye drain cleaner, and concrete powder (see Table V, Figure 3).

The leachate solutions are composed primarily of sulfate, Ca, and bicarbonate (several hundreds of ppm), with lesser Na, Cl, Mg, K, nitrate, Si, and Sr (generally several ppm to tens of ppm). Indoor samples generally produce leachates that are higher in sulfate, Ca, bicarbonate, Cl, K, and Sr, and lower in Mg than the outdoor dust samples. Total bicarbonate concentrations calculated from the alkalinities range from several tens of ppm in the outdoor dust leachates to 600-700 ppm in the indoor dust leachates. Of the trace metals and metalloids, the leachates are generally most enriched in Al, Mo, Cu, Zn, Ba, Sb, Cr, and Mn. Concentration ranges of Al, Ni, Pb, Se, W, Ga, and Hg in the indoor dust leachates generally fall at the upper end of or greater than their concentration ranges in the outdoor dust leachates. Mercury concentrations in the indoor dust leachates, while lower than those of most other metals, are nearly 10 times greater (120-150 ppt) than in the outdoor dust leachates (5-15 ppt).

Analyses of duplicate leachates for several outdoor and the two indoor samples show that there can be some variability in the determinations for a given sample (see Table IV and Appendix II), which likely results primarily from the heterogeneous nature of the dust samples. As with the bulk samples, compositions of the sample leachates show no consistent spatial variations between sample sites. Maps showing the spatial variations in leachate chemistry are presented elsewhere (*2*).

Table V. Alkalinity and pH of DIW Leachates from WTC Outdoor (2, 16, 18, 19, 25) and Indoor Dust Samples (20, 36) Compared to that of Some Common Household Products.

Sample	*Size*	*pH*	*Alkalinity*
20	bulk	12.3	460
36	bulk	12.4	560
2	bulk	10.1	40
16	bulk	9	20
25	bulk	9.37	49
18	bulk	12.04	—
18	>425μm, gr	11.5	257
18	425-150μm	11.51	245
18	150-75μm	11.73	347
18	75-38μm	11.73	336
18	38-20μm	11.47	222
18	<20μm	11.34	> 347
19	bulk	9.83	—
19	>850μm, gr	10.74	109
19	850-425μm, gr	10.93	147
19	425-150μm	9.23	95
19	150-75μm	9.18	97
19	75-38μm	9.09	103
19	38-20μm	9.15	98
19	<20μm	9.2	103
Liquid Drain Cleaner	—	12.2	3320
Cement	—	12.2	990
Baking Soda	—	8.5	26000
Milk of Magnesia	—	10.6	44
Lye Powder	—	13.5	59000
Liquid Bleach	—	11.8	1900
NaOH Granules	—	13.4	57000
Slag Wool	—	9.25	184

NOTE: Alkalinity is reported in ppm $CaCO_3$. Due to a malfunction during titration measurement, alkalinity for sample 18, <20μm size, should be considered as a minimum number. Samples ground after sieving are indicated by "gr".

A comparison of the leachate compositions with the bulk dust compositions shows that many of the elements are not leached from the settled dust deposits in proportion to their concentrations in the original bulk samples. The percentage of elements leached from the indoor and outdoor dust and girder coating samples is shown in Figure 4. In order to account for the 1:20 dilution of the sample in the leachate, the leachate concentrations in Figure 4 have been multiplied by 20.

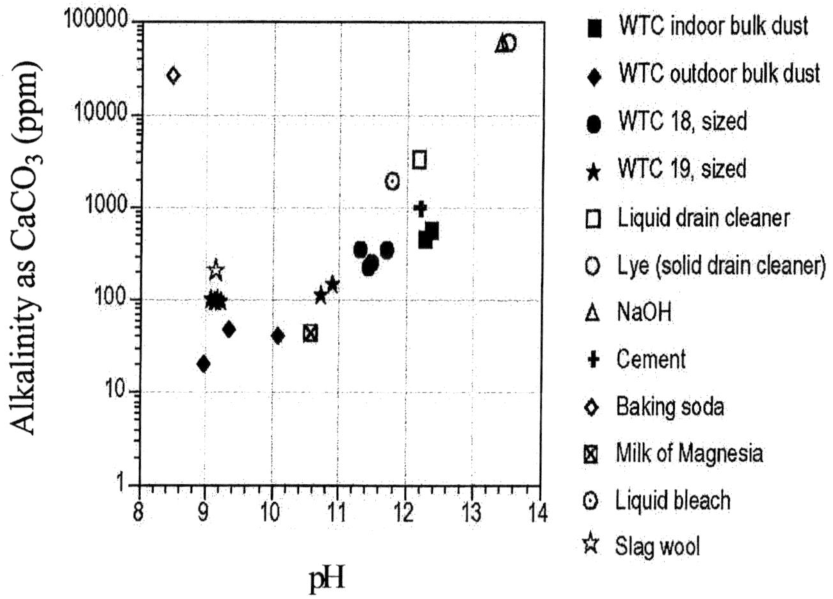

Figure 3. Alkalinity and pH of DIW leachates from bulk WTC dust samples and size fractions of samples 18 and 19, compared to that of some common household products.

Also, sulfur in the bulk sample has been converted to an equivalent weight of sulfate for direct comparison to the sulfate in the leachate. Sulfate is leached in the greatest proportions (ten to several tens of percent). Molybdenum, Ca, K, Sr, Na, Ni, As, Rb, Tl, Li, and Sb are leached in amounts of up to several percent. Silicon, Al, Pb, Ti, Mn, Ce, Y, La, Th, and U are leached in relatively small proportions compared to their bulk concentrations (generally less than 0.1 %). Most elements are leached in greater proportions from the indoor dust samples. However, elements such as Sb, V, Cd, Bi, Si, Mg, and Mn are leached in greater proportions from the outdoor dusts.

McGee *et al.* noted that the NIST urban particulate standard reference material SRM 1649 generated water leachates with substantially lower pH and higher levels of most heavy metals than the leachates generated by the WTC dusts (*17*). The girder coatings generated leachates having overall compositions that are generally similar to those of the settled dust samples, with several exceptions (see Figure 2 and Table IV). Specifically, the girder coatings generated leachates with an alkaline pH (near 10.8), substantially higher nitrate,

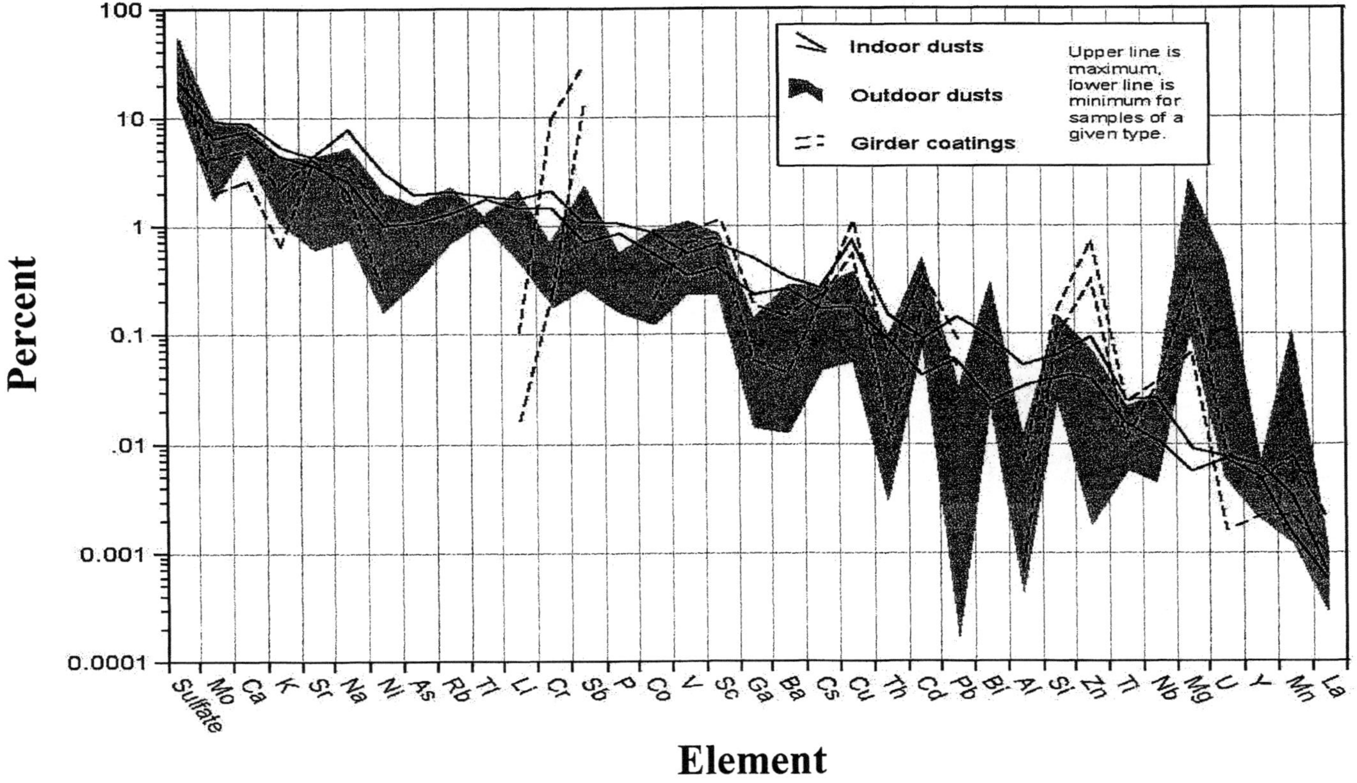

Figure 4. Percentage of elements leached from the indoor and outdoor dust and girder coating samples.

Cr, Zr, Ce, and La concentrations, but lower Na, K, Mo, Li, and V concentrations than the dust leachates.

Leachate pH and Alkalinity of Sized WTC Samples

The pH and alkalinity were measured on the different size fractions of samples 18 and 19, one of which (sample 18) generated the highest bulk leachate pH (12.04) of any of the outdoor samples (see Table V, Figure 3). There is for both samples a measureable decrease in leachate pH from the bulk and coarsest size fractions to that of the finest fraction (<20 μm). However, the leachate pH values of the finer size fractions of sample 18 are still very high (>11.34) compared to the bulk, coarse, and fine size fractions of sample 19.

The coarsest fractions of samples 18 and 19 were ground prior to leaching and analysis to examine whether grinding has any influence on leachate pH. Grinding resulted in higher leachate pH and alkalinity of the coarse fraction of sample 19 relative to the fine fractions, but resulted in a decreased leachate pH and alkalinity of the coarse fraction of sample 18 relative to the fine fractions.

*In Vitro L*each Tests Using Simulated Lung Fluids

Results of physiologically-based leach tests on indoor (sample 20) and outdoor (sample 16) WTC settled dust samples are given in Table VI and Figure 5. The results indicate that the hydroxyl alkalinity released from the dusts also produces pH shifts in the SLF, but that the pH shifts are somewhat lower than those obtained in DIW leachates due to the pH-buffering capacity of SLF components. Some metals in the dusts (Si, Al, Ba, Cu, Zn, Co, Ni) are more soluble in SLF than in DIW due to chelation by chloride, citrate, and glycine. Metals or metalloids that form oxyanion species in solution are not appreciably different in concentration between the SLF and DIW leaches. Phosphorous concentrations are substantially lower in the SLF-dust leachate solutions than in the SLF blank, suggesting that calcium dissolved from the dusts reacts with phosphate in the SLF to precipitate a calcium-phosphate solid.

Discussion

Factors Controlling the Chemical Composition of Bulk Settled Dust Samples

Based on the materials characterization studies of splits of the same samples described in Chapter 5 (*1*), it can be inferred that the major chemical

Table VI. Chemical Composition of SLF and DIW Leachates of WTC Outdoor (16) and Indoor (20) Dust Samples Compared to Procedural Blanks (Blk).

	SLF-20	*SLF-16*	*SLF-Blk*	*DIW-20*	*DIW-16*	*DIW-*
pH	9.54	7.59	7.85	11.45	8.22	6.2
Ca	932	1010	9	748	621	0.05
Mg	27	51	0.07	0.14	40	0.01
Na	>1000	>1000	>1000	24.6	9.4	1.5
K	33.5	25.2	0.5	21.4	14.6	0.2
Si	31.7	32.9	7.5	7.9	23.5	0.8
Sulfate	3075	2760	87.5	1460	1650	13.5
P	0.3	3.0	44.4	0.2	0.09	0.14
Sr	2860	2165	4.2	2410	2210	0.5
Cu	1385	799	4.0	61.8	28.4	0.5
Zn	368	324	12.1	19.2	40.6	3.3
Mo	204	72	6.9	202	66.6	1.5
Fe	190	169	236	54.5	75	104.5
Cr	187	92	71	228	37	6
Sb	169	74	0.3	11.2	57.7	0.1
Ba	95.1	65.5	4.1	115	47.6	1.3
Ni	80.3	70.9	0.8	43.1	36.3	0.1
Rb	79.5	82.0	0.1	36.9	24	0.01
V	62.4	45.3	27.0	10.2	20.6	9.7
Li	54.0	37.6	4.4	32.3	23.3	9.7
Ti	52.2	43.8	< 0.1	15.7	18.1	0.1
Se	37	14.5	19.3	28.3	2.4	1
Al	23.7	9.6	6.5	187	12.1	7.0
Sc	18.7	18.4	3.6	8.1	15.0	0.95
Cd	10.2	1.7	0.0	0.3	0.3	0.02
Co	9.0	4.0	0.2	2.4	1.8	0.02
As	8.6	10.0	9.6	7.3	—	1
Mn	3.9	72.9	0.8	0.6	15.4	0.2
Nb	3.6	1.9	0.5	2.2	1.3	0.2
Th	3.0	1.9	2.1	1.4	1.0	0.2
Zr	2.0	0.7	2.9	1.0	0.86	0.05
Bi	1.9	0.7	0.1	0.78	0.32	0.07
Pb	1.5	1.6	0.2	8.9	0.46	0.05
Cs	1.1	0.93	0.03	0.17	0.07	0.01
Tl	0.4	0.25	< 0.05	0.065	—	0.05
Ge	0.20	0.71	0.24	0.44	0.57	0.38
Ga	0.17	< 0.02	0.33	0.67	—	0.02
U	0.15	1.96	1.02	0.055	0.905	0.165

Continued on next page.

Table VI. Continued.

	SLF-20	*SLF-16*	*SLF-Blk*	*DI-20*	*DI-16*	*DI-Blk*
Y	0.05	0.02	0.11	0.02	0.06	0.01
Ce	0.02	< 0.01	0.045	< 0.01	0.03	0.01
La	0.02	0.02	0.035	< 0.01	0.015	0.01
Ag	< 3	< 3	< 3	< 3	-	3
Be	< 0.05	< 0.05	< 0.05	< 0.05	0.1	0.05

NOTE: Sample results are the average of two duplicate sample splits. Units for Ca through P are ppm. Units for Sr through Be are ppb.

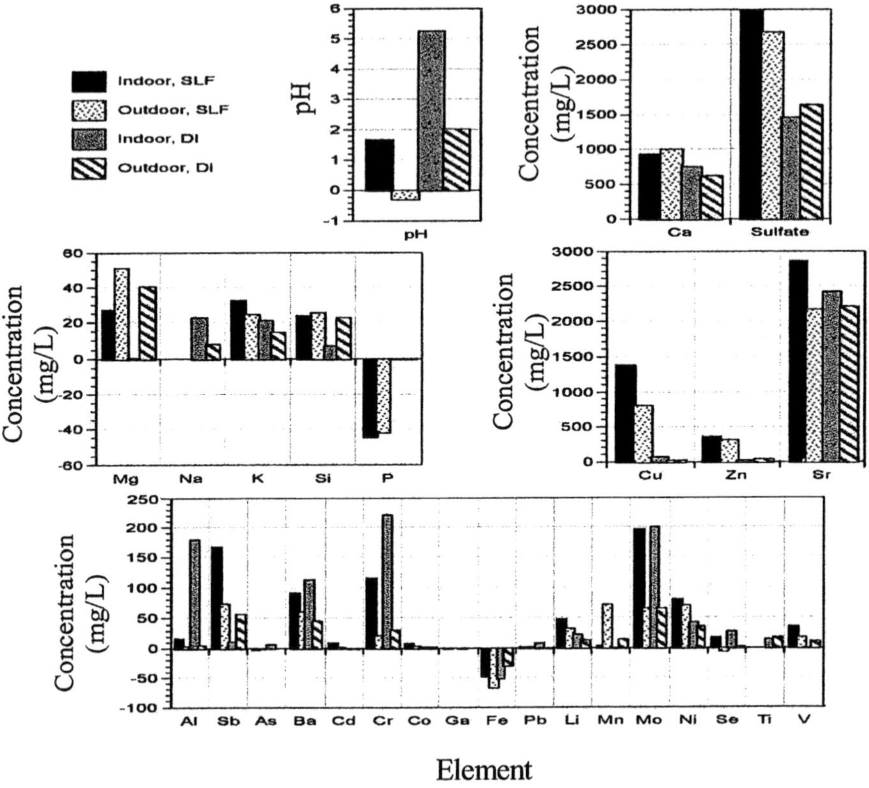

Figure 5. Chemical Composition of simulated SLF and DIW leaches on one indoor (20) and one outdoor (16) dust sampl. Concentrations are given as the difference between the dust leachate and the leaching fluid blank.

components of the settled dusts (Si, Ca, sulfate, organic carbon and other ignition-volatile compounds, Mg, Fe, Al, Na, and K) were contributed primarily by glass fibers, concrete, gypsum wallboard, window glass, paper, and other materials within the office buildings. A wide variety of metals, metalloids, and other elements present in lesser concentrations in the settled dusts (Ti, Sr, Zn, Mn, Ba, Pb, Cu, Ni, Sb, V, Mo, Li, Rb, etc.) were likely contributed by many different materials such as concrete, paints, light bulbs and fixtures, electrical wires, pipes, fire retardant materials, computer equipment, and electronics.

Chemical Reactions Between WTC Dust and Water

The compositional variability of the bulk dust samples results in part from the complex and heterogeneous makeup of the dusts observed in physical characterization studies described in Chapter 5. However, a variety of our results indicate that many of the outdoor dust samples analyzed in this study were modified chemically as a result of interactions with either rainwater or the water used in fire fighting or street washing, further enhancing the compositional variability. For example, differences in the composition ranges for many elements between the outdoor dust samples and indoor dust samples (see Figure 1) are consistent with the preferential leaching of soluble Ca and sulfate from outdoor dust samples by water prior to sample collection, with the accompanying enrichment of the less soluble components (such as Si, Fe, P, Ba, and Pb) in the outdoor dust samples relative to the indoor dust samples.

The alkaline pH of the dust leachate solutions, coupled with their high concentrations of Ca, sulfate, and carbonate, are consistent with an origin resulting primarily from the dissolution of soluble concrete and wallboard components. Lesser contributions likely came from partial dissolution of glass fibers and other less soluble materials in the dusts.

The extremely high leachate pH and alkalinity generated by the indoor samples and several of the outdoor samples require a source of soluble hydroxyl in the dusts, which most likely included portlandite (a calcium hydroxide) and possibly minor sodium and potassium hydroxides in the concrete particles. Portlandite is present in sufficient quantities to be detected by XRD in a number of the dust samples (see Chapter 5). Dissolution of gypsum (hydrous calcium sulfate) from the concrete and wallboard is likely the source of the high dissolved sulfate, and also, with portlandite, a source for the high calcium concentrations in the leachates.

Chemical speciation calculations on the WTC dust leachate compositions indicate that the leachates having pH from 11 to 12.4 are at or near saturation with respect to portlandite, but are supersaturated with respect to calcite (calcium carbonate). This suggests that, over the short duration of the leach experiments, calcite precipitation was inhibited kinetically but that, over time, calcite precipitation would likely occur as the leachate solutions equilibrated with carbonic acid, according to the following reaction:

$$Ca(OH)_{2(s)} + H_2CO_{3(aq)} \rightarrow CaCO_{3(s)} + 2H_2O \quad (1)$$

Hence, it is likely that accessible calcium hydroxide particles in the outdoor dusts reacted with carbonic acid in rain water or waters used in fire fighting or street cleaning, and were either replaced or armored by calcium carbonate. Such reactions help explain the lower pH values and alkalinities generated by the outdoor dust samples. Similar reactions may also have occurred eventually, albeit to a lesser and slower extent, in indoor dust deposits as fine particles of calcium hydroxide reacted with the moisture and carbon dioxide in the air.

The consistently higher conductivity of the indoor dust samples relative to the outdoor samples also indicates that the outdoor samples had already lost some of their soluble and reactive constituents by reactions with water prior to sample collection. The large variations in leachate pH, conductivity, alkalinity, and bulk chemistry between the different outdoor dust samples indicate, however, that the outdoor samples likely interacted with water to different extents prior to collection, most likely a function of the extent to which a given sampling location was exposed to water. Nonetheless, even the samples that generate a leachate pH near 8, which therefore likely had been substantially affected by water prior to collection, still release a wide variety of elements into solution.

The solubility of the various metals in the dust leachates is controlled by the solubility of the phases in which they reside in the dusts, coupled with their solubility and speciation in solution given the overall leachate compositions. The metals most efficiently leached from the dusts, especially the indoor samples, were those that form oxyanion species stable in solution at higher pH (Cr, Sb, Mo, As). In contrast, elements that form insoluble hydroxide precipitates at high pH were extracted much less efficiently (e.g. Mg, Fe, Mn). Lead was generally leached much less effectively from the dusts (as a percentage of original weight) than other less abundant metals such as antimony and molybdenum. Lead may occur in less soluble phases in the dusts, and saturation of lead sulfate or lead carbonate minerals in the leachates may have further limited the lead solubility.

Effects of Dust Particle Size

Particle size plays an important role in the extent of uptake by the human body, as described in Chapter 1. Particles less than approximately 500 microns are generally considered ingestible. Those less than 10-20 microns are inhalable at least into the upper respiratory tract but are trapped in mucus, removed by mucociliary action, and either expectorated or incidentally ingested. Particles less than 2-4 microns are respirable into the alveoli, where they are either exhaled or trapped. Respiratory problems can result if the trapped respired particles are chemically reactive or are not amenable (e.g. due to the number of particles present, their shape, and chemical composition) to clearance by the macrophages or dissolution in the lung fluids.

The lack of consistent compositional variations between different size fractions of two samples (18 and 19) suggests that there may not have been a consistent size fractionation of materials as the dust settled at a given location. Although materials characterization studies have identified many different metal-bearing particles that are of respirable to inhalable size (see Chapter 5), there is no observed chemical enrichment in these metals in the <20 μm size fractions of the dust. It may be that the fine metal-rich particles are adhered to and enclosed in larger particles, and/or that the metals reside as trace constituents in many different materials.

General similarities in the compositional ranges, element concentration profiles, and leachate pH between size fractions of sample 18 and the indoor bulk dusts, and between size fractions of sample 19 and the outdoor dusts, is consistent with sample 18 having interacted to a greater extent with water prior to collection than sample 19. However, the effects of water interactions apparently did not completely mask the differences in the amounts of portlandite originally present between the two samples, which was detected by x-ray diffraction in all size fractions of sample 18 (see Chapter 5). Sample 18 generated consistently higher leachate pH and alkalinity across all size fractions than sample 19, including the ground coarse fractions where grinding should have exposed all encapsulated portlandite. It further appears that water-induced armoring or replacement of portlandite by calcite was incomplete in all the analyzed size fractions of sample 18. Due to sample size limitations, we were unable to determine whether calcium hydroxide is present in this sample in respirable size fractions. McGee *et al.* *(17)* found, at the sampling sites they studied, that leachate pH values were less than 10 in the PM-2.5 fraction, suggesting that calcium hydroxide was either not present or not accessible in their samples of respirable sizes.

Implications for Assessing Human Health Effects Resulting from Dust Exposure

The high alkalinity of the WTC dusts has been linked to chemical-induced irritation of the eyes, mouth, throat, upper respiratory tract, and gastrointestinal tract in persons exposed to the dust (*18*). Our results confirm that the indoor settled dust deposits and at least some outdoor settled dust deposits contained sufficient portlandite, even in particle size fractions less than 20μm, to generate caustic alkalinity that could cause chemically induced irritations. Further work would be needed to determine whether the dust deposits having high levels of calcium hydroxide in coarser size fractions also had sufficient calcium hydroxide in the PM-2.5 fraction to trigger alkalinity-related irritation of the deeper respiratory system.

The ability of the dusts (especially outdoor dusts) to generate caustic alkalinity and alkalinity-induced health effects may have diminished over time as calcium hydroxides reacted with water and atmospheric carbon dioxide to form calcium carbonate (calcite) coatings on the coarser particles. However, calcite-armored calcium hydroxide particles still could be a source for alkalinity if physical disturbance of the settled dusts was sufficient to break up the particles and expose fresh calcium hydroxide.

Comparison of the WTC dusts with the NIST urban dust standards suggests that the WTC settled dusts did not have as high concentrations of many potentially toxic and bioaccessible heavy metals as particulates to which urban populations are regularly exposed on a daily basis. However, the substantially greater intensity but shorter duration of exposure to the WTC dusts must be factored in to any interpretation of metal exposures beyond urban baselines. Our SLF leach test results suggest that some metals and metalloids in the WTC dusts could be bioaccessible in the fluids lining the respiratory tract, including those elements that form strong aqueous complexes with chloride and organic acids (e.g. Zn, Cu, Ni, Mn, Hg), and those that form aqueous oxyanion species stable in near-neutral to alkaline fluids (e.g. Sb, Mo, Cr, V). Lead, although one of the elements of concern in the dusts (*19*), may have been somewhat less bioaccessible in respiratory fluids than other metals. Further work is needed to assess the potential bioaccesibility of metals such as Pb in the WTC dusts via ingestion exposure and such tests are currently under way. Results of exposure assessments and other health monitoring studies are required to determine whether any among this suite of potentially bioaccessible metals in the dusts were actually absorbed *in vivo* and, more importantly, had any deleterious health effects. For example, although biomonitoring of blood and urine in firefighters who responded to the WTC fire and collapse found statistically higher levels of urine Sb and blood Pb compared to controls, the levels were below workplace maximum exposure guidelines (*20*). However, not all of the potentially bioaccessible metals noted above were monitored.

A broad spectrum of minerals and materials are found in the dusts in inhalable to respirable size ranges (see Chapter 5), including particles of calcite, portlandite, gypsum, glass fibers, chrysotile asbestos, and metal-rich particles. At least some of these materials have been noted to be of health concern for their potential to trigger irritation of the respiratory tract (*17*) and other health effects (*19*). The SLF leach tests indicate that a variety of secondary mineral phases might also form through reactions of the dusts with respiratory tract fluids, including calcium carbonates (replacing calcium hydroxide) and calcium phosphates (through reactions of calcium released from calcium hydroxide and gypsum with phosphate in the fluids). Chemical speciation calculations indicate the SLF-WTC dust leachates are thermodynamically supersaturated with a wide variety of silicate minerals such as zeolites, amphiboles, and chrysotile. This raises the possibility that some silicate phases present in the dusts may not be amenable to clearance by particle dissolution in the respiratory tract as long as other more soluble and reactive particles remain, and that some secondary silicate phases might actually precipitate through reactions with respiratory tract fluids. Mineralogical characterization of particles in sputum and lung tissue samples would help determine if such mineral transformations and diminished dissolution rates actually occur *in vivo* in response to dust inhalation.

Conclusions

Chemical characterization studies provide insights into the chemical composition and chemical reactivity of dusts and debris left around lower Manhattan by the the World Trade Center collapse, as well as insights into the chemical processes by which the dusts reacted afterward with rainfall and water used in fire fighting or street cleaning. The complex makeup and heterogeneous nature of the dusts resulted in a general lack of consistent discernible variations in their chemical composition and chemical reactivity, either as a function of location around lower Manhattan or as a function of particle size. In contrast, there do appear to be differences in chemical composition, alkalinity, and soluble components between indoor dust samples and most outdoor dust samples that resulted from the variable interactions of the outdoor dust samples with water prior to sample collection. Such interactions reduced the alkalinity and reactivity of the dust (by leaching reactive calcium hydroxide and precipitating calcium carbonate), and partially removed soluble components of the dusts such as calcium sulfate and some metals. Dusts generated during the initial collapse and dusts subsequently resuspended from indoor deposits and some outdoor deposits may have also been chemically reactive as they came into contact with water-based respiratory tract and gastrointestinal fluids via inhalation and ingestion exposure routes. These studies identify mineralogical and chemical characteristics of potential concern that can be evaluated by health workers as they continue to assess health effects resulting from short-term exposure to the dust clouds generated by the collapse or longer-term exposure via resuspension of settled dusts.

References

1. Meeker, G. P.; Sutley, S. J.; Brownfield, I. K.; Lowers, H. A.; Bern, A. M.; Swayze, G. A.; Hoefen, T. M.; Plumlee, G. S.; Clark, R. N.; and Gent, C. A., Paper presented in ACS Symposium: *Urban Aerosols and their Impact: Lessons Learned from The World Trade Center Tragedy*, New York City, Sept. 10, 2003. *ACS Division of Environmental Chemistry Preprints of Extended Abstracts, paper 133,* **2003,** *43 (2)*, 1354.
2. Clark, R.N.; Green, R.O.; Swayze, G.A.; Meeker, G.; Sutley, S.; Hoefen, T.; Livo, K.E.; Plumlee, G.; Pavri, B; Sarture, C.; Wilson, S.; Hageman, P.; Lamothe, P.; Vance, J.S.; Boardman, J.; Brownfield, I.; Gent, C.; Morath, L.C.; Taggart, J.; Theodorakos, P.; Adams, M. *Environmental studies of the World Trade Center area after the September 11, 2001 attack;* Open-File Report 01-0429; U.S. Geological Survey, **2001,** [http://pubs.usgs.gov/of/2001/ofr-01-0429/].
3. Clark, R.N.; Swayze, G.; Hofen, T.; Livo, E.; Sutley, S.; Meeker, G.; Plumlee, G.; Brownfield, I.; Lamothe, P.J.; Gent, C.; Morath, L.; Taggart, J.; Theodorakos, P.; Adams, M.; Green, R.; Pavri, B.; Sarture, C.; Vance, S.; Boardman, Paper presented in ACS Symposium: *Urban Aerosols and their Impact: Lessons Learned from The World Trade Center Tragedy*, New York City, Sept. 10, 2003. *ACS Division of Environmental Chemistry Preprints of Extended Abstracts, paper 110,* **2003,** *43 (2)*, 1324.
4. *Analytical methods for chemical analysis of geologic and other materials;* Taggart, J.E., Jr., Ed.; Open-file Report 02-0223, version 5.0; U.S. Geological Survey [http://pubs.usgs.gov/of/2002/ofr-02-0223/].
5. Hageman, P.L.; Briggs, P. H. *Proceedings from the Fifth International Conference on Acid Rock Drainage, Denver, Colorado, May 21-24, 2000;* Soc. Mining, Metallurgy, and Exploration Inc., **2000,** 1463-1475.
6. EPA *Method 1312, Synthetic precipitation leaching procedure.* **1994,** [http://www.epa.gov/SW-846/1_series.htm].
7. Ficklin, W.H.; Mosier, E.L. In *The Environmental Geochemistry of Mineral Deposits, Part A: Processes, Techniques, and Health Issues;* Rev. Econ. Geol.; Soc. Econ. Geol.: Littleton, CO, **1999,** *6a*, 249-265.
8. *Alkalinity calculator;* U.S. Geological Survey [http://oregon.usgs.gov/alk/]
9. Bethke, C.M. *Geochemical Reaction Modeling: Concept and Applications;* Oxford University Press: Oxford, Great Britain, **1996**; 397 pp.
10. Plumlee, G.S.; Ziegler, T.L. *Treatise on Geochemistry;* Elsevier: Amsterdam, **2003**, *9*, 263-310.

11. Bauer, J.; Mattson, S.M.; Eastes, W. *In-vitro acellular method for determining fiber durability in simulated lung fluid,* **1997,** [http://my.ohio.voyager.net/~eastes/fibsci/papers.html].
12. Chatfield, E.J.; Kominsky, J.R. *Summary report: characterization of particulate found in apartments after destruction of the World Trade Center,* **2001,** 42 p.
13. Lioy, P.; Weisel, C.P.; Millette, J.R.; Eisenreich, S.; Vallero, D.; Offenberg, J.; Buckley, B.; Turpin, B.; Zhong, M.; Cohen, M.D.; Prophete, C.; Yang, I.; Stiles, R.; Chee, G.; Johnson, W.; Porcja, R.; Alimokhtari, S.; Hale, R.C.; Weschler, C.; Chen, L.C. *Env. Health Persp.* **2002,** *110,* 703-714.
14. Shacklette, H.T.; Boerngen, J.G. *Element concentrations in soils and other surficial materials of the conterminous United States;* Professional Paper 1270; U.S. Geological Survey, **1984,** 105 p.
15. NIST, *Certificates of analysis, Standard Reference Material SRM 1648,* [http://patapsco.nist.gov/srmcatalog/common/view_detail.cfm?srm=1648].
16. NIST, *Certificate of analysis, Standard Reference Material SRM-1649a,* [http://patapsco.nist.gov/srmcatalog/common/view_detail.cfm?srm=1649a].
17. McGee, J.K.; Chen, L.C.; Cohen, M.D.; Chee, G.R.; Prophete, C.M.; Haykal-Coates, N.; Wasson, S.J.; Conner, T.L.; Costa, D.L.; Gavett, S.H. *Env. Health Persp.* **2003,** *111,* 972-980.
18. Landrigan, P.J.; Lioy, P.J.; Thurston, G.; Ng, S.P. *Env. Health Persp.* **2004,** *112,* 731-739 [http://ehp.niehs.nih.gov/docs/2004/112-6/toc.html].
19. NYCDHMH-ATSDR, *Final technical report of the public health investigation to assess potential exposures to airborne and settled surface dust in residential areas of lower Manhattan;* New York City Dept. Health and Mental Hygiene, Agency for Toxic Substances Disease Registry, **2002,** 127 p.
20. Edelman, P.; Osterloh, J.; Pirkle, J.; Caudill, S.P.; Grainger, J.; Jones, R.; Blount, B.; Calafat, A.; Turner, W.; Feldman, D.; Baron, S.; Bernard, B.; Lushniak, B.D.; Kelley, K.; Prezant, D. *Env. Health Persp.* **2003,** *111,* 1906 – 1911.

Appendix I

Analytical Results for WTC Bulk Outdoor Dust (2-7, 12, 14-18, 21, 22, 24, 25, 27, 28, 30, 32, 34, and 35), Bulk Indoor Dust (20 and 36) and Girder Coating (8 and 9) Samples.

Element	*2*	*3*	*4*	*4dup*	*5*	*6*	*7*	*12*
Si	21.2	26.3	-	-	11.4	11.4	-	-
Ca	15.0	9.6	20.8	21.0	20.9	20.6	21.1	20.3
LoI	8.0	13.6	-	-	19.6	19.6	-	-
Mg	3.1	2.2	2.6	2.5	2.7	2.7	2.4	2.9
S	1.3	0.9	-	-	-	-	-	-
Fe	4.1	2.2	1.4	1.4	1.4	1.4	1.4	2.0
Al	4.1	2.7	2.8	2.7	2.7	2.7	2.6	3.2
Org C	1.0	3.6	-	-	-	-	-	-
Cbt C	1.2	1.6	-	-	-	-	-	-
Na	0.82	0.76	0.59	0.53	0.50	0.50	0.45	0.43
K	0.63	0.69	0.55	0.54	0.46	0.47	0.50	0.55
Ti	0.39	0.25	0.27	0.25	0.24	0.24	0.25	0.34
Mn	0.15	0.08	0.11	0.10	0.10	0.10	0.10	0.15
P	305	524	415	498	305	305	450	362
Ba	765	376	402	383	-	-	360	439
Sr	1000	409	687	733	-	-	720	739
Zn	2990	1200	1460	1540	-	-	1260	2780
Pb	710	176	203	268	-	-	183	277
Cu	438	142	167	212	-	-	118	388
Ce	108	51	91	85	-	-	82	108
Y	59	30	53	50	-	-	50	63
Cr	224	98	116	116	-	-	114	123
Ni	88	31	22	22	-	-	22	30
La	51	26	48	43	-	-	42	54
Sb	52	26	35	41	-	-	28	69
V	39	43	33	33	-	-	30	35
Mo	25	15	18	18	-	-	15	29
Li	27	17	24	23	-	-	24	28
Th	11	6	8	8	-	-	8	10
Rb	21	24	20	20	-	-	20	22
Co	13.9	8.4	5.8	5.8	-	-	5.2	7.2
Nb	11.0	7.8	1.2	< 0.1	-	-	< 0.1	3.0
Sc	8.8	6.6	7.2	7.0	-	-	6.9	7.9
U	3.9	2.0	3.1	3.0	-	-	3.0	3.7
Cd	7.3	3.2	4.7	5.2	-	-	3.8	5.5
As	6.8	3.7	4.2	4.3	-	-	4.1	5.0
Ga	6.0	5.4	4.3	4.3	-	-	4.0	4.9
Be	3.7	2.2	2.8	2.7	-	-	2.7	3.5
Ag	1.2	3.8	< 2	< 2	-	-	< 2	4.4
Cs	0.73	0.76	0.69	0.71	-	-	0.73	0.78
Bi	0.50	0.68	0.58	0.73	-	-	0.38	0.78
Tl								

Element	*14*	*15*	*16*	*16dup*	*17*	*17dup*	*18*	*21*
Si	15.3	13.6	17.0	-	16.0	-	-	12.8
Ca	17.7	18.6	13.4	16.1	17.0	17.0	21.4	18.9
LoI	18.1	17.3	22.8	-	15.9	-	-	21.2
Mg	2.8	2.6	1.8	2.1	2.1	1.7	2.1	2.7
S	4.3	5.4	3.7	-	-	-	-	5.1
Fe	1.9	1.9	1.9	1.8	1.7	1.8	1.4	1.5
Al	2.9	2.6	2.3	2.3	2.3	2.3	2.3	2.7
Org C	3.1	2.3	2.5	-	-	-	-	4.0
Cbt C	1.5	1.5	1.5	-	-	-	-	1.4
Na	0.59	0.66	0.87	0.61	0.93	0.57	0.42	0.50
K	0.56	0.49	0.69	0.47	0.54	0.50	0.46	0.50
Ti	0.31	0.25	0.26	0.22	0.25	0.20	0.21	0.24
Mn	0.12	0.10	0.07	0.09	0.07	0.07	0.08	0.12
P	218	262	218	578	218	248	454	262
Ba	461	405	3670	298	-	260	317	460
Sr	643	736	3130	563	-	645	805	787
Zn	1570	1110	1410	1280	-	1150	1190	1500
Pb	276	152	208	109	-	271	259	278
Cu	242	367	307	133	-	189	133	153
Ce	69	65	132	71	-	59	76	77
Y	47	46	31	41	-	31	43	55
Cr	116	129	95	96	-	80	106	104
Ni	29	33	31	20	-	19	21	31
La	35	33	70	37	-	30	40	39
Sb	40	30	148	35	-	596	29	33
V	31	27	25	29	-	29	27	28
Mo	19	12	10	16	-	17	15	9
Li	23	22	18	20	-	18	21	23
Th	8	7	5	5	-	6	7	8
Rb	25	22	22	19	-	20	18	21
Co	7.1	6.5	6.5	5.2	-	5.9	4.9	5.3
Nb	9.1	7.6	6.6	< 0.1	-	< 0.1	< 0.1	9.0
Sc	6.1	5.9	4.4	5.6	-	5.1	5.8	6.2
U	2.9	2.7	2.3	2.5	-	2.1	2.7	3.2
Cd	3.4	4.0	3.0	3.6	-	3.4	5.9	4.6
As	5.1	4.0	4.3	5.0	-	6.0	3.8	3.6
Ga	4.1	3.9	4.3	4.0	-	4.1	3.8	3.9
Be	2.9	2.4	1.8	2.2	-	2.0	2.2	2.9
Ag	1.2	1.4	1.5	< 2	-	3.4	< 2	2.4
Cs	0.88	0.78	0.87	0.70	-	0.84	0.69	0.76
Bi	0.56	0.25	0.28	0.25	-	1.87	0.40	0.50
Tl	0.11	0.11	0.12	0.10	-	0.13	0.09	0.10

Element	*22*	*24*	*24dup*	*25*	*27*	*28*	*30*	*30dup*
Si	17.0	-	-	13.2	15.2	13.8	15.1	-
Ca	16.8	20.9	21.5	20.4	19.5	19.7	19.7	20.2
LoI	15.3	-	-	17.5	14.4	16.7	17.5	-
Mg	2.8	2.7	2.7	3.3	3.0	2.8	3.5	3.3
S	3.7	-	-	4.0	4.3	4.6	-	-
Fe	2.8	1.5	1.5	1.3	1.7	1.8	1.8	2.0
Al	2.8	2.9	2.8	3.3	3.0	2.9	3.6	3.4
Org C	2.6	-	-	2.9	2.0	2.4	-	-
Cbt C	1.3	-	-	1.9	1.8	1.7	-	-
Na	0.83	0.45	0.45	0.62	0.62	0.76	0.71	0.59
K	0.52	0.56	0.56	0.56	0.50	0.54	0.56	0.56
Ti	0.29	0.28	0.29	0.29	0.29	0.26	0.29	0.32
Mn	0.12	0.12	0.13	0.15	0.12	0.12	0.14	0.14
P	262	594	498	262	262	218	393	545
Ba	452	439	444	624	470	491	-	501
Sr	710	726	737	695	701	711	-	868
Zn	1380	2350	2360	1910	1650	1720	-	2370
Pb	452	355	365	756	204	234	-	242
Cu	130	232	238	251	188	218	-	222
Ce	72	88	92	85	78	75	-	117
Y	48	53	54	62	55	54	-	68
Cr	111	121	137	134	126	106	-	142
Ni	31	24	25	39	39	26	-	35
La	35	44	46	44	40	38	-	59
Sb	28	55	61	66	50	52	-	44
V	30	34	38	31	30	29	-	40
Mo	7	30	32	31	27	42	-	17
Li	23	25	25	29	25	25	-	30
Th	9	8	8	10	9	8	-	11
Rb	21	23	24	24	22	23	-	22
Co	6.3	9.9	7.5	7.4	6.2	5.9	-	7.2
Nb	9.2	< 0.1	< 0.1	11.0	11.0	10.0	-	< 0.1
Sc	6.2	6.9	7.0	7.1	6.6	6.2	-	9.3
U	3.1	3.2	3.2	3.8	3.4	3.3	-	4.1
Cd	3.8	8.0	13.0	7.5	5.0	5.2	-	7.0
As	6.6	4.4	4.9	4.2	5.0	4.8	-	4.8
Ga	4.0	4.5	4.6	4.3	4.3	4.1	-	5.2
Be	2.9	2.9	3.0	3.6	3.2	3.1	-	3.8
Ag	1.4	2.8	< 2	1.4	1.4	1.7	-	2.9
Cs	0.76	0.80	0.84	0.83	0.77	0.76	-	0.79
Bi	0.43	0.69	1.04	0.67	0.40	0.48	-	1.28
Tl	0.10	0.11	0.10	0.10	0.09	0.11	-	0.11

Element	*32*	*34*	*34dup*	*35*	*20*	*36*	*8*	*9*
Si	-	12.2	-	-	14.2	11.7	15.0	15.5
Ca	22.0	20.5	21.0	21.4	19.4	21.3	20.7	26.0
LoI	-	18.5	-	-	15.7	16.9	15.8	13.0
Mg	2.5	3.0	2.8	2.6	2.6	2.9	6.9	3.2
S	-	-	-	-	5.5	5.8	1.4	1.2
Fe	1.6	1.4	1.5	1.6	1.3	1.4	1.3	0.6
Al	2.6	3.0	2.8	2.8	2.5	2.9	2.9	3.6
Org C	-	-	-	-	2.7	2.3	2.5	2.5
Cbt C	-	-	-	-	1.3	1.5	1.9	1.9
Na	0.40	0.50	0.44	0.40	1.16	0.58	0.12	0.16
K	0.48	0.51	0.50	0.48	0.46	0.46	0.28	0.32
Ti	0.25	0.25	0.28	0.28	0.25	0.23	0.21	0.28
Mn	0.11	0.12	0.12	0.12	0.10	0.11	0.14	0.19
P	295	305	432	361	175	218	87	87
Ba	376	-	382	376	390	438	317	472
Sr	764	-	799	785	706	823	444	378
Zn	1510	-	1790	1830	1330	1400	57	101
Pb	137	-	309	253	153	159	9	12
Cu	117	-	132	194	176	95	10	13
Ce	91	-	95	96	62	70	202	356
Y	53	-	55	55	44	53	134	243
Cr	123	-	114	106	94	107	153	87
Ni	23	-	25	34	30	29	202	23
La	46	-	48	49	31	36	102	175
Sb	30	-	40	40	39	34	1	1
V	31	-	32	31	25	29	31	40
Mo	10	-	20	18	19	16	1	1
Li	24	-	26	25	22	25	25	36
Th	9	-	9	9	7	9	18	31
Rb	19	-	21	20	19	21	8	8
Co	5.2	-	5.9	5.5	5.0	5.3	12.3	1.7
Nb	< 0.1	-	< 0.1	< 0.1	8.0	9.0	4.4	6.3
Sc	7.2	-	7.4	7.3	5.4	6.4	9.2	9.8
U	3.2	-	3.4	3.4	2.7	3.2	4.7	7.6
Cd	7.3	-	6.2	5.2	4.2	5.8	0.1	0.2
As	4.0	-	4.0	4.3	3.5	3.8	< 2	< 2
Ga	3.9	-	4.3	4.2	3.6	4.0	2.8	4.2
Be	2.9	-	3.1	3.2	2.5	3.1	4.0	4.2
Ag	2.6	-	2.4	2.2	3.5	1.6	1.8	1.0
Cs	0.68	-	0.73	0.69	0.72	0.78	0.18	0.22
Bi	0.48	-	1.17	0.51	0.64	0.82	0.01	0.01
Tl	0.10	-	0.10	0.09	0.09	0.09	0.02	0.02

NOTE: Units for Si through Mn are percent dry weight. Units for P through Tl are ppm. LoI is loss on ignition, (-) indicates parameter not measured, dup is duplicate. Org C is organic carbon, Cbt C is carbonate carbon.

Appendix II

Analytical Results for DIW Leachates of WTC Bulk Outdoor Dust (2-7, 12, 14-18, 21, 22, 24, 25, 27, 28, 30, 32, 34, and 35), Bulk Indoor Dust (20 and 36) and Girder Coating (8 and 9) Samples.

	2	*2dup*	*3*	*4*	*5*	*6*	*10*	*11*
pH	10.1	10.1	9.51	8.56	9.9	9.65	8.66	9.16
S Cond	1.58	1.5	1.31	2.17	1.9	2.01	1.96	1.96
Alkalinity	-	40	-	-	-	-	-	-
Sulfate	834	890	694	1400	1210	1040	1390	1360
Ca	388	385	314	584	577	523	546	538
Bicarbonate	-	48	-	-	-	-	-	-
Na	6.1	4.69	2.84	33.2	7.69	5.76	11.5	6.69
Cl	7.8	0.4	3.7	-	-	-	-	-
Mg	1.75	1.77	2.83	2.98	3.2	3.65	3.35	2.79
K	6	4.2	3.8	12.5	7.71	6.33	2.84	5.58
Nitrate	1.5	1.1	0.5	-	-	-	-	-
Si	5.8	4.2	4.5	3.5	8.1	5.9	2.7	2.9
F	<.8	-	<.8	-	-	-	-	-
Sr	834	782	561	1190	1150	1100	890	879
Al	111	43.2	44.6	13.2	24.3	26	7.8	11.1
P	50	40	100	100	40	30	20	40
Fe	< 50	< 50	< 50	< 50	< 50	< 50	< 50	< 50
Mo	56.8	43.7	14	57.9	45.7	42.2	8.9	12.4
Cu	19.2	16.4	19.8	110	22.4	13.5	19.8	16.4
Zn	10.7	2.7	7.7	54.2	15.6	20.9	104	11.3
Ba	36.5	44.6	23.4	49.4	38.3	36	56.6	39.6
Sb	33.1	26.2	22.9	22.7	46.3	42	18.6	17.5
Cr	25.9	19.7	9	20.3	25.1	18.2	16	17.8
Mn	1	0.9	3.2	2.1	2	3.8	1.8	2.3
Ni	18.1	6.9	14.4	< 0.4	21.4	19.4	< 0.4	< 0.4
Li	11.2	8.4	4.1	11.6	11.2	9.4	8.6	8
Ti	17.9	10.2	13.4	16.4	18.9	18.7	11.7	13.8
Rb	12.6	7.7	8.1	10.7	12.4	12.9	3.6	11.7
V	6.2	4.4	6	11.8	11.8	8.8	8.2	9.3
Pb	0.64	0.06	0.5	0.4	0.5	0.51	0.2	0.3
Se	2.5	1.7	1	8.1	< 5	< 5	1.6	1
As	1	2	1	6.8	< 3	< 3	2	4
Sc	1.8	1	1.3	2.4	2.2	1.9	2.4	2.7
Co	1.2	0.83	0.72	2.0	1.0	1.02	< 0.02	< 0.02
Cd	0.44	0.44	0.26	0.62	1.1	0.82	0.98	0.85
Nb	< 0.02	0.1	0.03	1.2	0.1	0.1	1	0.9
Zr	0.07	< 0.05	0.1	0.97	0.5	0.3	0.3	0.3
Th	0.04	<0.005	0.09	0.26	0.8	0.37	< 0.2	< 0.2
W	-	0.6	-	0.67	-	-	0.65	0.58
Ge	0.07	0.09	0.09	0.65	0.1	0.08	0.45	0.45
U	0.03	0.03	0.15	0.31	0.08	0.09	0.14	0.14
Ga	0.23	0.1	0.1	< 0.05	0.1	0.1	< 0.05	< 0.05
Bi	<0.005	0.03	0.01	0.28	<0.005	<0 .005	< 0.2	< 0.2
Tl	< 0.05	0.06	< 0.05	0.2	0.2	0.06	< 0.1	< 0.1
Y	< 0.01	0.06	< 0.01	0.05	0.08	0.08	0.03	0.02
Be	< 0.05	< 0.05	< 0.05	< 0.05	< 0.05	< 0.05	< 0.05	< 0.05
Cs	0.08	0.04	0.05	0.04	0.04	0.04	< 0.02	0.08
Ce	< 0.01	< 0.01	0.03	0.04	0.02	0.02	0.01	0.02
La	< 0.01	< 0.01	0.01	0.02	< 0.01	0.01	< 0.01	< 0.01
Hg	-	0.005	-	-	0.018	0.007	-	-
Ag	< 3	< 3	< 3	< 3	-	-	< 3	< 3

	12	*13*	*14*	*15*	*16*	*16dup*	*17*	*18*
pH	9.85	11.16	9.68	10	8.22	9	9.47	12.04
S Cond	1.95	2.15	2.03	2.01	2.08	2.02	1.96	2.79
Alkalinity	-	-	-	-	-	20	-	-
Sulfate	1360	1310	1250	1230	1350	1460	1110	1440
Ca	547	548	544	528	526	565	517	649
Bicarbonate	-	-	-	-	-	24	-	-
Na	2.51	11.5	3.05	2.65	5.09	7.36	4.81	24.6
Cl	-	-	5.1	3.4	8.5	<.08	-	-
Mg	2.82	1.46	3.52	1.71	20.2	22.4	2.54	0.4
K	3.37	8.07	6.9	5.9	9.2	6.8	4.83	8.39
Nitrate	-	-	1.4	1.5	<1.6	<.08	-	-
Si	6.4	2.8	6.4	4.9	4.3	3.2	2	3.3
F	-	-	<1.6	<1.6	<1.6	-	-	-
Sr	985	1140	1230	1060	999	926	1000	1130
Al	21.7	238	30.3	53.9	6.33	12	50.6	505
P	< 10	50	50	20	30	60	< 10	60
Fe	< 50	< 50	< 50	< 50	< 50	200	< 50	< 50
Mo	21.8	26.3	30.8	10.6	46.3	31.7	35.5	44.6
Cu	14.4	18.6	11.4	10.2	15.6	36.6	14.4	25.1
Zn	6.2	15	11.6	10.6	24.1	48.9	12.7	23.7
Ba	40.5	36.2	45.1	28.9	23.2	52.6	17.5	33.7
Sb	33.6	11	35.9	15.3	28.6	19.2	11.2	14.6
Cr	21	37.6	31.4	42	20.8	17.6	17.4	48.2
Mn	2.1	1.7	2.3	1.2	35.1	7.7	1.7	0.5
Ni	< 0.4	< 0.4	25.2	22.2	25	12.6	21.9	< 0.4
Li	9	16.5	9.8	6.4	11.2	8	6.9	13.5
Ti	13.6	13.6	25.7	24.8	25.1	18.3	19.4	16.9
Rb	5.4	9.5	14.1	12.4	14.1	7.6	8.9	9.3
V	13.7	10.5	9.7	6.6	6.5	6.2	2.7	11.7
Pb	0.4	0.5	0.97	1.5	0.4	0.3	0.3	1.5
Se	1.6	5.4	1.9	< 1	3	5.1	< 5	4.1
As	2	4.9	1	< 1	2	2	< 3	2
Sc	3.9	2.5	1.9	1.5	1.2	1	0.8	2.8
Co	< 0.02	0.1	1.15	1.0	1.3	0.81	1.0	0.04
Cd	0.94	0.51	0.37	0.55	0.39	0.31	0.47	0.27
Nb	0.78	0.82	0.02	0.02	0.02	0.09	0.07	0.8
Zr	0.3	0.2	0.08	0.2	0.2	< 0.05	0.09	0.4
Th	< 0.2	< 0.2	0.06	0.06	0.24	0.008	0.2	< 0.2
W	0.71	0.64	-	-	-	0.22	-	0.58
Ge	0.43	0.45	0.07	0.06	0.1	0.05	0.05	0.4
U	< 0.1	< 0.1	0.06	0.03	0.52	0.1	0.01	< 0.1
Ga	0.2	0.2	0.1	0.1	0.05	0.03	0.08	0.34
Bi	< 0.2	< 0.2	0.01	0.01	0.01	0.04	<0.005	< 0.2
Tl	< 0.1	< 0.1	< 0.05	< 0.05	< 0.05	< 0.05	< 0.05	< 0.1
Y	0.02	0.02	0.11	0.1	0.1	0.09	0.05	0.03
Be	< 0.05	< 0.05	< 0.05	< 0.05	< 0.05	< 0.05	< 0.05	< 0.05
Cs	0.03	0.04	0.06	0.05	0.05	0.02	0.03	0.02
Ce	< 0.01	0.02	0.01	0.01	0.02	0.02	0.02	0.01
La	< 0.01	< 0.01	< 0.01	< 0.01	0.01	0.02	< 0.01	< 0.01
Hg	-	-	-	-	-	0.015	17	18
Ag	< 3	< 3	< 3	< 3	< 3	< 3	9.5	12.0

	19	*21*	*22*	*23*	*24*	*25*	*25dup*	*26*
pH	9.83	9.98	10.4	9.65	10.94	9.37	-	9.0
S Cond	2.31	2.02	2.02	1.98	2.31	2.16	-	2.1
Alkalinity	-	-	-	-	-	-	49	-
Sulfate	1350	1270	1170	1370	1400	1240	1600	1400
Ca	553	549	529	541	596	558	624	555
Bicarbonate	-	-	-	-	-	-	59	-
Na	55	4.11	5.69	3.44	20	12.9	16.3	20.1
Cl	-	7.8	8.1	-	-	37	<.08	-
Mg	1.84	2.61	2.12	3.21	3.54	6.15	8.06	4.7
K	10.1	7.7	5.2	4.36	9.37	11.7	11.2	14.2
Nitrate	-	2.4	1.5	-	-	11	9.3	-
Si	1.8	5.8	5.4	3.2	6.3	8.1	7	4.8
F	-	<1.6	<1.6	-	-	<1.6	-	-
Sr	916	1020	943	1110	1340	1240	1180	1050
Al	26.8	53.6	153	7.7	77.6	23.8	7.1	5.7
P	40	40	50	< 10	60	60	60	200
Fe	< 50	< 50	< 50	< 50	< 50	< 50	< 50	< 50
Mo	30.9	10.7	7.4	23.6	124	140	116	35.1
Cu	28.8	6.2	9.6	31.2	33.4	39	44.9	110
Zn	8.6	9.6	6.5	11.9	45	11	8.7	21.2
Ba	35.9	33.9	32.2	52.5	57.6	58.4	83.7	26.9
Sb	9.9	21.2	17	29.6	43.1	73.6	75.6	35.9
Cr	13.1	19.3	27.7	16.9	55.8	24.4	24.7	21.2
Mn	0.4	1.4	1	1.7	3.8	4.9	5.3	3.2
Ni	< 0.4	24.6	24.8	< 0.4	< 0.4	32.1	16.7	< 0.4
Li	8.3	7.4	7.8	11.3	24.5	29.7	26.4	13.2
Ti	17.2	25.9	24	14.1	17	25.5	19.1	15.8
Rb	10.4	14.1	10	7.8	10.6	19.3	13.7	13.2
V	7	8	5.5	12	16.8	13.2	11.5	16.3
Pb	0.65	1.1	0.68	0.2	0.55	11.5	9.8	0.4
Se	4.2	2.2	1.6	3.4	10.9	7.4	5.6	3.2
As	2	1	< 1	3.1	4.6	3.2	3	3.6
Sc	2.4	1.7	1.5	2.8	3.8	2.2	1.8	3.6
Co	0.31	1.2	0.98	< 0.02	0.16	3.2	2.7	1.3
Cd	0.91	0.25	0.16	0.63	1.72	1.6	1.5	2.0
Nb	0.87	0.04	0.04	0.56	0.65	0.03	0.1	0.57
Zr	0.62	0.2	0.2	0.2	0.5	0.2	0.1	0.3
Th	< 0.2	0.1	0.17	< 0.2	< 0.2	0.13	<0.005	< 0.2
W	< 0.5	-	-	< 0.5	0.74	-	0.63	0.67
Ge	0.47	0.05	0.04	0.45	0.48	0.2	0.2	0.6
U	< 0.1	0.02	0.02	0.21	< 0.1	0.13	0.17	0.24
Ga	0.06	0.2	0.27	< 0.05	0.41	0.1	0.1	< 0.05
Bi	< 0.2	0.006	0.01	< 0.2	< 0.2	0.01	0.02	< 0.2
Tl	< 0.1	< 0.05	< 0.05	< 0.1	< 0.1	< 0.05	< 0.05	< 0.1
Y	0.02	0.1	0.07	0.01	0.04	0.11	0.08	0.02
Be	< 0.05	< 0.05	< 0.05	< 0.05	< 0.05	< 0.05	< 0.05	< 0.05
Cs	0.04	0.06	0.05	0.03	0.03	0.08	0.05	0.06
Ce	0.03	0.01	0.02	0.01	0.03	0.02	0.01	0.01
La	< 0.01	< 0.01	< 0.01	< 0.01	< 0.01	0.01	< 0.01	< 0.01
Hg	19	21	22	23	24	-	-	-
Ag	9.8	10.0	10.4	9.7	10.9	< 3	< 3	< 3

	27	*28*	*29*	*30*	*31*	*32*	*33*	*34*
pH	10	9.93	9.67	9.63	8.61	10.07	9.72	9.8
S Cond	2.31	2.02	2.08	1.9	2.0	2.02	1.99	2.02
Alkalinity	-	-	-	-	-	-	-	-
Sulfate	1240	1250	1370	986	1410	1380	1340	1180
Ca	568	553	562	461	533	554	549	524
Bicarbonate	-	-	-	-	-	-	-	-
Na	12.7	5.57	8.87	4.28	3.56	8.25	2.48	2.76
Cl	52	12	-	-	-	-	-	-
Mg	2.01	2.85	3.54	5.27	17.6	2	2.38	3.2
K	9.7	11.3	5.77	3.22	4.35	6.94	3.82	5.06
Nitrate	3	3.2	-	-	-	-	-	-
Si	7.2	8.6	6.5	5	3.4	5.4	3.8	4.2
F	<1.6	<1.6	-	-	-	-	-	-
Sr	1440	1160	1090	1540	1120	1070	885	1070
Al	33.4	45	16.4	22.6	5.3	10.4	13.4	27.8
P	40	40	40	20	30	30	100	20
Fe	< 50	< 50	< 50	< 50	< 50	< 50	< 50	< 50
Mo	126	50.4	65.8	30.6	21.8	15.2	14.2	27.9
Cu	21.5	9	24.2	14	31.4	14	26.3	10.6
Zn	8.4	12.1	8.1	5.3	20.6	48.6	9.4	12.2
Ba	38.6	43.5	46	53.9	65.5	81.3	45.4	32.4
Sb	25.5	43.6	37.6	35.5	13.6	22.1	16.2	33.5
Cr	15.7	34.5	26.3	26.1	15.1	27	16.7	16.2
Mn	1	2	1.9	3.3	15.6	< 0.2	1.7	1.8
Ni	27	25.9	< 0.4	18.1	< 0.4	< 0.4	< 0.4	20.7
Li	24.3	11.2	11.6	9.6	12.5	9.8	4.7	7.9
Ti	25	26.3	16.1	16.5	14.4	16.9	15	18.9
Rb	14.9	25	8.1	9.3	6	7.6	4.6	10.8
V	16.1	12.2	16	7.2	7.5	13.4	11.4	7
Pb	0.4	0.83	0.4	0.2	0.07	0.56	0.4	0.5
Se	8.8	3.5	1.6	< 5	1.2	2.4	1.3	< 5
As	3	2	3	< 3	2	< 1	1	< 3
Sc	2.1	2.5	3.8	1.6	2.5	3.2	2.6	1.4
Co	1.2	1.3	< 0.02	0.72	< 0.02	< 0.02	< 0.02	0.87
Cd	0.38	0.54	1.1	1.1	0.32	0.54	0.77	1.0
Nb	0.06	0.03	0.57	0.07	0.48	0.43	0.4	0.06
Zr	0.2	0.2	0.3	0.2	< 0.2	0.2	< 0.2	0.1
Th	0.16	0.08	< 0.2	0.12	< 0.2	< 0.2	< 0.2	0.1
W	-	-	0.75	-	< 0.5	< 0.5	< 0.5	-
Ge	0.05	0.08	0.47	0.09	0.55	0.46	0.55	0.08
U	0.008	0.04	0.11	0.09	0.34	< 0.1	0.22	0.03
Ga	0.2	0.2	0.1	0.1	< 0.05	0.09	< 0.05	0.1
Bi	0.01	0.007	< 0.2	< 0.005	< 0.2	< 0.2	< 0.2	< 0.005
Tl	< 0.05	< 0.05	< 0.1	< 0.05	< 0.1	< 0.1	< 0.1	< 0.05
Y	0.09	0.12	0.01	0.08	0.02	0.03	0.01	0.07
Be	< 0.05	< 0.05	0.1	< 0.05	< 0.05	< 0.05	< 0.05	< 0.05
Cs	0.05	0.1	0.04	0.04	0.02	0.02	0.02	0.03
Ce	0.02	0.02	< 0.01	0.02	0.01	0.02	0.02	0.02
La	< 0.01	0.02	< 0.01	0.01	0.02	0.01	0.01	< 0.01
Hg	-	-	-	0.012	-	-	-	-
Ag	< 3	< 3	< 3	-	< 3	< 3	< 3	-

	35	*20*	*20dup*	*36*	*36dup*	*8*	*9*
pH	9.69	11.8	12.3	11.8	12.4	INS	10.8
S Cond	2.05	3.41	3.51	3.4	3.9	INS	1.43
Alkalinity	-	-	460		560	-	-
Sulfate	1380	1320	1600	1640	1840	1090	674
Ca	554	718	835	888	913	528	336
Bicarbonate	-	-	552	-	671	-	-
Na	3.32	15.3	15.3	18.3	22.1	2.1	1.54
Cl	-	45	<.08	40	<.08	16	3
Mg	3.07	0.11	0.1	0.08	0.08	10.3	1.1
K	4.48	10.9	10.2	12.3	12.2	3	1
Nitrate	-	9.1	8	17	14	62	4.1
Si	4.6	3.4	3.2	3.2	3.5	6.7	11.3
F	-	<1.6	-	<1.6	-	<.8	<.8
Sr	935	1420	1420	1690	1490	990	758
Al	11.2	611	501	702	480	10.8	121
P	30	90	90	90	100	< 10	< 10
Fe	< 50	< 50	< 50	< 50	< 50	< 50	< 50
Mo	19.5	73.8	60.4	72.9	63.2	1.7	1.2
Cu	12	15.1	18.6	33.6	34.2	5.6	3.5
Zn	8	28.4	26.2	61.8	61.4	20.1	15.8
Ba	39.5	61.7	56.7	57.2	63.6	22.8	10.4
Sb	30.8	20.8	13.6	17.1	13.7	8.7	8.0
Cr	13.6	69.4	70.2	109	110	18	408
Mn	< 0.2	1.3	0.9	1.7	0.8	5.5	2.1
Ni	< 0.4	36.2	14.5	42.6	16.9	24.9	16.6
Li	7.2	18.5	16.3	19.5	18.2	1.3	0.3
Ti	15.1	25.5	19.3	28.4	21.7	24.9	15.3
Rb	6.77	17.7	12.3	20.8	14.2	3.5	1.4
V	10.8	6.5	4.8	7.8	4.8	13.8	14.4
Pb	0.3	5.8	4.3	10.9	10.2	0.4	0.3
Se	2.1	10.5	6.4	10.3	6.8	< 5	< 5
As	3	3.3	2	3.3	2	< 3	< 3
Sc	3.1	1.2	1.1	2.1	1.3	3.6	5.5
Co	< 0.02	1.8	1.4	2.2	1.5	1.3	0.75
Cd	0.69	0.18	0.12	0.18	0.12	0.02	0.02
Nb	0.4	0.08	0.1	0.05	0.1	0.08	< 0.02
Zr	0.3	0.4	0.1	0.4	0.1	3.7	0.2
Th	< 0.2	0.51	< 0.005	0.38	< 0.005	0.52	0.18
W	< 0.5	-	0.5	-	0.71	-	-
Ge	0.58	0.05	0.04	0.07	0.04	0.1	< 0.02
U	0.16	0.01	< 0.005	< 0.005	< 0.005	0.02	0.006
Ga	< 0.05	0.59	0.48	0.97	0.46	0.08	0.38
Bi	< 0.2	0.02	0.03	< 0.005	0.01	< 0.005	< 0.005
Tl	< 0.1	0.08	< 0.05	< 0.05	< 0.05	< 0.05	< 0.05
Y	0.03	0.13	0.11	0.16	0.13	0.31	0.27
Be	0.06	< 0.05	< 0.05	< 0.05	< 0.05	< 0.05	< 0.05
Cs	< 0.02	0.09	0.06	0.08	0.07	0.02	< 0.01
Ce	< 0.01	< 0.01	< 0.01	0.01	< 0.01	0.26	0.4
La	0.01	0.01	< 0.01	0.01	< 0.01	0.05	0.18
Hg	-	0.13	0.045	0.12	0.06	-	-
Ag	< 3	< 3	< 3	< 3	< 3	< 3	< 3

NOTE: Units for Alkalinity, sulfate, bicarbonate and Ca through F are ppm. Units for Sr through Cs are ppb. S Cond is specific conductance, dup is duplicate, (-) indicates parameter not measured or all measurements were below detection limits. Bicarbonate concentrations were calculated based on the measured alkalinity. Samples 2-6, 10-19,and 21-35 are outdoor dusts, 20 and 36 are indoor dusts, and 8 and 9 are girder coatings.

Chapter 13

World Trade Center Environmental Contaminant Database: A Publicly Available Air Quality Dataset for the New York City Area

Steven N. Chillrud[1], Alison S. Geyh[2], Diane K. Levy[3], Elsie M. Chettiar[3], and Damon A. Chaky[1]

[1]Lamont-Doherty Earth Observatory of Columbia University, 61 Route 9W, Palisades, NY 10964
[2]Johns Hopkins Bloomberg School of Public Health, 615 North Wolfe Street, Baltimore, MD 21205
[3]Statistical Analysis Center, Department of Biostatistics, 722 West 168th Street, 6th Floor, Columbia University Mailman School of Public Health, New York, NY 10032

Environmental data collected in response to the World Trade Center attack on September 11, 2001 and air quality data collected between from 1970 to December 31, 2003 in New York City and New Jersey are now available on the World Wide Web via the World Trade Center Environmental Contaminant Database (*1*). This database was developed by the Mailman School of Public Health at Columbia University and Bloomberg School of Public Health at John's Hopkins and is intended for use by individuals interested in having access to raw data (i.e., not summary statistics or averaged data) in order to explore the exposures related to the WTC disaster as well as general air quality issues in New York City and New Jersey. This website provides an easily accessible format to tabular data as well as various help files and sampling site maps to facilitate ease of use. Within the next year, the WTCECD should provide links to other (offsite) datasets relevant to WTC exposures, and will incorporate additional datasets provided by researchers at the National Institute of Environmental Health (NIEHS) Centers for Environmental Health.

Introduction

As a result of the World Trade Center (WTC) attack on September 11, 2001, a strong interest in environmental contaminant data for the New York Metropolitan area has emerged. There are currently multiple independent sources for these data, including:

- various federal, state and local agencies that have collected New York City air quality data in direct response to the disaster (*2-5*);
- the U.S. EPA's Aerometric Information Retrieval System (AIRS) dataset (*5*), which includes air quality readings for much of the New York metropolitan area dating from the present back to the 1970s.

While data from these various sources are public, they are not easily accessible to interested individuals. They reside in physically separate locations, in a variety of unfamiliar formats that do not lend themselves to being easily searched. In order to simplify the process of acquiring relevant air quality data, the World Trade Center Environmental Contaminant Database (WTCECD) has been developed by the Mailman School of Public Health at Columbia University and the Bloomberg School of Public Health at John's Hopkins.

The purpose of the WTCECD is to provide pre- and post-9/11 environmental data from the New York City area and the state of New Jersey in an easily accessible format at one location (*1*). This database is intended for use by individuals interested in having access to raw data (i.e., not summary statistics or averaged data) in order to explore exposures related to the WTC disaster and/or general air quality issues in New York City and New Jersey.

Discussion

The WTCECD currently provides data collected by the United States Environmental Protection Agency (U.S. EPA), New York State Department of Environmental Conservation (NYSDEC) and New York State Department of Health (NYSDOH) on samples of outdoor air, outdoor bulk dust, indoor air, and indoor dust wipes specifically in response to the WTC disaster. Figure 1 shows the location of the monitoring sites established in the New York City area by these agencies in response to the WTC disaster.

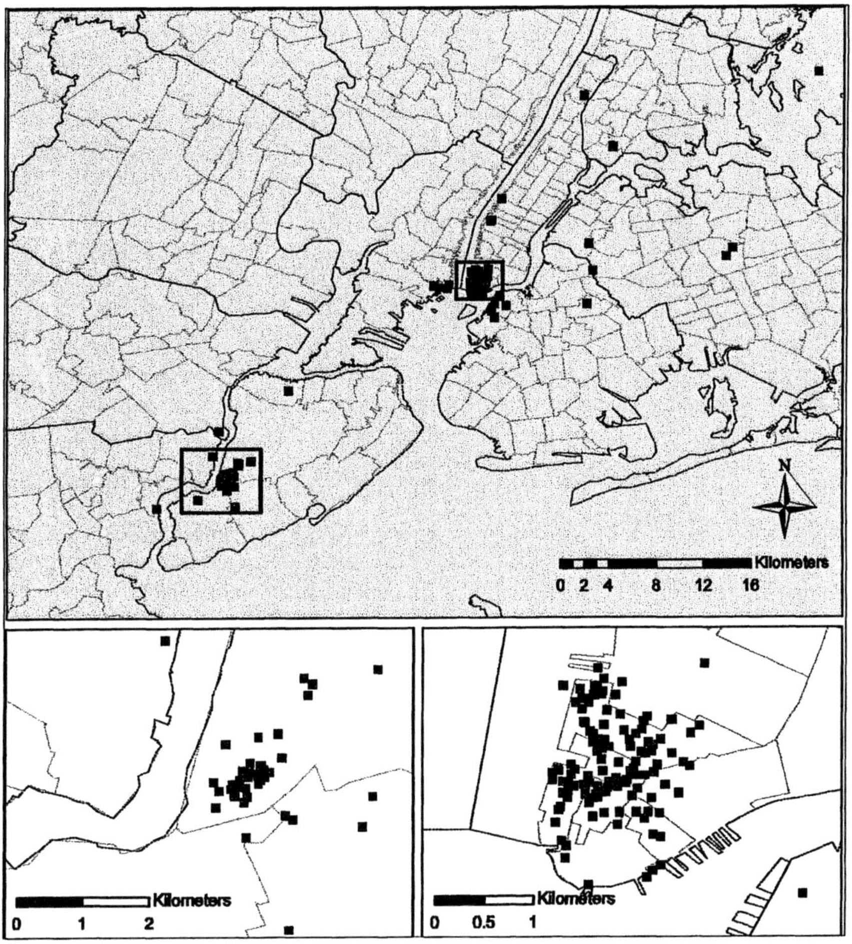

Figure 1. Monitoring sites established in the New York City area by the U.S. Environmental Protection Agency, the New York State Department of Environmental Conservation and the New York State Department of Health in response to the WTC disaster. Insets are Staten Island (lower left), and Lower Manhatten (lower right).

Approximately 250 parameters are included in the WTCECD including fine and coarse particulate matter mass (PM-2.5, PM-10), volatile organic compounds (VOCs), dioxins, polychlorinated biphenyls (PCBs), metals, and asbestos. The sampling frequencies and sampling periods for these different parameters varied greatly depending both on the parameter and/or the sampling site, ranging from hourly integrations, to 24-hr integrated samples that occurred daily to weekly, to muliple day integrations (2 to 10 days) that occurred with the same frequency as the sampling period, to one time grab samples.

The WTCECD also provides air quality data reported to the U.S. EPA's Aerometric Information Retrieval System (AIRS) from fixed-site air quality monitoring stations operated by local, state and federal environmental agencies from 1970 to December 31, 2003. The location of these sites are shown in Figures 2 and 3. The list of pollutants reported include ozone, sulfur dioxide, nitrogen dioxide, and carbon monoxide, on an hourly basis, lead every 3 to 6 days, total suspended particulates (TSP) and fine and coarse particulate matter mass (PM-10, PM-2.5), which were sampled at frequencies from hourly to daily to every 6th day, depending upon the time period and size cut. Also available are parameters measured on an irregular basis, which include a suite of VOCs, dioxins, PCBs, particle-associated metals, and some limited meteorological data.

The web-based interface allows users to search the WTCECD for one or more contaminant parameters from one or more specific sites. Users are guided through the search process by labels which enumerate each step (see Plate 1). Context sensitive help is available as the user moves through the process. Based on the criteria entered in each step, the system determines the valid criteria for successive steps. For example, when the user selects a specific site, the system responds by displaying only those pollutants that are measured at that site. This interactive search process saves the user from receiving messages from the system that no records are available to satisfy the search criteria.

All quality assurance/quality control (QA/QC) documentation resides with the reporting agencies. The WTCECD conducted no additional QA/QC on the data sets other than analyzing the AIRS and post 09/11 data for data discrepancies before loading it into the WTCECD. Any discrepancies identified were corrected with agency input. Examples of discrepancies identified include duplicate sample identifications and end sample date being before the begin sample date. Any questions on QA/QC issues should be directed to the reporting agencies listed for that data set.

A typical WTCECD query automatically provides sampling location, sampling date and other fields helpful for data interpretation (e.g. agency provided data quality flags). Additional fields are available and may be selected.

Query results are provided for download as either comma or tab-delimited text files, and may be date-limited prior to download.

Within the next year, the WTCECD should provide links to other (offsite) datasets relevant to WTC exposures, and will incorporate additional datasets provided by researchers at the National Institute of Environmental Health (NIEHS) Centers for Environmental Health.

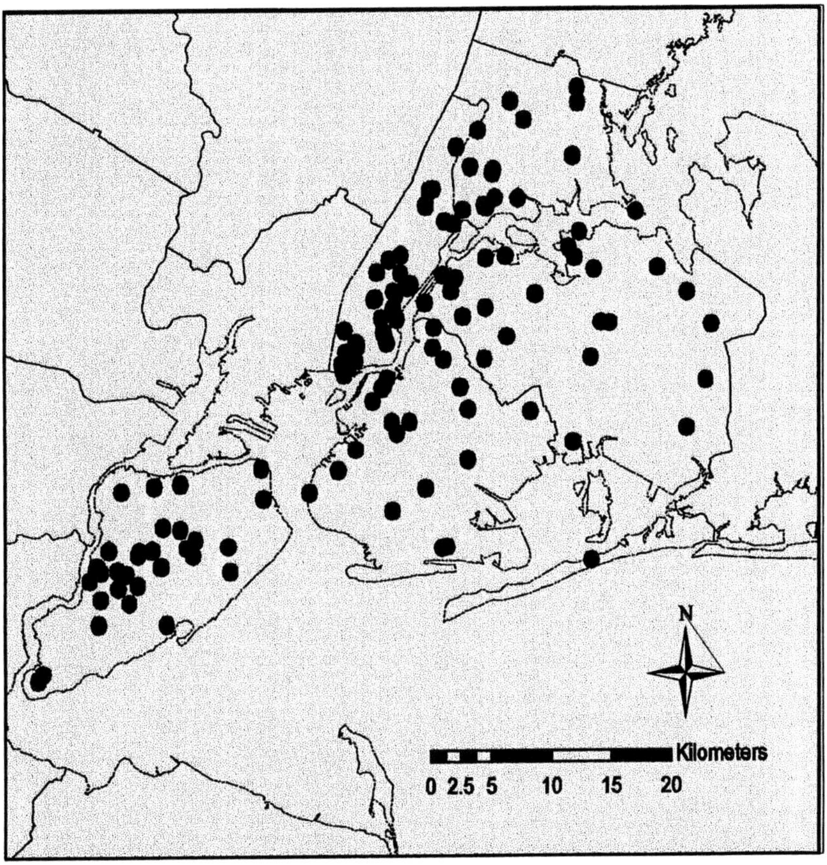

Figure 2. The U.S. Environmental Protection Agency's Aerometric Information Retrieval System (AIRS) monitoring sites in New York City.

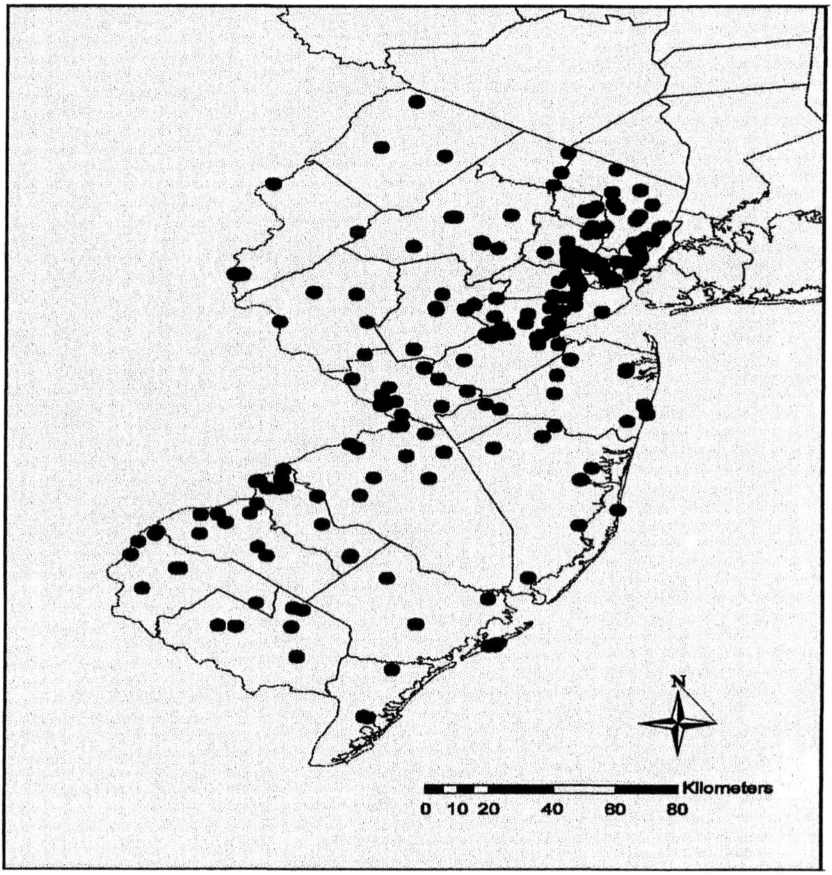

Figure 3. The U.S. Environmental Protection Agency's Aerometric Information Retrieval System (AIRS) monitoring sites in New Jersey.

Conclusions

The WTCECD provides pre- and post-9/11 environmental data from the New York City area and the state of New Jersey in an easily accessible format at one location (*1*). This database is intended for use by individuals interested in having access to raw data (i.e., not averaged data or summary statistics) in order to explore exposures related to the WTC disaster as well as general air quality issues in New York City and New Jersey.

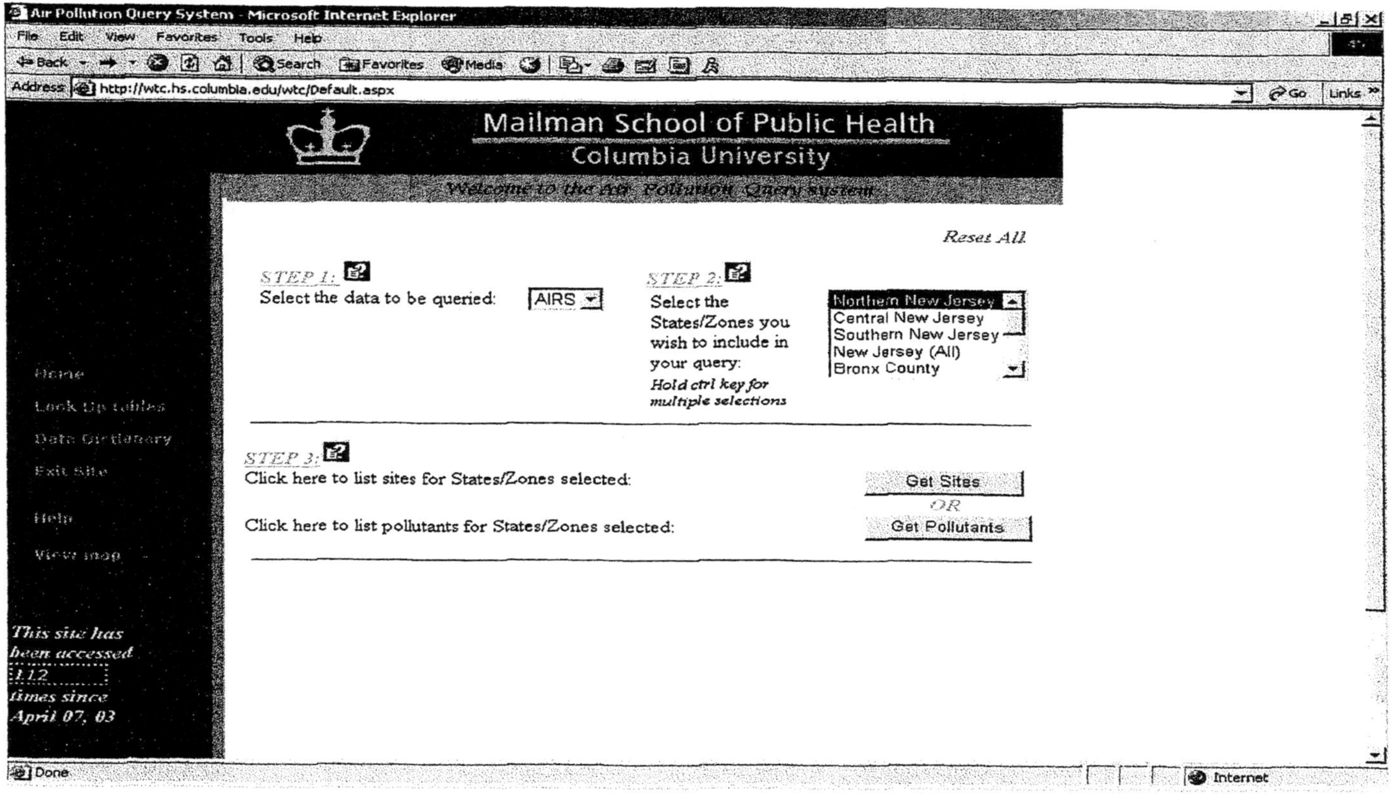

Plate 1. Example of the enumerated steps for searching the World Trade Center Environmental Contaminant Database.
(See page 30 of color inserts.)

Acknowledgements

The WTCECD was developed with funding from the National Institute of Environmental Health Sciences via their Centers for Environmental Health Programs at the Mailman School of Public Health at Columbia University and the Bloomberg School of Public Health at Johns Hopkins (NIEHS grant numbers P30 ES09089 and P30 ES03819). We also gratefully acknowledge all the help we received from numerous staff at The U.S. EPA and NYSDEC. This is LDEO contribution 6710.

References

1. "World Trade Center Environmental Contaminant Database (WTCECD)" [http://wtc.hs.columbia.edu/wtc/] Mailman School of Public Health, Columbia University, New York, NY.
2. *"EPA Response to September 11"* [http//www.epa.gov/wtc] U.S. Environmental Protection Agency.
3. U.S. Geological Survey (2002). Fact sheet 0050-02.
4. "OSHA WTC Sampling Results" [http://www.osha.gov/nyc-disaster/summary.html] U.S. Department of Labor, Occupational Safety and Health Administration.
5. "Air Monitoring in Lower Manhatten" [http://www.nyc.gov/html/dep/html/airmonit.html] New York City Department of Environmental Protection.
6. "EPA-Air Facility System (AFS)" [http://www.epa.gov/Compliance/planning/data/air/afssystem.html] U.S. Environmental Protection Agency.

Aerosol Transport Issues

Chapter 14

The Importance of the Chemical and Physical Properties of Aerosols in Determining Their Transport and Residence Times in the Troposphere

Jeffrey S. Gaffney and Nancy A. Marley

Environmental Research Division, Argonne National Laboratory, 9700 South Cass Avenue, Argonne, IL 60439

Tropospheric aerosols are now clearly identified as one of the key "missing links" in many aspects of climate and radiative forcing processes, as well as in the health effects of air quality. This chapter examines some basic chemical and physical properties of aerosols that determine their diffusional growth, atmospheric removal by gravitational settling, and uptake of water leading to increased size and subsequent removal by both wet and dry deposition. The importance of aerosol size and chemical compositions in determining their atmospheric lifetimes is stressed, with particular attention to combustion-derived carbonaceous soots. The atmospheric oxidation of "oily" soot surfaces to make these organic aerosols sufficiently hygroscopic for particle growth and atmospheric removal is expected to require a substantial time period. Mean residence times for submicron aerosols, as determined by using natural radionuclides, support the observation that a significant fraction of fine aerosols (0.1–1 μm) have considerably longer atmospheric lifetimes than anticipated from normal washout and dry depositional processes of hygroscopic aerosols. It is concluded that the surface and chemical properties of aerosols and particulate matter must be accounted for in both climate effects and assessments of health impacts of atmospheric aerosols (*18*).

Introduction

During the World Trade Center tragedy, two types of aerosols were produced in two time intervals (see Chapter 2). First, the initial collapse of the buildings generated large amounts of mechanically produced dusts. These dusts were of a sufficiently large size to settle on surfaces before they could be transported far from the site. Subsequent dust-generating events, occurring during the post-attack rescue and cleanup efforts, produced dust particles that were much smaller in size. A second type of aerosol also produced during the event was combustion aerosols from fires that ignited and continued to smolder long after the initial attack. These smoldering fires produced much finer aerosols that were of concern because of their respirable size and their longer lifetimes in the atmosphere.

Urban aerosols smaller than 2.5 µm have recently been implicated in the health effects of urban air pollution. Aerosols in the size range 0.1–1.0 µm are of particular interest, because they are respirable and travel deep into the lungs. In contrast, larger particles are typically removed in the upper respiratory track by impaction onto surfaces. Studies have indicated a relationship between lower levels of exposure to atmospheric particulates and increases in the incidence of cardiopulmonary illnesses (*1,2*) and cancer (*3*), as well as in mortality rates in general (*4*). These effects are particularly strong in individuals with preexisting conditions. Evidence suggests that aerosol mass concentration is not the most appropriate measure of potential health effects. Rather, aerosol characteristics such as size, surface area, particle number, chemical composition, and morphology are important measurements to consider.

Fine tropospheric aerosols (0.1–1.0 µm) can also play important roles in radiative balance and climate change. Aerosols in this fine size range scatter light most effectively and are therefore most important in radiative cooling and visibility reduction. The fine aerosols can cool the atmosphere both directly by scattering incoming solar radiation and indirectly by serving as cloud condensation nuclei (CCN). Fine aerosols, particularly carbonaceous soots formed in combustion processes, can also warm the atmosphere by absorbing incoming solar radiation (*5,6*). The same properties that determine human health impacts (i.e., size, chemical composition, morphology) also determine the aerosol radiative effects.

Determining the consequences of fine aerosols in the atmosphere requires an understanding of their sources and their residence times, or lifetimes, in the troposphere. The effects of aerosols on human health and climate depend on their physical and chemical properties. Their sizes and affinities for water determine their removal rates from the atmosphere by wet and dry deposition, as well as the extent to which they can be inhaled. Also important to radiative issues are their optical properties as a function of size, which in turn depend on the chemical composition of the aerosol. Specifically, the optical constants (real and imaginary components of the refractive index) that determine the contributions of the aerosol materials to light extinction are needed to calculate their effects on incoming and outgoing radiation as a function of wavelength. An empirical method has been developed for obtaining this type of information, particularly for black carbonaceous soots from various sources (*7*). However, much work remains in determining the radiative properties and health impacts of fine atmospheric aerosols.

This chapter deals with the chemical and physical properties of fine aerosols that are key to understanding their human health, air quality, and regional climate impacts. The discussion focuses on key aspects of fine aerosols that relate to their atmospheric transport and removal processes (i.e., their atmospheric lifetimes).

Aerosol Atmospheric Lifetimes

Here we examine data addressing the lifetimes, or mean residence times in the atmosphere, of fine aerosols in the size range 0.1–1.0 µm. Natural radioactive isotopes attached to these fine aerosols have been used successfully as tracers of their transport and atmospheric lifetimes (*8–11*). The dominant radioactive isotopes in fine aerosols are ^{7}Be and ^{210}Pb and its daughters, ^{210}Bi and ^{210}Po (*8,9*). Cosmogenic particles hitting the atmosphere lead to the production of ^{7}Be in the lower stratosphere and upper troposphere. Uranium-238, which is common throughout the Earth's crust, decays to the inert gas ^{222}Rn, which escapes into the atmosphere. Radon gas emitted into the troposphere decays to form ^{210}Pb and its daughters. The sources, radioactive lifetimes, and decay schemes of these radioactive isotopes are shown in Plate 1.

Once ^{7}Be is produced in the upper troposphere and lower stratosphere, it attaches itself to available aerosol surfaces with a mean diameter of 0.3 µm (*9–13*). Thus, the measurement of the ^{7}Be content of fine aerosols can give a measure of the transport of this material from the upper troposphere/lower stratosphere, where attacment occurred, to the surface, where it is sampled. Past work suggested that this might be a useful way to identify stratospheric intrusions into the troposphere and also estimate whether stratospheric ozone is

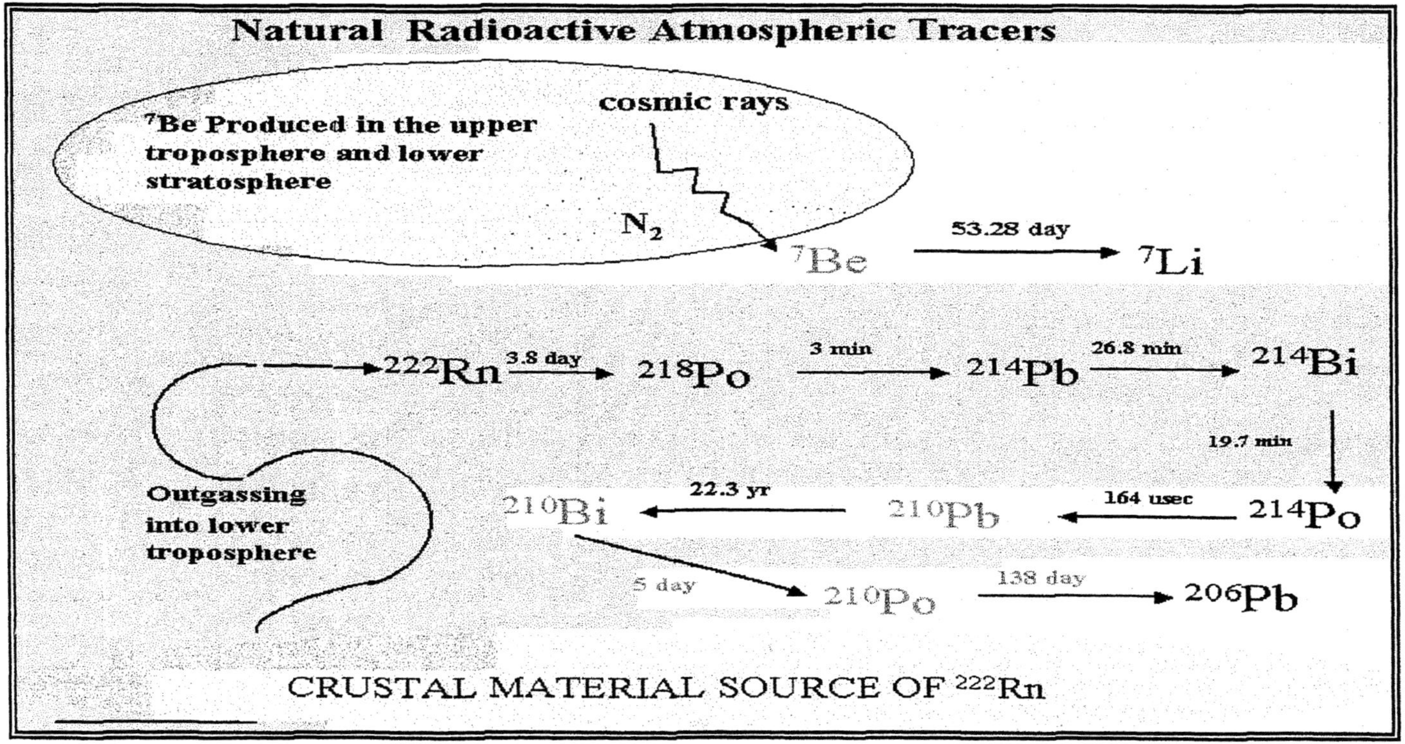

Plate 1. Sources of natural radioactivity in the atmosphere for ^{7}Be and ^{210}Pb and its daughters, ^{210}Bi and ^{210}Po (shown in red). Very short lived or unstable isotopic species are shown in black. *(See page 31 of color inserts.)*

affecting tropospheric sites during these events. Indeed, ^{7}Be studies have been informative as a means of exploring aerosol removal from aloft during thunderstorms and hail events (*8*).

The decay of ^{222}Rn leads to the formation of ^{218}Po, which rapidly becomes attached to preexisting fine aerosols in the troposphere and decays to form the long-lived ^{210}Pb (*8*). The ^{210}Pb then decays to form the daughters ^{210}Bi and ^{210}Po (see Plate 1). Although the parent ^{210}Pb has a relatively long lifetime (half-life of 22.3 years) the daughters have shorter lifetimes (half-life of ^{210}Po is 138 days, ^{210}Bi is 5 days) which make them useful for the determination of aerosol residence times. The ratios of the daughters to the parent can then be used to estimate residence times of size-fractionated aerosols.

The size of the ^{218}Po-attached particles in the boundary layer are somewhat larger than that for ^{7}Be, with a median diameter of 0.4 µm upon initial attachment. This larger diameter is a consequence of the slightly greater median diameter of particles in the planetary boundary layer versus the free troposphere (*9–13*). The attached aerosols are removed from the atmosphere by wet and dry deposition, with precipitation (wet deposition) being the dominant path. In contrast to ^{7}Be with its 53-day half-life, ^{210}Pb is present long enough in soils where it is deposited to be measurable in wind-blown dusts. Contamination of aerosol samples with resuspended surface materials results in excess ^{210}Po and subsequently extremely high activity ratios and very long apparent residence times. Therefore, resuspended dusts can act as contamination in studies of aerosol residence times. However, wind-blown dusts are usually larger than 1–2 µm, while the diameter of the initial ^{218}Po-attached aerosols is typically 0.4 µm. Determinations made on whole aerosol samples can generate results that are more or less equilibrated with regard to the $^{210}Bi/^{210}Po/^{210}Pb$ in soils. However, this problem can be avoided by collecting size-fractionated samples and separating the larger wind-blown dust from the fine aerosols.

In the absence of contamination by wind-blown soil or other sources of ^{210}Bi or ^{210}Po, the disequilibria in $^{210}Bi/^{210}Pb$ and $^{210}Po/^{210}Pb$ ratios are useful estimates of atmospheric residence times (*9*). Key to this measurement is the ability to readily separate and determine ^{210}Pb and its daughters (*10*). A methodology has been developed for easily determining the activities of these species in air samples that have been sized by using cascade impactors (*9*). This method uses anion-exchange-impregnated filter media to collect ^{210}Bi and ^{210}Po as the chloride adducts effectively separating the ^{210}Bi and ^{210}Po from the parent ^{210}Pb.

Data obtained in this way for precipitation samples yielded typical values of 10 days for the mean residence times of aerosols in rain, snow, and hail (*8*). Use of the same methods on size-fractionated dry aerosols in air determined apparent atmospheric residence times of 10–40 days for aerosols < 1 µm, results that were very consistent from site to site (*11*). The 10-day lifetimes observed for precipitation samples are consistent with the concept that washout is the predominant removal mechanism for soluble aerosols such as sulfate and nitrate. The longer lifetimes observed for the air samples are thought to be due to carbonaceous soots or other fine particulate matter that was not water soluble and therefore was not readily washed out by precipitation.

The observed ages reported here are not absolute but are representative of the total aerosol sample. For example, a sample containing a mixture of 90% of a 5-day-old aerosol and 10% of a 100-day-old aerosol would have an apparent age of 14.5 days by this method. Similarly, a 50:50 mixture of a 5-day-old aerosol and a 100-day-old aerosol would have an apparent residence time of 52.5 days while a mixture of 90% of a very young aerosol (1-day aerosol lifetime) and 10% of an older aerosol (100-day aerosol lifetime) would have a 1.9-day apparent lifetime. The mean residence time should therefore be viewed as more of a probability distribution function than an absolute lifetime. The results of isotopic disequilibria studies of aerosol samples from many locations indicate that a significant fraction of the submicron aerosol has a fairly long lifetime, typically 25–30 days or more. This apparent residence time is therefore a composite value for all of the aerosols containing attached ^{210}Pb in the sample and should be viewed as a mean residence time.

While the atmospheric lifetimes of hygroscopic aerosols found in precipitation samples are relatively short (5–10 days), air particulate samples appear to have a significant fraction of much older aerosols. Clearly, the data obtained for the radionuclides on size-fractionated aerosol samples indicate that a significant portion of fine atmospheric aerosols are transported for more than 20 days, as do recent results obtained with similar methods (*12*). This means that regional if not global control strategies are needed to maintain adequate air quality and to reduce the impacts of elevated levels of fine aerosols on our health, weather, and climate.

Aerosol Removal Processes

The observations of the radionuclide tracers presented above strongly indicate the importance of combustion-derived carbonaceous soots entrained in air for longer time periods than sulfate or nitrate aerosols that are readily washed out in precipitation. Here we review the physical and chemical removal processes of atmospheric aerosols and present a simple chemical model based on the hypothesis that the surface characteristics of carbonaceous soot particles determine their atmospheric lifetimes. The combustion conditions probably determine these surface characteristics and consequently the hydrophilic or hydrophobic natures of soots released into the atmosphere as primary aerosol pollutants. These same surface characteristics will also determine how reactive the soot aerosols are in oxidative chemical reactions with hydroxyl radical (OH), nitrate radical (NO_3), etc. Such reactions, in turn, affect the aerosol surface aging and wet removal processes.

The atmospheric removal of other types of aerosols, like those observed during the World Trade Center collapse and the post-collapse cleanup period, also depends on the surface characteristics and sizes of the aerosol. Thus, much of the subsequent discussion, though focused on carbonaceous soots, is relevant to aerosols in the fine size ranges produced in the World Trade Center events. The main removal processes for atmospheric aerosols (i.e., particles and fine droplets) are diffusion to surfaces, gravitational settling, and hygroscopic growth to form larger droplets that settle out or act as CCN. Figure 1 shows the relative rates of diffusion (dashed line) and gravitational settling (solid line) for a model particle system consisting of spherical particles of unit density, as a function of particle size. This graph clearly indicates that small particles (< 0.1 μm) diffuse rapidly to existing particle surfaces (diffusional loss). Particles larger than 2–3 μm have settling velocities sufficiently large to rule out their contribution to long aerosol lifetimes in the troposphere. For example, the very large aerosols from the initial collapse of the World Trade Center were observed to be very large in size (many >10 μm) and did not travel far from the site of the

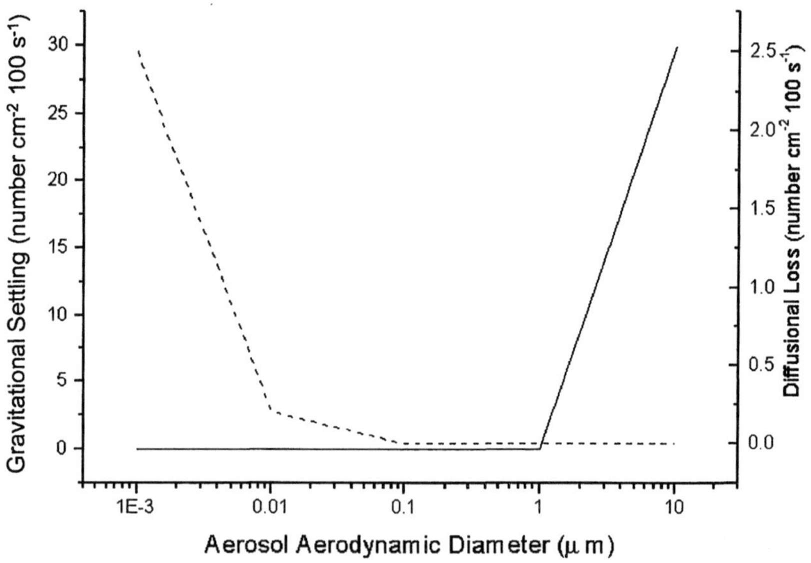

Figure 1. Diffusional (---) and gravitational settling (—) loss rates for aerosol particles as a function of size.

event. Figure 1 clearly indicates that a stable region exists for particles with diameters of 0.1–1 µm that do not diffuse readily to existing particle surfaces or settle out because of gravitational forces. This stable region leads to the accumulation of particles of this size in the atmosphere and is known as the “accumulation mode”.

Removal of these fine particles from the atmosphere depends on their ability to grow to a size large enough to be deposited by gravitational settling. Alternatively, these same particles can take up enough water to act as CCN and be incorporated into aqueous cloud droplets. The surface chemistry of particles in the stable region for particle removal (0.1–1.0 µm) is therefore very important in determining the lifetimes of aerosols in the accumulation mode or of fine aerosols in the atmosphere. Hygroscopic aerosols will actively equilibrate with the available water vapor in the troposphere. A good example is sulfuric acid. If the water vapor content is sufficiently high, fine hygroscopic sulfuric acid droplets actively add water and grow in size. The sulfate particles that grow to sizes greater than 2–3 µm can be dry deposited or settle out on surfaces. They can also act as CCN and grow large enough to rain out or be wet deposited, again leading to their atmospheric removal and shorter lifetimes.

Inorganic aerosols that are very hygroscopic include secondary acidic species such as sulfuric acid, ammonium sulfates, and ammonium nitrates. The washout of these species is strongly dependent on water vapor content. In most cases, the cycling time of water in the atmosphere is about 10 days. As noted above, the lifetimes of 10–11 days observed for many aerosol species through the use of radionuclide clocks are consistent with the preferential washout of inorganic sulfates and nitrates (*8*).

However, findings for fine dry aerosols indicate a significant background of older particles in many locations (*8–12*). We hypothesize that these older particles were initially hydrophobic oily carbonaceous soots that required time for surface oxidation. With sufficient surface oxidation, oily soots become more hygroscopic, take on water, and are removed from the atmosphere. Note that the processes involved in the atmospheric growth and removal of fine particles also occur during inhalation by humans. Thus, hygroscopic aerosols will take up water and grow in size in the humid environment of the human respiratory system. Hygroscopic aerosols therefore are more readily removed before they travel deep into the lungs, whereas hydrophobic aerosols are more likely to be carried along the airways without increasing in size.

A Simple Model

Let us hypothesize that most carbonaceous soots have a thin coating of either aromatic or paraffinic organic material upon formation, depending on the source and the combustion temperatures and conditions under which the primary soots were formed. For this simple model we envision a paraffinic coating. Thus, the soot surface initially looks s omething like the surface shown in Figure 2A. After formation, daytime oxidation reactions are initiated in the atmosphere. The

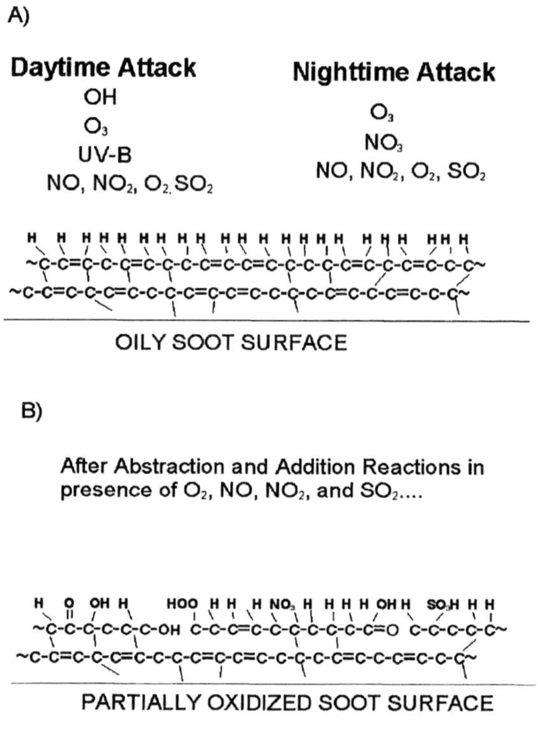

Figure 2. A) Hypothetical "oily" soot surface, showing reaction sites and likely atmospheric oxidants for day and night. B) Soot surface after oxidation and reaction with atmospheric species. OH = hydroxyl radical; O3 = ozone; NO = nitrogen oxide; NO2 = nitrogen dioxide; O2 = oxygen; SO2 = sulfur dioxide; NO3 = nitrate radical; UV-B = ultraviolet B radiation.

primary reaction is the abstraction of a hydrogen from the surface by OH to form water. A radical site created on the surface by this reaction will rapidly add oxygen. Atmospheric ozone (O_3) is another possible initiator of oxidation at surface sites on organic soots where olefinic (alkene) groups or unsaturated bonds are exposed. The addition of OH to olefinic positions also occurs,leading to the production of hydroxy-substituted radicals that again add oxygen. The oxygen-containing organic radicals in turn undergo further reactions, leading to the formation of keto-peroxy, peroxy-hydroxy, and carboxylic acid-peracid functionalities on the surfaces. Thus, the surface reactions are of the sort in equations 1–3:

$$\text{OH} + \text{alkane site}{\sim} \rightarrow H_2O + \text{alkyl radical}{\sim} \qquad (1)$$

$$O_2 + \text{alkyl radical site}{\sim} \rightarrow {}^{\cdot}\text{O-O-alkyl}{\sim} \qquad (2)$$

$${}^{\cdot}\text{O-O-alkyl}{\sim} + NO + O_2 \rightarrow \text{CHO-alkyl}{\sim} + HO_2 + NO_2 \text{ or } NO_3\text{-alkyl}{\sim} \qquad (3)$$

Here ~ indicates attachment of the group to the soot surface. Reactions with alkene positions yield similar products but have OH addition as a predominant pathway, as in equations 4–6:

$$\text{OH} + \text{alkene site}{\sim} \rightarrow \text{HO-CR-C-R}{\sim} \qquad (4)$$

$$\text{HO-CR-C-R}{\sim} + O_2 \rightarrow \text{HO-CR-}\overset{/\text{O-O}^{\cdot}}{\text{C}}\text{-R}{\sim} \qquad (5)$$

$$NO + O_2 + \text{HO-CR-}\overset{/\text{O-O}}{\text{C}}\text{-R}{\sim} \rightarrow \text{HO-CR-}\overset{//\text{O}}{\text{C}}\text{-R}{\sim} + NO_2 \text{ or HO-CR-}\overset{/ONO_2}{\text{C}}\text{-R}{\sim} \qquad (6)$$

When the nitrogen oxide (NO) concentration is low, the peroxy radicals formed from OH (or O_3) reactions will likely react with hydroperoxyl radical (HO_2) to form hydroperoxy functional groups on the surfaces of the soot particles. Thus, reactions of this type on soot surfaces have mechanisms similar to those of gas-phase atmospheric peroxyradical reactions. The organohydroperoxide groups (-OOH) would be able to react with atmospheric sulfur dioxide (SO_2) to form organosulfate products on the surface, similar to the reactions of SO_2 with organic peroxides (R-OOH) and hydrogen peroxide (H_2O_2) that yield sulfate in aqueous aerosol solutions. The hydrogen shift reactions observed for the larger organic alkoxyradical species should also be possible, leading to the formation of hydroxyketones in the case of NO reactions (*13*).

During the nighttime hours, if sufficient nitrogen dioxide (NO_2) and O_3 are present, NO_3 will also oxidize the surface by similar mechanisms, forming nitric acid and radical sites on the initially oily surface. Again, NO_3 will abstract hydrogen at the alkyl sites, and/or add to alkene sites on the carbonaceous surfaces. These reactions form products similar to those of the oxidations initiated by O_3 or OH. Photolysis of some nitrated or carbonyl products (i.e., aldehydes) would lead to the formation of carboxylic acids and other oxidized products on the surface as in Figure 2B.

The initial surface reactions are fairly fast. However, once the reactions begin, hydrogens on the surface near the oxidized sites are slower to undergo abstraction with OH or NO_3 radicals than those on an unoxidized surface, because the strongly electronegative oxygen atoms already bound to the surface are electron-withdrawing agents, and OH, O_3, and NO_3 are electrophilic agents. The result is a strong competition for electrons (i.e., hydrogen atoms) and a decrease in the reactivity of the surface as it begins to undergo oxidation.

In Figure 2B the olefinic sites, having the highest reactivity with OH, NO_3, and O_3, have reacted to form oxidized products (*13,14*). After this initial oxidation of the unsaturated sites, the surface of the oily soot is partially oxidized, and it will be slower to react by addition or abstraction. The oxidized areas decrease the reactivity of the surface to further oxidation, as noted above (*13,14*). Although this idealized surface has been "aged," it still is not substantially hygroscopic. The key argument here is that although the surface is still fairly hydrophobic, it is now less reactive to oxidation reactions. Indeed, the surface will eventually react sufficiently to form organic acid sites, resulting in a hydrophilic surface that allows water uptake and particle growth. The use of reaction rate prediction tools based on OH reactivities yields an estimate that this may take 40–80 days (*14*), not inconsistent with total mean residence times determined for fine aerosols through use of radionuclide isotopes (*9–12*).

Recently, as part of the Mexico City Metropolitan Area/Mexico City Megacity 2003 collaborative air quality study organized by the Massachusetts Institute of Technology, an aethalometer (Anderson, seven channel) was used to obtain black carbon measurements with 1- to 2-min time resolution in the Mexico City urban center, which has a high black carbon content. Black carbon levels were observed to be on the order of 1–5 $\mu m\ m^{-3}$ for a 24-hour average during April 2003. On a number of occasions during that study, black carbon levels were observed to be unaffected by rain events, though hygroscopic inorganic aerosols were removed rapidly, and their concentrations reduced significantly during the same rain events. Some aging of the black carbon aerosols was observed, but this did not modify their hydrophobic nature sufficiently to result in removal of particles in the size range 0.1–1.0 µm through increased humidity and precipitation (*15*). These observations are consistent with an oily hydrophobic surface for carbonaceous soots. In addition, soot particles 0.1–1 μm in aerodynamic diameter are not removed by impaction by falling raindrops (hydrometeors), because the fine particles tend to flow around the falling drops and are not physically captured as the raindrops fall through the gas to the ground. That is, the impaction of the fine aerosols by the drops is not significant.

Conclusions and Connections to the World Trade Center Tragedy

The simple model presented above for black carbon and the observations of longer mean residence times for fine atmospheric aerosols have important consequences for both human health and climate. The fact that fine carbonaceous soot produced by combustion processes, such as diesel soots, are unlikely to be hydrated readily indicates that these soots will not grow sufficiently in size when exposed to high humidity to be dry deposited or to be removed from the atmosphere by wet deposition during rain events. This conclusion is noteworthy, because it explains the limited removal of this type of aerosol material from the environment by wet and dry deposition and also from human respiratory organs by the normal defense system.

The upper respiratory tract (trachea, nasal passages, and bronchi) is designed to remove particles > 1 μm by impaction on cilia and surfaces during inhalation. The respiratory system is very high in humidity, so that hygroscopic aerosols will grow to sufficient size to be removed in the upper respiratory tract before they reach the deep lung. Very fine aerosols will be removed by diffusional processes very rapidly or, if hygroscopic, will quickly grow to a size that can be removed by impaction in the upper respiratory tract. Thus, particles in the size range 0.1–1.0 μm that are not hygroscopic, such as black carbon, are the ones that can reach the deep lung and are important in exposures leading to chronic or longer-term medical problems (asthma, lung cancer, etc.). Note also that combustion-generated black carbon soots are known to contain polycyclic aromatic hydrocarbons (PAHs), suspected carcinogens that can travel to the deep lung along with the soot particles (*13*).

Once fine hydrophobic aerosols reach the deep lung, the respiratory system has two means of removing them. The first is oxidative biochemistry, which oxidizes and "dissolves" the materials so that they can be removed or processed by the body. The second is the removal of the more resistant particles by macrophages or white blood cells that ingest them and effectively make them "larger" for removal by the lymphatic system or by being coughed out as expectorate or sputum. Black carbon is known to reach the deep lung and, once there, to be quite resistant to biochemical oxidation and to dissolution in the body (*16*). Black pulmonary pigment obtained from autopsy tissue samples has been shown through $^{13}C/^{12}C$ measurements to consist of particulate material generated from fossil fuel combustion (*16*). These observations are consistent with the model proposed above for oxidation processes on oily soot surfaces.

Although the smoldering fires of the World Trade Center did not likely contribute large amounts of black carbon directly to the total atmospheric levels in New York City, contributions from the use of heavy-duty diesel engines during the cleanup activities after the event added to the normal urban burden of black carbonaceous soots. The larger dust particles and more hygroscopic materials released at the site all probably contributed to upper respiratory exposures, as noted by the development of the "World Trade Center cough."

This chapter has briefly reviewed available data indicating that fine aerosols have longer lifetimes than are currently assumed in global climate models, which typically use mean residence times of 2–7 days for atmospheric aerosols and treat all aerosols similarly (*17,18*). The discussion here of the simple soot surface model and aerosol removal rates clearly indicates that the size and chemistry of the fine aerosols are important factors to be considered in the estimation of aerosol lifetimes. It is also clear that the more hydrophobic aerosols in the size range 0.1–1.0 µm, such as black carbon, are likely to have the longest lifetimes and the greatest impacts on both climate and health.

Submicron aerosol levels are of concern in all urban areas with regard to their human health effects and these same urban centers are major sources of fine aerosols transported over regional and global areas, leading to radiative effects (both direct and indirect) of importance to global climate change. We conclude that control strategies for carbonaceous aerosols, with their longer lifetimes than sulfate or nitrate aerosols and their radiative importance and health effects, merit more attention (*17,18*). Moreover, the surface and chemical properties of aerosols and particulate matter must be accounted for in both climate models and assessments of health impacts of atmospheric aerosols (*18*).

Acknowledgement

The authors' work is supported by the U.S. Department of Energy, Office of Science, Office of Biological and Environmental Research, Atmospheric Science Program, under contract W-31-109-Eng-38.

References

1. Pope, C.A. *Aerosol Sci. Technol.* **2000,** *32*, 4-14.

2. Schwartz, J.; Ballester, F.; Saez, M.; Perez-Hoyos, S.; Bellido, J.; Cambra, K.; Arribas, F.; Canada, A.; Perez-Boillos, M.J.; Sunyer, J. *Environ. Health Perspect.* **2001,** *109,* 1001-1006.
3. Beeson, W.L.; Abbey, D.E.; Knutsen, S.F. *Environ. Health Perspect.* **1998,** *106,* 813-822.
4. Lipfert, F.W.; Morris, S.C.; Wyzga, R.E. *J. Air Waste Manage. Assoc.* **2000,** *50,* 1501-1513.
5. Marley, N.A.; Gaffney, J.S.; Cunningham, M.M. *Environ. Sci. Technol.* **1993,** *27,* 2864-2869.
6. Gaffney, J.S.; Marley, N.A.. *Atmos. Environ.* **1998,** *32,* 2873-2874.
7. Marley, N.A.; Gaffney, J.S.; Baird, J.C.; Blazer, C.A.; Drayton, P.J; Frederick, J.E. *Aerosol Sci. Technol.* **2001,** *34,* 535-549.
8. Gaffney, J.S.; Orlandini, K.A.; Marley, N.A.; Popp, C.J. *J. Appl. Meteorol.* **1994,** *33,* 869-873.
9. Marley, N.A.; Gaffney, J.S.; Cunningham, M.M.; Orlandini, K.A.; Paode, R.; Drayton, P.J. *Aerosol Sci. Technol.* **2000,** *32,* 569-583.
10. Marley, N.A.; Gaffney, J.S.; Orlandini, K.A.; Drayton, P.J.; Cunningham, M.M. *Radiochim. Acta* **1999,** *85,* 71-78.
11. Gaffney, J.S.; Marley, N.A.; Cunningham, M.M. *Atmos. Environ.* **2004,** *38,* 3191-3200.
12. Dueñas, C.; Fernándeza, M.C.; Carreterob, J.; Ligerb, E.; Cañetea, S. *Atmos. Environ.* **2004,** *38,* 1291-1301.
13. Finlayson-Pitts, B.J.; Pitts, J.N., Jr. *Atmospheric Chemistry of the Upper and Lower Atmosphere.* Academic Press, San Diego, 2000.
14. Gaffney, J.S.; Bull, K. In *Chemical Kinetics of Small Organic Radicals*; Z. Alfassi, Ed.; CRC Press: Boca Raton, 1988; Vol. II, Chapter 8.
15. Gaffney, J.S. et al., *Science* **2004,** in preparation.
16. Slatkin, D.N.; Friedman, L.; Irsa, A.P.; Gaffney, J.S. *Human Pathol.* **1978,** *9,* 259-267.
17. Chung, S.H.;. Seinfeld, J.H. *J. Geophys. Res.* **2002,** *107(D19),* 4407.
18. Jacobson, M.Z. *J. Geophys. Res.* **2002,** *107(D19),* 4410.

Chapter 15

$^{210}Po/^{210}Pb$ in Outdoor–Indoor PM-2.5, and PM-1.0 in Prague, Wintertime 2003

Jan Hovorka[1], Robert F. Holub[2], Martin Braniš[1], and Bruce D. Honeyman[2]

[1]Institute for Environmental Studies, Faculty of Science, Charles University in Prague, Benátská 2, 128 01 Prague 2, Czech Republic
[2]Environmental Science and Engineering Division, Colorado School of Mines, Coolbaugh Hall, Golden, CO 80401

After the large aerosol releases that occurred during the WTC collapse there was a significant amount of particulates that were transported inside the nearby buildings. This type of transport is occurring in many urban areas and is particularly significant for fine aerosols in the sub micron size ranges. The ^{210}Pb net alpha activities and the mass concentrations of 24-hour and 15-min averages of PM-2.5 outdoor/indoor and PM-1.0 indoor aerosols were employed to determine penetration factors of outdoor PM-2.5 into a naturally ventilated flat on the third floor in Prague center. The PM-2.5 penetration factors were in a range of 0.42 - 0.54. The factor values determined from alpha activities were higher than the factors derived from mass concentrations. The discrepancies can be explained by size selective aerosol penetration through the building envelope, which favors aerosol particles of smaller sizes having a higher specific alpha activity than larger ones. Both the alpha activities and the 24-hour mass concentrations of PM-2.5 did not reflect personal activities indoors unlike the 15-min mass concentrations. The 15-min averages also exhibited the highest temporal variation. Nevertheless, all three variables gave similar penetration factors. Both the alpha activities and 24-hour mass concentrations have shown indoor PM-1.0 to comprise about 70% of the indoor PM-2.5. The penetration factors and ratio of indoor PM-1.0 to indoor PM-2.5 were constant during the sampling campaign. Indoor aerosol concentrations were affected predominantly by the outdoor aerosol concentration.

Introduction

After the large aerosol releases that occurred during the World Trade Center collapse there was a significant amount of particulate matter (PM) that was transported inside the buildings in lower Manhattan. This type of transport is occurring in many urban areas and is particularly significant for fine aerosols that are in the submicron size ranges. The work presented here describes the use of natural radionuclides that are attached to these fine aerosols to determine the extent of this type of aerosol infiltration into buildings.

The long-lived ^{222}Rn decay products ^{210}Po and ^{210}Pb present unique tracers for the study of atmospheric aerosol transport (*1*). These decay products can also be used as markers of outdoor aerosols penetrating into indoor environments (*2, 3*). This marker method is based on a well-justified assumption that there is no notable source of ^{210}Pb indoors. The ^{210}Pb net alpha activities and the mass concentrations of 24-hour and 15-min averages of fine (PM-2.5) outdoor/indoor and ultrafine (PM-1.0) indoor aerosols were employed to determine the penetration factors of outdoor PM-2.5 into a naturally ventilated flat on the third floor in the Prague city center. The indoor/outdoor penetration factors ($^{15m}C_{mass}$, $^{24h}C_{mass}$, and C_{alpha}) of atmospheric aerosols were determined from the slopes of the linear regressions of indoor against outdoor values of the 15-min, and 24-hour mass concentrations and alpha activities, respectively (*4*).

Experimental Methods

The PM-2.5 indoor/outdoor and PM-1.0 indoor aerosols were sampled using a Harvard Impactor at an airflow rate of 10.0 l/min and 23 l/min, respectively. Samples were collected for 24 hours on a single polytetrafluorethylene (PTFE) membrane. The volumetric ($\mu Bq/m^3$) and specific (Bq/g) activities of $^{210}Pb/^{210}Po$ were measured by means of a α-spectrometer equipped with a multichannel analyzer (Canberra MCA) by using the ^{210}Po alpha 5.3 MeV peak. Correction was made for additional alpha activity due to ^{210}Pb decay and the net activity of ^{210}Po was calculated. The ratio of $^{210}Po/^{210}Pb$, based on repetitive measurements of aerosols sampled by the same technique in 2002, was estimated to be 0.2. The 15-min averages of PM-2.5 indoor/outdoor mass concentrations were determined by a laser photometer (TSI 8520 DustTrak) and values were corrected for changes in humidity and aerosol particle size distribution (*5*). Also, 24-hour concentrations of the gaseous atmospheric constituents nitrogen oxide (NO), nitrogen dioxide (NO_2), and ozone (O_3) were calculated from 15-min averages simultaneously measured indoors and outdoors (Horiba; APNA-360, APOA-360). Hourly averages of personal activities (e.g. number of people present, cooking activities, vacuum cleaning, ventilation, etc.) were recorded in a diary. The sampling campaign was conducted on daily basis from February 7 – March 13, 2003 in the living room of a three room flat on the third floor of an apartment house situated in the Prague city center. A five-person family occupies this no-smoking flat.

Results and Discussion

One of the main tasks in indoor/outdoor air quality studies is the estimation of how much outdoor aerosols penetrate indoors. Our results have shown that the $^{24h}C_{mass}$ is constant over a broad aerosol concentration range. According to the 24-hour averages of indoor PM-1.0 and PM-2.5 and outdoor PM-2.5 aerosol mass concentration shown in Figure 1, 46% and 33% of outdoor PM-

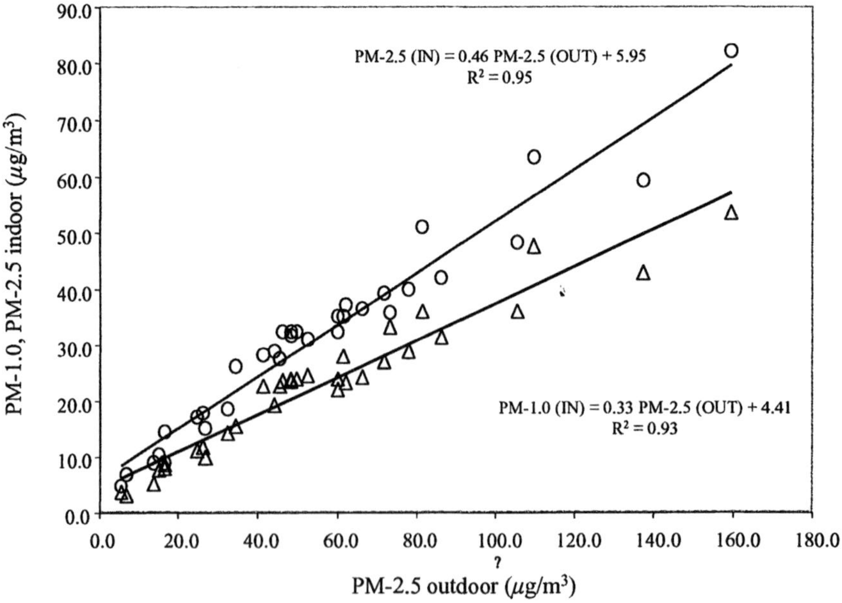

Figure 1. Average 24-hour mass concentrations of PM-2.5 (○) and PM-1.0 (△) indoors (IN) versus PM-2.5 outdoors (OUT).

2.5 contributes to indoor PM-2.5 and PM-1.0, respectively. Also, 71% of the indoor PM-2.5 mass is due to PM-1.0, as determined from the slope of the regression line shown in Figure 2. The ${}^{15m}C_{mass}$ values shown in Figure 3 are higher than ${}^{24h}C_{mass}$ of Figure 1. The higher temporal variations of 15-min mass concentrations, probably reflecting personal activity indoors, result in a lower correlation in Figure 3 ($R^2 = 0.64$) than obtained in the case of 24-hour mass concentrations ($R^2 = 0.95$).

The volumetric ${}^{210}Po$ net activity in PM-2.5, shown in Figure 4, is usually higher outdoors than indoors because of the higher outdoor mass concentration of PM-2.5. In contrast, the ${}^{210}Po$ specific activity in PM-2.5 is higher indoors than outdoors. This mainly reflects the particle size-selective penetration through the building envelope that favors particles of the 200-500 nm sizes which carry more ${}^{210}Po$ and ${}^{210}Pb$ due to their larger surface area per unit mass than particles of the 500-2500 nm size range (*2, 3*). Also, the increase of indoor specific activity, but to a lesser extent compared with the penetration effect,

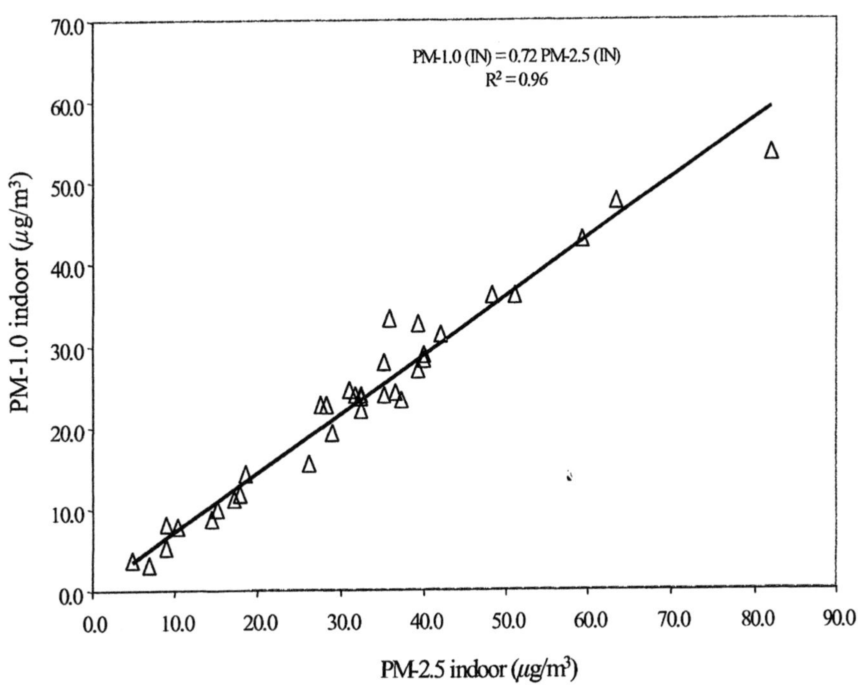

Figure 2. Average 24-hour mass concentrations of PM-1.0 indoors (IN) versus PM-2.5 indoors.

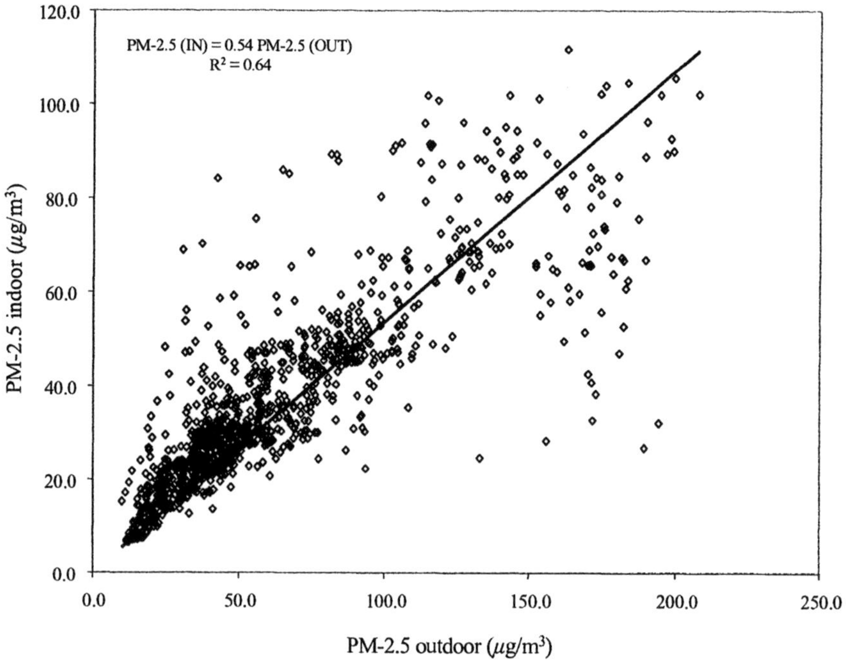

Figure 3. Average 15-minute mass concentrations of PM-2.5 indoors(IN) versus PM-2.5 outdoors (OUT).

can also be caused by the lower values of indoor plate-out rates for the 200-500 nm sized particles.

The C_{alpha} (0.57, R^2= 0.73), obtained from Figure 4, is higher than the ${}^{24h}C_{mass}$ (0.46, R^2= 0.95), but very close to the ${}^{15m}C_{mass}$ (0.54, R^2= 0.64) values (see Figures 1, and 3). Similar to the 15-minutes mass values, the regression line of the alpha activities does not have as tight a fit as do the 24-hour values due to the experimental error of the activity measurements which is on average 11%. The proportionality of indoor PM-1.0 to indoor PM-2.5, a value of 71% (R^2= 0.76) obtained from the alpha measurements shown in Figure 5, agrees well with the value of 72% (R^2= 0.96) estimated from the 24-hour mass concentrations (Figure 2).

The building ventilation rate was calculated to be ~0.36 ± 0.12 h^{-1}. This rate was determined on the basis of the disappearance rate of NO produced indoors. A non-ventilated gas stove in the kitchen is a strong source of NO indoors. Such a source significantly prevails over the outdoor NO sources and NO production indoors is clearly associated with the intensity of cooking. Typical time

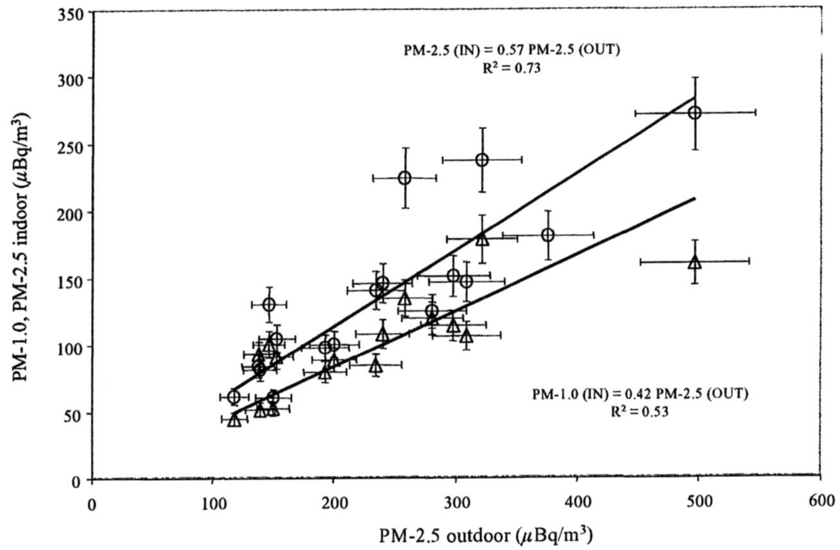

Figure 4. Volumetric ^{210}Po net activity in PM-2.5 (○) and PM-1.0 (△) indoors (IN) versus PM-2.5 outdoors (OUT).

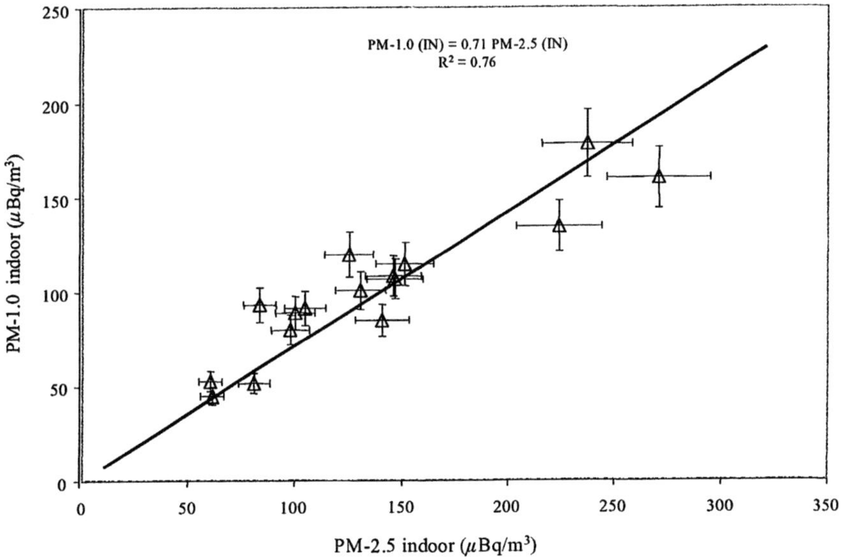

Figure 5. Volumetric ^{210}Po net activity in PM-1.0 indoors (IN) versus PM-2.5 indoors

evolution of NO concentration caused by cooking is depicted in Figure 6. Cooking activities initiated between six and seven o'clock cause a steep growth followed by an exponential decrease of NO concentration in the living room. Nevertheless, while NO produced in the kitchen is easily spread throughout the whole flat causing occasional peak values around 0.30 mg/m^3 in the living room, our results of aerosol measurements clearly have shown that no significant mass of aerosol produced by cooking in the kitchen penetrates into the living room where the measurements were conducted. The reason for such a behavior is the high plate-out rates of cooking-associated particles with median diameters well below 100 nm (6, 7).

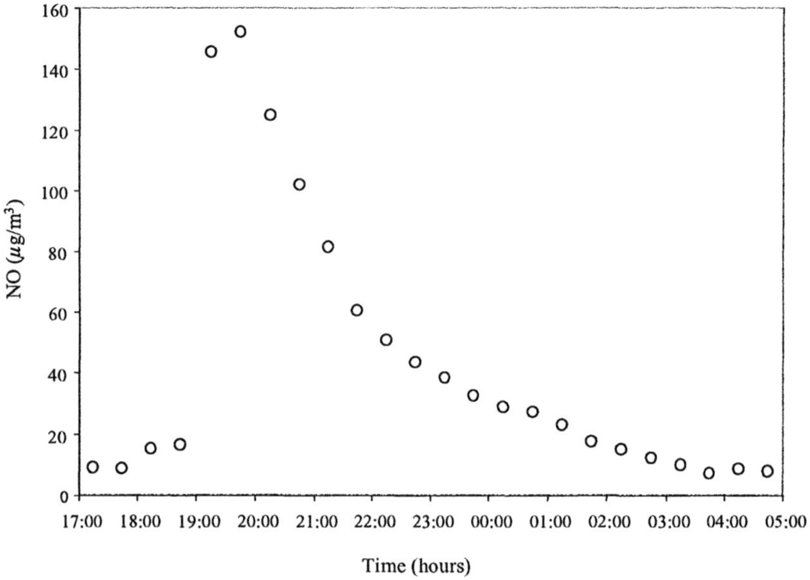

Figure 6. Typical time evolution of NO concentration indoors recorded in the living room on March 4, 2003.

Conclusion

The measurement of ^{210}Po alpha activity in low-volume aerosol samples allows for the determination of indoor and outdoor aerosols present in the indoor air and for a reliable estimation of penetration factors of outdoor aerosols into buildings. Such factors are key in assessing indoor/outdoor air quality relationships. Results of this study also have shown that losses of outdoor aerosol particles during the infiltration through the building envelope were the most important process governing indoor aerosol concentrations.

Acknowledgment

The sampling campaign went within the framework of cooperative research performed in several European urban areas funded by EU as the URBAN-AEROSOL project (Characterization of Urban Air Quality – Indoor/Outdoor Particulate Matter Chemical Characteristics And Source-to-Inhaled Dose Relationships) under grant EVK4-2000-00541. Careful sampling of all the PM's by M.Domasová and P.Řezáčová is gratefully acknowledged. Alpha counting was done in the LAER laboratory of the Colorado School of Mines and supported by grant of Charles University in Prague 206/2001 B GEO, PrF.

References

1. Marley, N.A.; Gaffney, J.S.; Drayton, P.J.; Cunnigham, M.M.; Orlandini, K.J.; Paoda, R. *Aerosol Sci.Technol.* **2000,** *32,* 569.
2. Mosley, R.B.; Greenwell, D.J.; Sparks, L.E.; Guo, Z.; Tucker, W.G.; Fortmann, R.; Whitfield, C. *Aerosol Sci.Technol.* **2001**, *34*, 127.
3. Harley, N.H.; Chittaporn, P.; Fisenne, I.M.; Perry, P. *J.Environ. Radioactiv.* **2000,** *51*, 27.
4. Ní Riain, C.M.; Mark, D.; Davies, M.; Harrison, R.M.; Byrne, M.A. *Atmos. Environ.* **2003,** *37*, 4121.
5. Ramachandran, G.; Adgate, J.L.; Pratt, G.C.; Sexton, K. *Aerosol Sci.Technol.* **2003,** *37,* 33.
6. Long, C.M.; Suh, H.H.; Koutrakis, P. *J.Air&Waste Manage Assoc.* **2000,** *50,* 1236.
7. Li, C.S.; Lin, W.H.; Jenq, F.T. *Environ. Int.* **1993,** *19*, 147.

Chapter 16

Estimates of the Vertical Transport of Urban Aerosol Particles

Edward E. Hindman

Earth and Atmospheric Sciences Department and NOAA–Cooperative Remote Sensing Science and Technology Center, The City College of New York, 138th Street and Convent Avenue, New York, NY 10031

Diurnal measurements obtained in urban Kathmandu, Nepal and rural Steamboat Springs, Colorado are presented to illustrate the role of convection in the vertical transport of aerosol particles in the atmosphere. At both locations, CN concentrations were a maximum during the morning and evening rush periods and a minimum during early afternoon. The vertical transport of aerosols was estimated for Steamboat Springs based on elementary meteorological principles and a method is presented for determining the verticle transport that makes use of archived meteorologocal data and analysis tools available from the National Oceanic and Atmospheric Administration, Air Resources Laboratory. The boundary layer depth was seen to peak each day at about 21 UTC. The vertical extent of the plume from the collapse of the World Trade Center on September 11, 2001 was estimated using this method and results indicated that the plume could have risen to between 1.7 and 2 km above sea level, consistent with the observed depth. The method presented here can also be used to forecast the depth and estimate vertical mixing in the atmospheric boundary layer.

Introduction

The diurnal-cycle of solar heating and nighttime cooling is confined to a shallow layer in contact with the ground called the atmospheric boundary layer (ABL) (*1*). As a result of the heating and cooling, the ABL experiences variations in temperature, humidity, wind speed, air pollutants and depth. In the absence of major storm systems, the ABL is shallow, stable and most polluted in the early morning and is deep, unstable and least polluted in the afternoon. The increasing depth of the ABL results in vertical transport and, hence, dilution of the pollutants. Diurnal pollution and meteorological measurements are presented here to illustrated this behavior of the ABL

In the absence of major storm systems, vertical transport is accomplished largely by convection. Air in contact with the ground heats more rapidly than the air above and it becomes positively buoyant. As this convective "bubble" rises, it expands and cools. When the "bubble" cools to the temperature of the surrounding environment, the upward motion ceases and the "bubble" mixes with the surrounding air. Thus, heat, moisture, aerosol particles and gases from the surface are transported aloft by convection.

Convection typically reaches a maximum in the early afternoon when the maximum amout of heat has been transferred from the surface to the air in contact with the surface (*1*). This fact is illustrated by the diurnal variation of pollutants mesured by Hindman and Upadhyay in urban Kathmandu, Nepal (*2*) and by Hindman in rural Steamboat Springs, Colorado (*3*).

In this chapter, diurnal measurements obtained in urban Kathmandu, Nepal and rural Steamboat Springs, Colorado are presented to illustrate the role of convection in the vertical transport of aerosol particles in the atmosphere. The vertical transport of aerosols is estimated based on elementary meteorological principles and a method is presented for determining the verticle transport that makes use of archived meteorologocal data and analysis tools available from the National Oceanic and Atmospheric Administration (NOAA) , Air Resources Laboratory (ARL). The depth of the ABL on September 11, 2001 over Manhattan, New York is estimated by using this method and this estimated depth of is consistent with the observed depth. Thus, the method appears useful for estimating the vertical transport of air pollutants

Experimental Methods

Diurnal measurements obtained in urban Kathmandu, Nepal and rural Steamboat Springs, Colorado are presented here to illustrate the role of convection in the vertical transport of aerosol particles in the atmosphere. The Steamboat Springs site was located near the floor of the Yampa River Valley (40.5N, 106.72W) near a major roadway. At both the Kathmandu and Steampoat Springs sites, the concentrations of aerosol particles with diameters around 10 nm, called condensation nuclei (CN), were continuously measured during the diurnal-cycle using standard instrumentation (TSI Incorporated). Motorized vehicles were a significant source of CN in both of these mountain valley locations. Consequently, the number of moving vehicles were counted as a function of time-of-day at both sites. Finally, the surface meteorological conditions were continuously measured.

Results

Figures 1 and 2, illustrate, respectively, the diurnal pattern of CN concentrations in Kathmandu between October 18-23, 1995 and in Steamboat Springs between January 11-25, 2001. At both locations, the CN concentrations were a maximum during the morning and evening rush-periods when the maximum vehicle movements occurred and a minimum during early afternoon. The CN concentrations were larger in Steamboat Springs due to the shallow wintertime ABL.

The corresponding Kathmandu and Steamboat Springs temperature and moisture measurements (dew-point and relative humidity) are illustrated, respectively, in Figures 3 and 4. The diurnal variation in temperature and moisture is clearly illustrated at both locations. The ABL was moist at both locations. Kathmandu was moist due to the recent ending of the summer monsoon and Steamboat Springs was moist due to snow cover.

The diurnal variations of wind speed and direction at Kathmandu and Steamboat Springs are illustrated in Figures 5 and 6. The air flow was controlled by the topography at both locations. At Kathmandu, in the early morning, a low-speed flow from widely varying directions was caused by "drainage" from the mountains surrounding the valley while in the afternoon, high-speed flow from 270 degrees was a component of the up-slope flow along the Himalayas to the north and east as recently modeled by Regmi, et al. (*4*). In contrast, the around-the-clock, low-speed flow from approximately 90 degrees in Steamboat Springs was "drainage" from the snow-covered Park Range to the east. There were, however, higher speed winds around noontime with a corresponding shift in the wind direction to 180 degrees indicating some weak up-slope flow along the Range.

The diurnal CN and temperature patterns measured in the Kathmandu and Steamboat Springs urban areas, while much smaller than the New York City (NYC) area, are expected to resemble the NYC area in fair weather conditions.

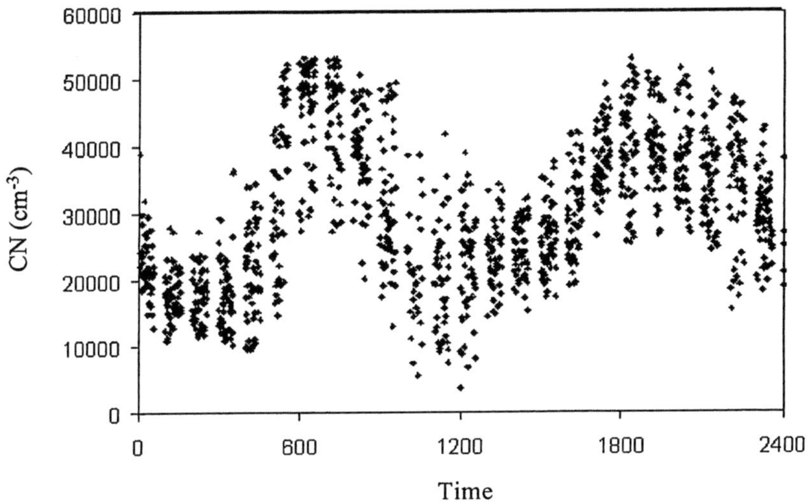

Figure 1. Condensation nucleus (CN) concentrations in Kathmandu, Nepal between October 18-23, 1995.

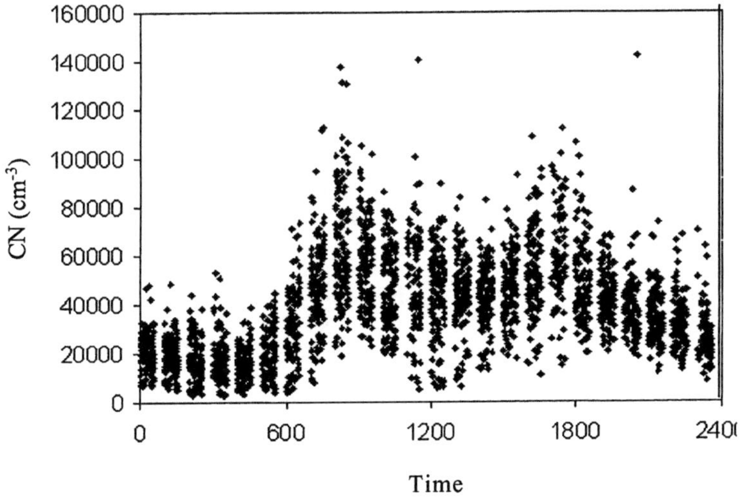

Figure 2. Condensation nucleus (CN) concentrations in Steamboat Springs, Colordo between January 11-25, 2001.

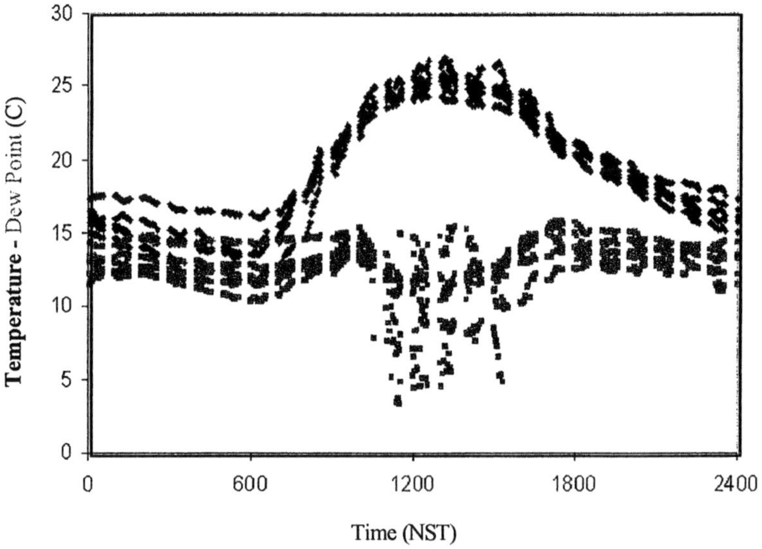

Figure 3. Temperature (black) and dew point (gray) values in Kathmandu, Nepal between October 18-23, 1995.

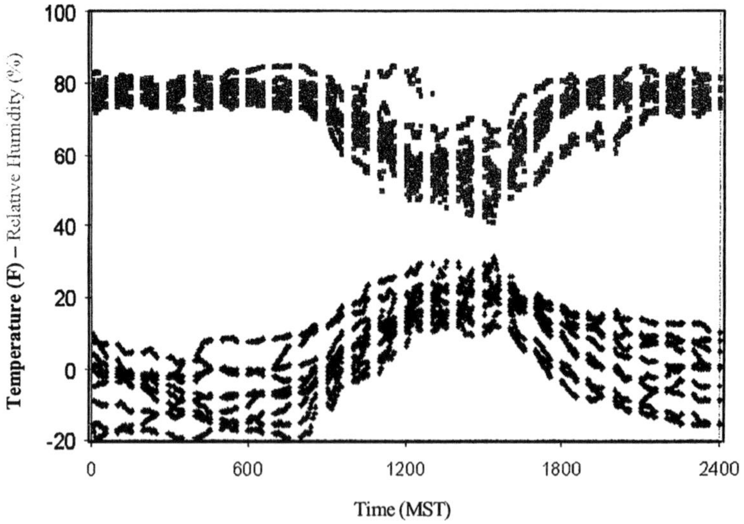

Figure 4. Temperature (black) and relative humidity (gray) values in Steamboat Springs, Colordo between January11-25, 2001.

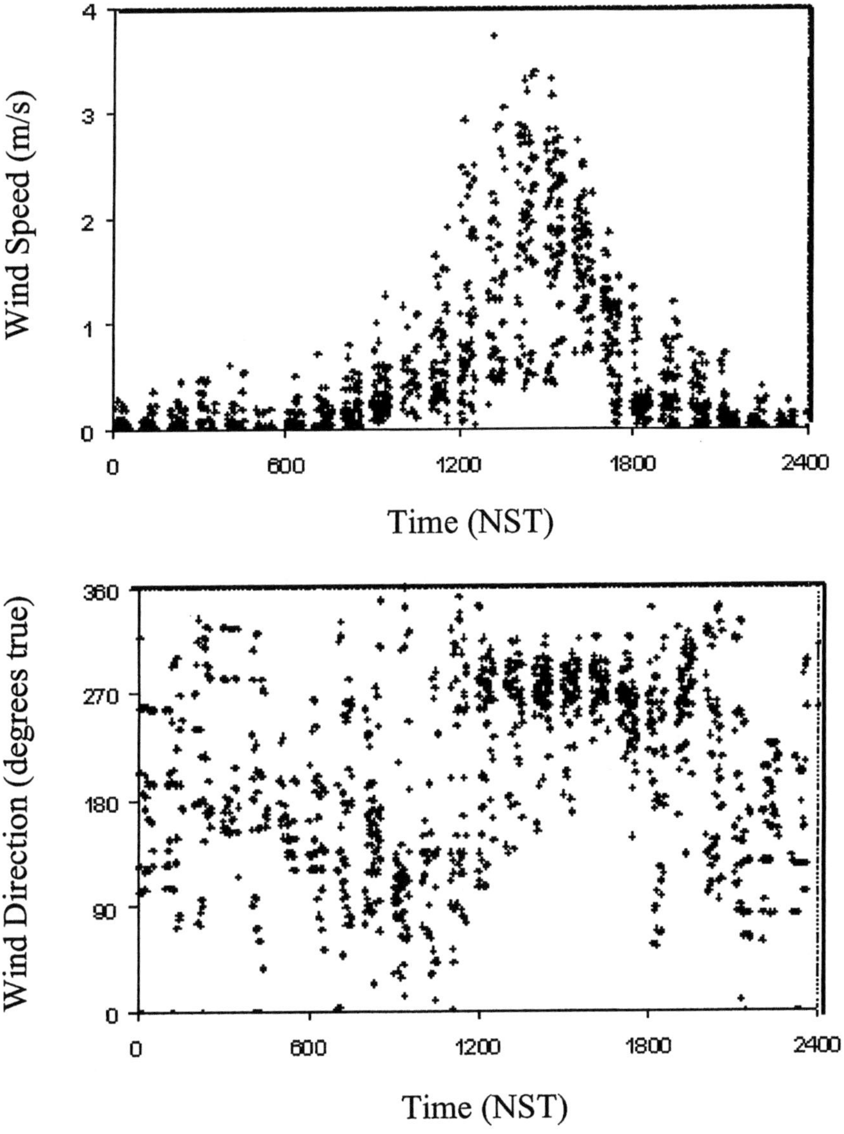

Figure 5. Wind speed (top) and direction (bottom) in Kathmandu, Nepal between October 18-23, 1995.

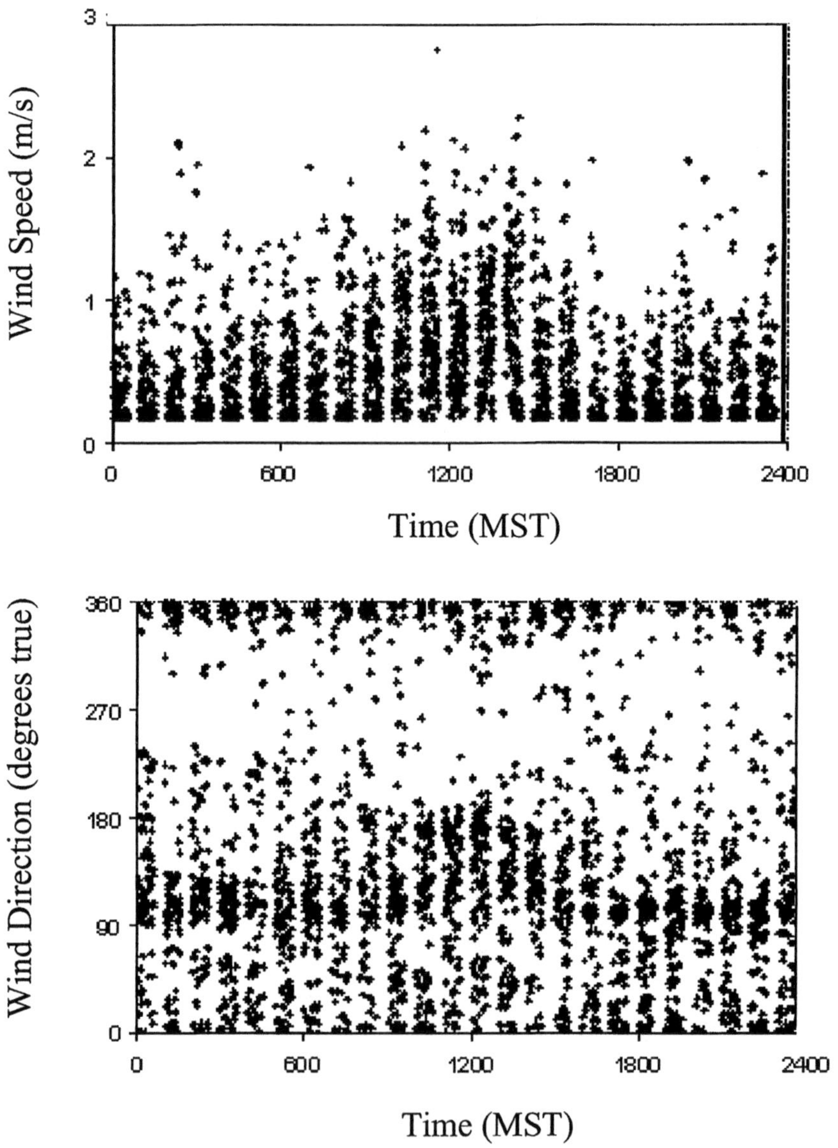

Figure 6. Wind speed (top) and direction (bottom) in Steamboat Springs, Colorado, January 11-25, 2001.

The mountain-vallley flows in Kathmandu and in Steamboat Springs are replaced by the land-sea breeze flows in NYC (*1*).

The depth of the ABL over Steamboat Springs for the period January 11-25, 2001 was estimated using archived meteorological data available from the NOAA ARL Real-time Environmental Applications and Display System (READY) web site (*5*). The latitude and longitude for the site was entered and the section on "Stability Time-series, EDAS 80km" was interrogated using first the "edas.subgrid.jan01.001" data (EDAS = Eta Data Assimilation System). The month (01), day (11) and hour (15) and plot duration of 72 h was entered. The resulting plot for January 11, 12, and 13 is shown in Figure 7. The boundary layer depth is seen to peak each day at about 21 UTC (15 MDT) near the time of maximum convection. The remaining days of the period were analyzed using the "edas.subgrid.jan01.002" data. The average maximum depth of the ABL for the entire period was 558 m, double the 250 m minimum value. This result indicates a weak convection occurred over the snow covered Yampa River Valley during the period and is consistant with the modest mid-day reduction in CN concentrations shown in Figure 2. The reductions were due to the weak convection becasue the maximum vehicle movements occurred about noontime.

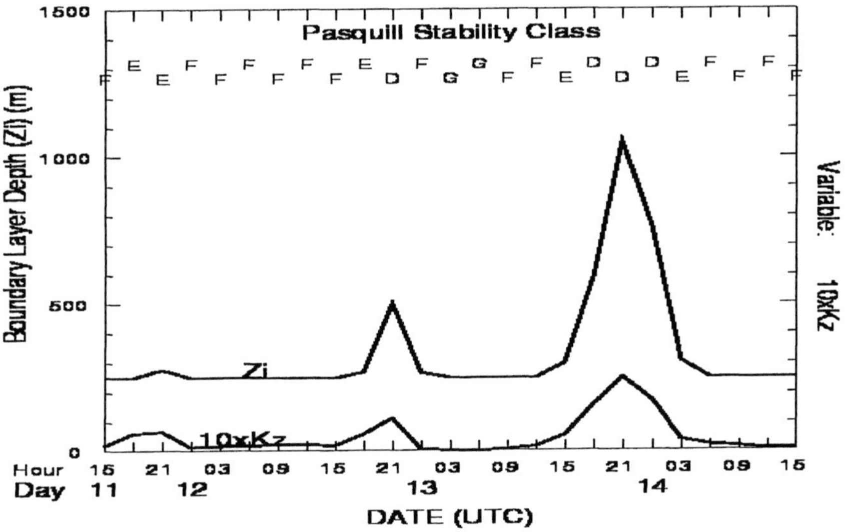

Figure 7. The Atmospheric Boundary Layer Depth and Stability Parameters (D = neutral, E = slightly stable, F = moderately stable, G = extremely stable)for Steamboat Springs, CO , January 11-13, 2001.

Esitmates of vertical transport in New York City on 9/11/01

The depth of the ABL on September 11, 2001 over NYC was estimated from archived meteorological data available from the NOAA-ARL-READY web site (*4*). First, the latitude and longitude of Central Park was entered (40.78N, 73.97W) and then the section "Meteogram, EDAS 80km" was interrogated. The data "edas.subgrid.sep01.001" was selected and the month (09), day (11), hour (12) and plot duration of 12 h were specified. The meteorological parameters chosen were temperature, relative humidity, wind speed and direction, sea-level pressure and precipitation. The resulting temporal display of the measurements is shown in Figure 8. The period displayed is from 12 UTC (0800EDT) September 11, 2001 to 12 UTC the next day. The figure illustrates the fine weather; warm temperatures with low relative humidities, wind from the north with an early morning shower left over from the deluge the previous evening.

Next, the section on "Soundings, EDAS 80km" was interrogated also using the latitude and longitude of Central Park to produce an atmospheric sounding every three hours between 12Z (0800EDT), September 11, 2001and and 12Z the next day. Plate 1 shows the 21Z (17EDT) sounding that illustrates the maximum depth of the ABL. This depth was reached at the time of the maximum surface air temperature of 28C (from Figure 8). The dashed line in Plate 1 represents the parcel of surface air rising and cooling. The parcel rose until it cooled to the environmental temperature denoted by the solid red line. The dotted line is the corresponding cooling of the dew point temperture as the parcel rose. The two lines almost intersect indicating the parcel nearly became saturated. Indeed, on the afternoon of September 11, there were no cumulus clouds observed by the author. The parcel rose to nearly 800 mb or almost 2 km above the surface indicating that the ABL was nearly 2 km deep on the afternoon of September 11.

Finally, the READY section on "Stability Time-series, EDAS 80km" was once again interrogated using the latitude and longitude of Central Park and using the period between 12Z (0800EDT), 11 September 2001 and and 12Z the next day. The atmospheric stability time-series and ABL depth that were produced are illustrated in Figure 9. It can be seen in Figure 9 that the early-morning stable air (Pasquill stability class E) with its shallow ABL was quickly replaced by slightly unstable air (Pasquill stability class C) with a rapidly deepening ABL. After sunset about 00Z (20EDT), the ABL quickly became shallow as the air became stable. The ABL was 1700 m deep according to this analysis, which is consistant with the nearly 2 km deep ABL from the earlier analysis of atmospheric soundings.

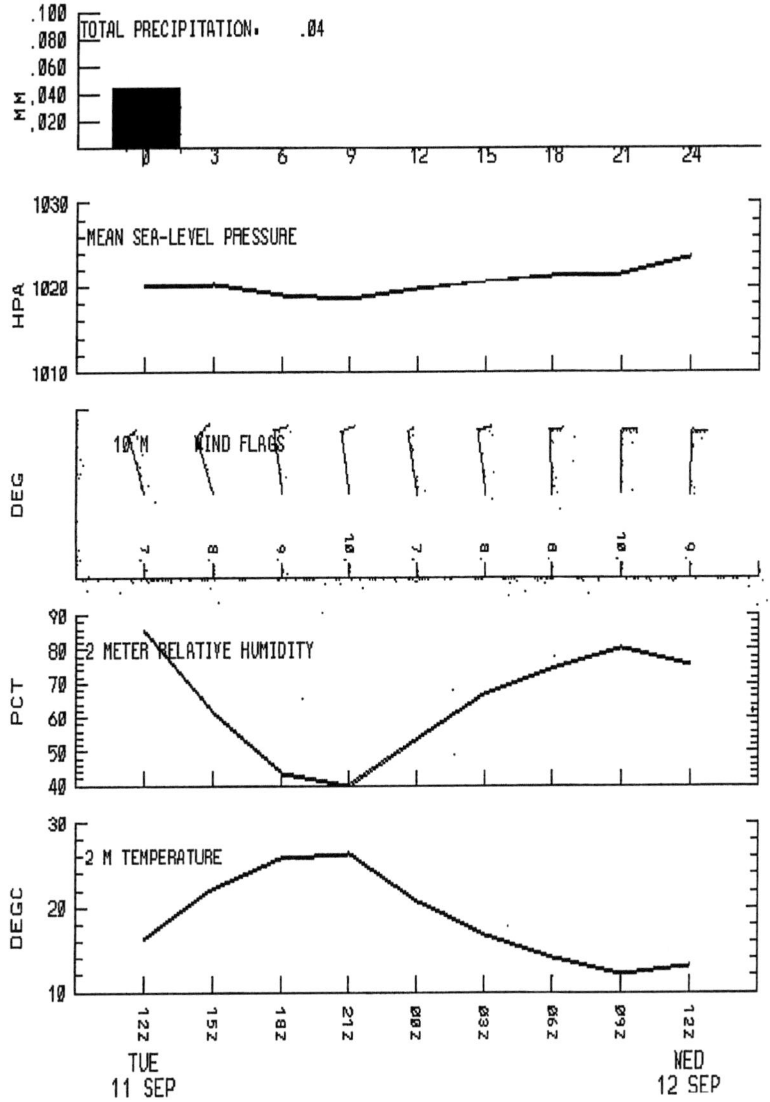

Figure 8. Meteorological conditions (total precipitation, wind direction, relative humidity, temperature and pressure) for Central Park, NYCon September 11-12, 2001.

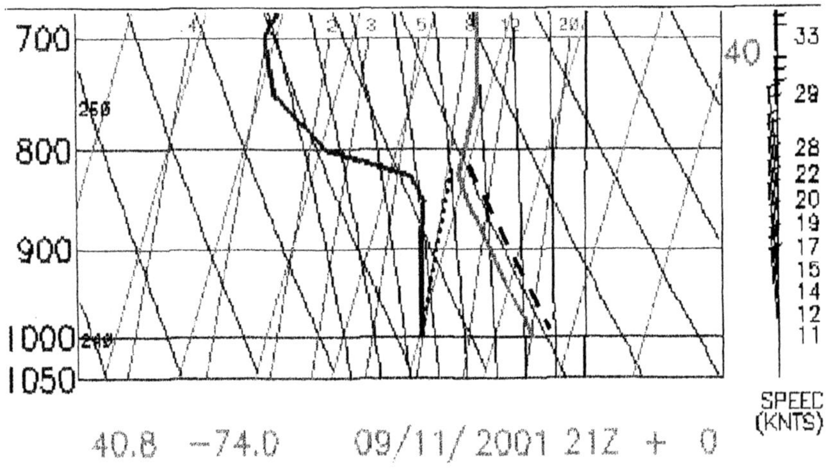

Plate 1. The environmental sounding for Central Park, NYC; September 11; 21Z (17EDT) (See page 32 of color inserts.)

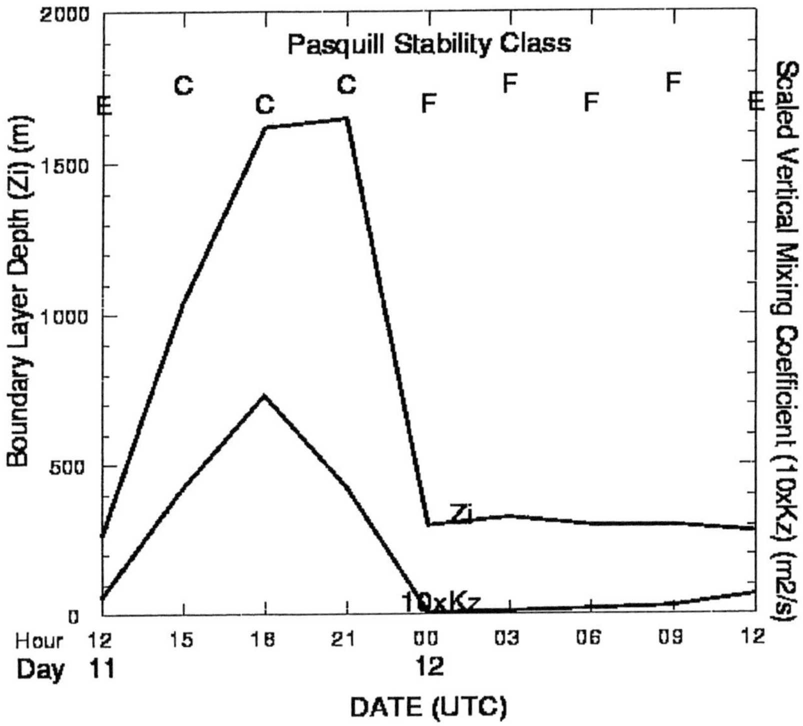

Figure 9. . The Atmospheric Boundary Layer Depth and Stability Parameters (D = neutral, E = slightly stable, F = moderately stable, G = extremely stable) forCentral Park, NYC for September 11-12, 2001.

The depth of the ABL on the afternoon of September 11, 2001 was estimated from studying photographs of the smoke rising from the WTC site (see Chapter 11, Figures 2 and 4). The smoke rose to well above the original height of the Twin-Towers which were about 500 m tall. Some of the plume-rise was due to the intense heat of the fires. Further, the author observed the plume on the afternoon of September 12 as he left Manhatten when the George Washington Bridge was reopened to bicycle traffic. Again, the plume rose to well above the original height of the Towers. Although these estimates are qualitative, they indicate the plume propably rose to the top of the ABL which was estimated to be between 1.7 and 2 km above sea level. It is fortunate that the ABL was so deep on the 11th and 12th to effectively disperse the potentially toxic plume, perhaps, saving more lives.

The most detailed transport and dispersion estimates from the WTC tragedy may have been provided by Georgopopoulos (*6*). The extremely high time and space resolved Colorado State University, Regional Atmospheric Modeling System (RAMS) was used to reproduce the WTC plume in this work.

Forecasting vertical transport

The method outlined here for estimating the depth of the ABL can also be used to forecast the depth of the ABL and estimate the vertical mixing of particulate matter and other pollutants in the atmosphere. The forecast data are available on the NOAA-ARL-READY web site under "Current meteorology" (*5*). The procedures used with the forecast data are identical to those used with the archived data.

Conclusions

The method presented here to estimate the vertical extent of the plume from the collapse of the Workd Trade Center indicated that the plume could have risen to between 1.7 and 2 km above sea level. The nearly 2 km depth of the layer rapidly diluted the plume and perhaps prevented further loss of life. This method also can be used to forecast the depth of the atmospheric boundary layer.

References

1. Stull, R. B. *Meteorology for Scientists and Engineers(2nd Ed.)* Brooks/Cole Publishing Co., St. Paul, MN, 2000, pp. 65-76.
2. Hindman, E. E. and Upadhyay, B. P. *Atmos. Environ.* **2002**, *36*, 727-739.
3. Hindman, E.E. unpublished results.
4. Regmi, R., Kitada, T., Kurata, G. *J. Appl. Meteor.* **2003**, *42*, 389–403.
5. "NOAA ARL Real-time Environmental Applications and Display sYstem - Archived Meteorology" [http://*www.arl.noaa.gov/ready/amet.html*] National Oceanic and Atmospheric Administration, Air Resources Laboratory, 7/26/2004.
6. Georgopopoulos, P. G. Paper presented in ACS Symposium: Urban Aerosols and their Impact: Lessons Learned from The World Trade Center Tragedy, New York City, Sept. 10, 2003. ACS Division of Environmental Chemistry Preprints of Extended Abstracts, paper 135, **2003,** 43 (2), 1363.

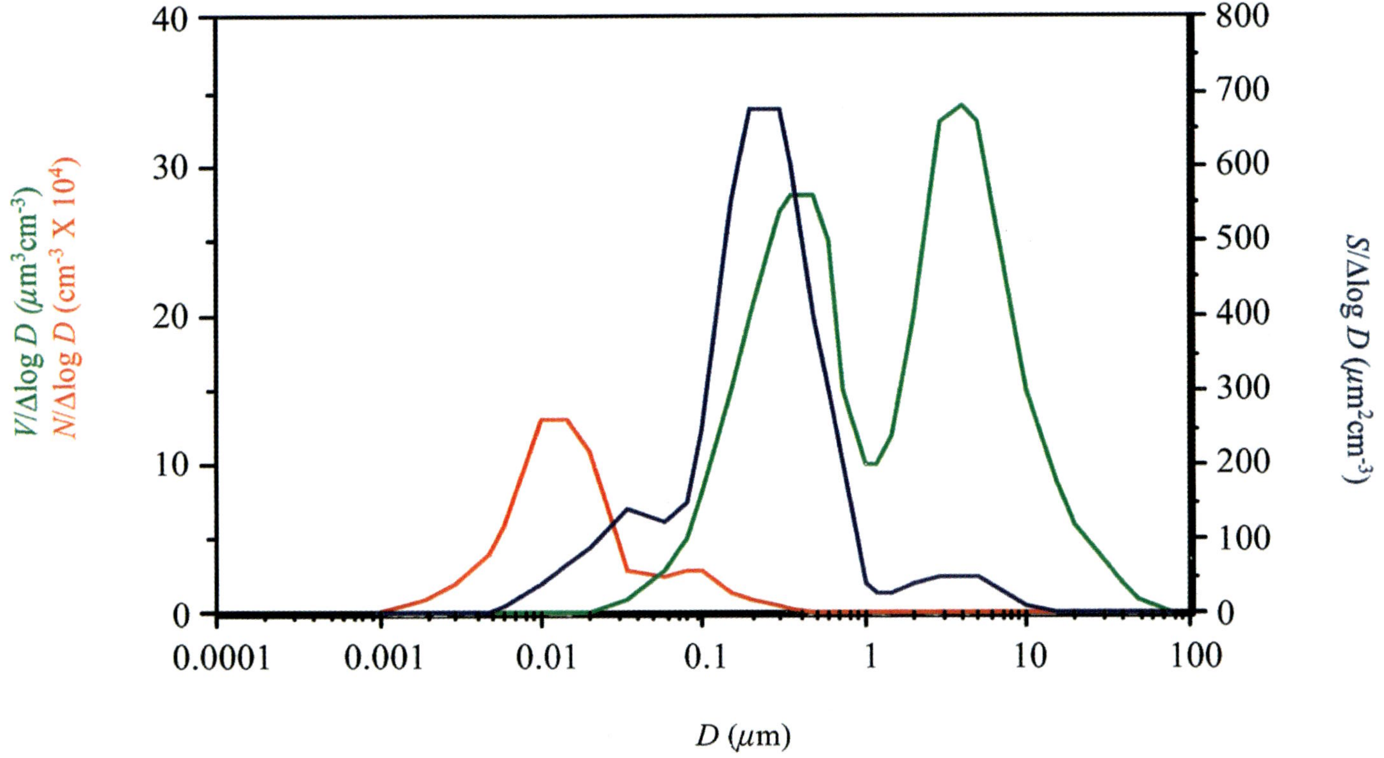

Plate1. Number (——), volume (——), and surface (——) distributions for a model urban aerosol(1).

Plate 1.2. Aerial photograph of Mexico City taken in April 2003, showing the early afternoon "haze" over this megacity.

Plate 1.3. Photograph from a commercial aircraft, taken in the early afternoon in April 2003, showing the Mexico City megacity plume of aerosols leaving the urban area and its regional-scale impact.

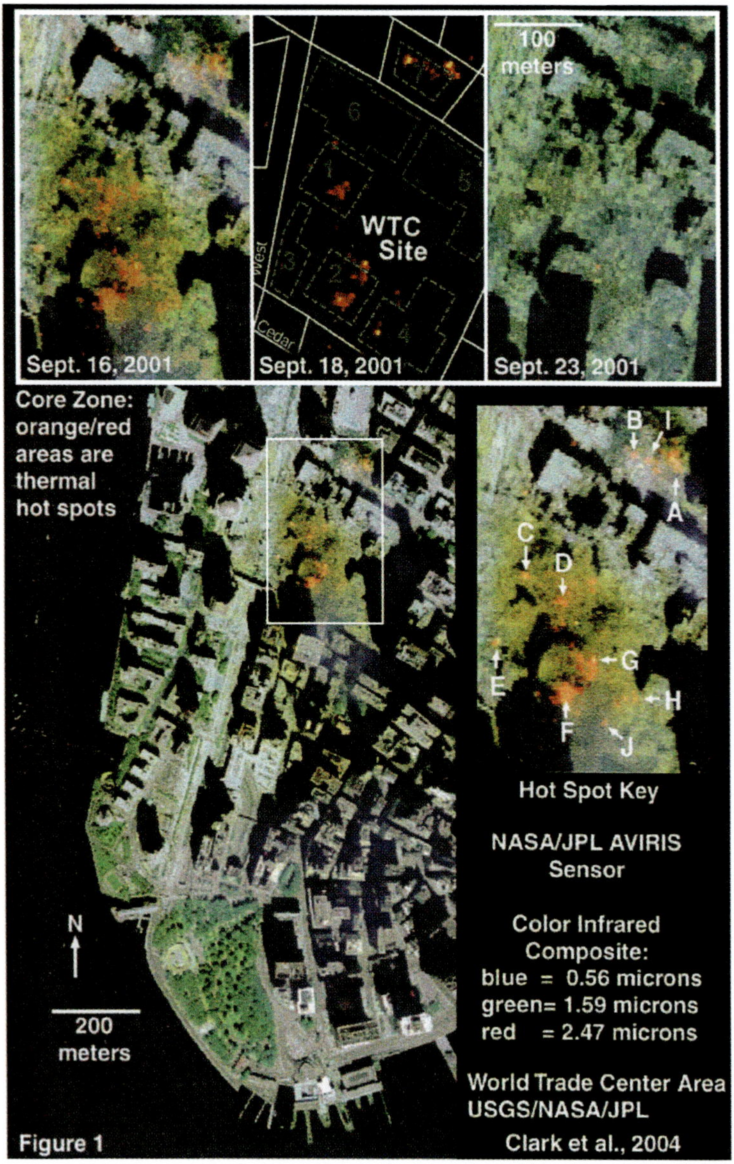

Plate 4.1. False color images of the core affected area around the WTC extending from 5 to 12 days after the collapse. Hot spots appear orange and yellow. The key shown at the right corresponds to the hot spot locations listed in Table 1.

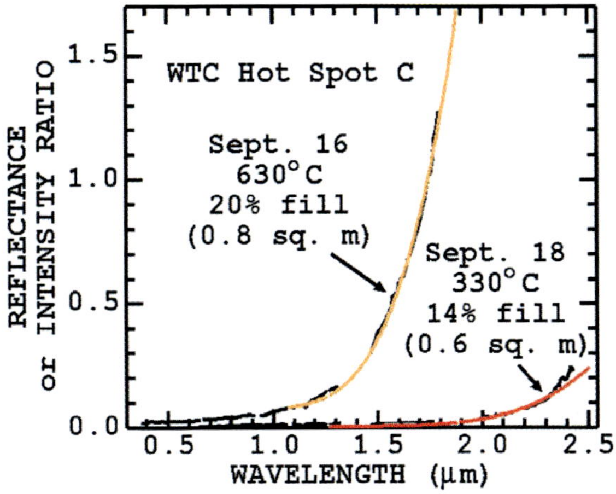

Plate 4.2. AVIRIS spectra (black) for thermal hot spot C on September 16 and 18 shown with the fitted thermal response (orange, red). The spectral shape is used to constrain temperature and the intensity constrains area.

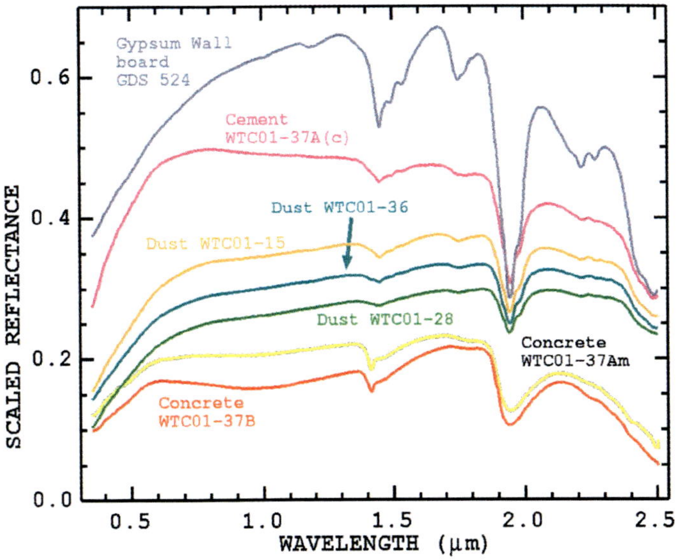

Plate 4.3. Spectra of the WTC field samples used to map the WTC plume.

Plate 4.4. Map showing the WTC dust and debris plume as detected by AVIRIS indicating an asymmetry in the dust and debris distribution, with more iron bearing materials to the south by southeast.

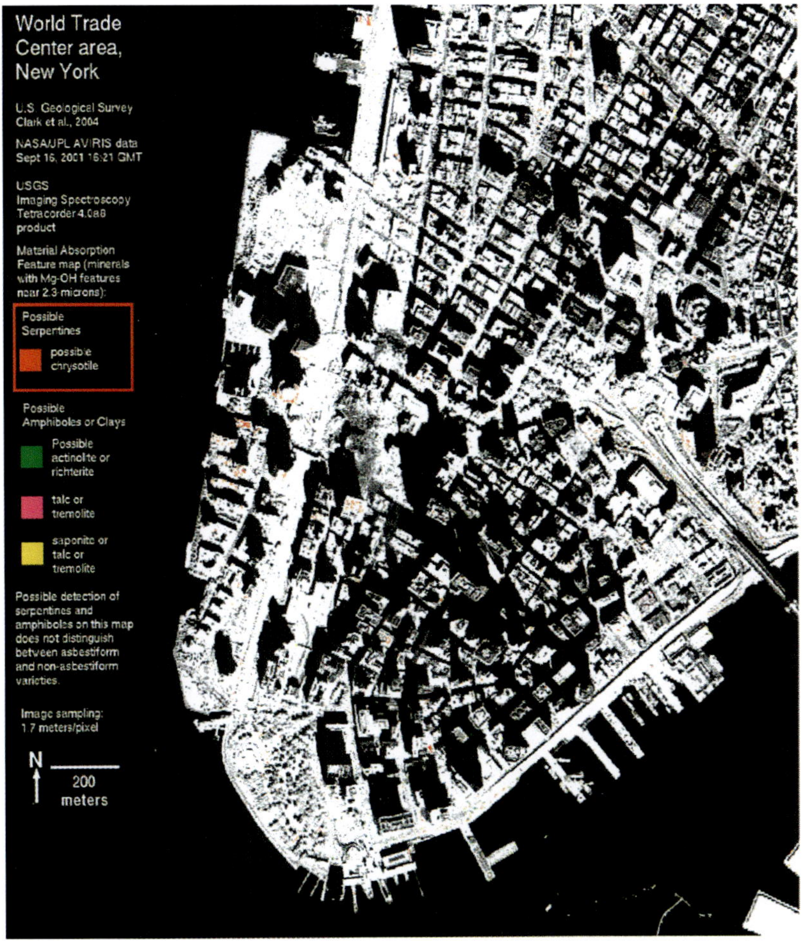

Plate 4.5. Spectral reflectance map keyed to minerals that can have asbestiform morphology showing only scattered possible occurrences at the surface around the WTC area.

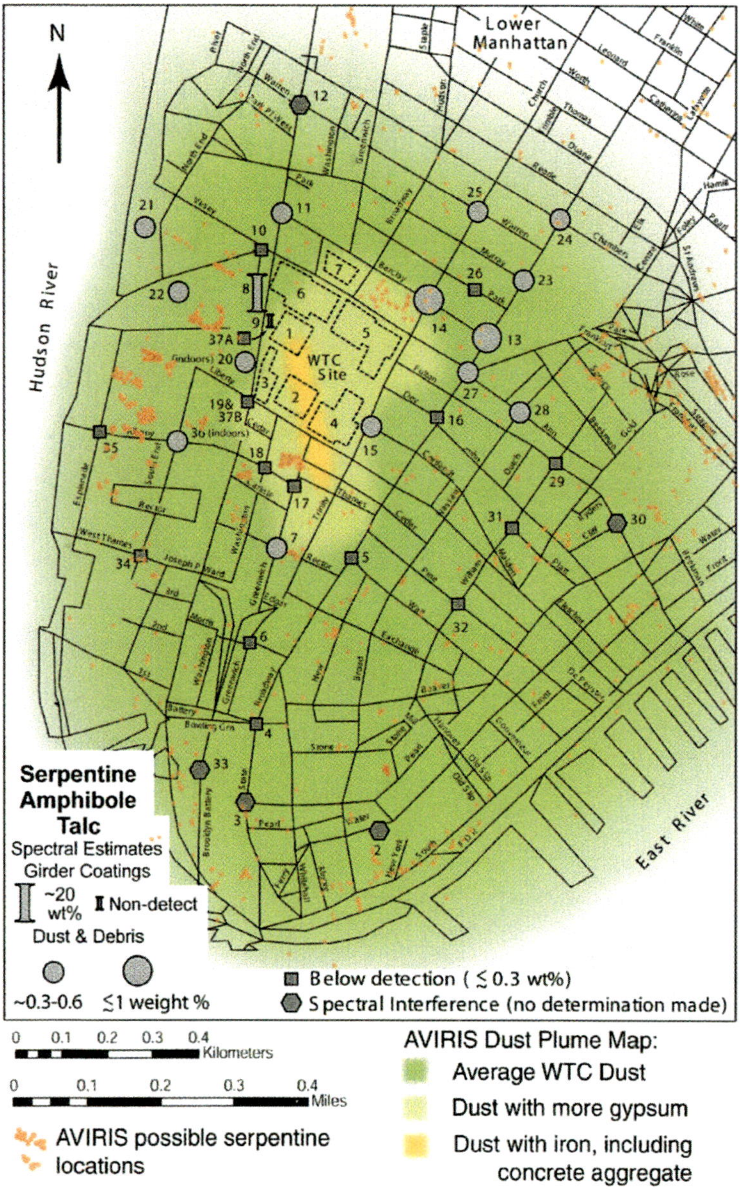

Plate 4.6. The combination of AVIRIS asbestiform mineral spectral signatures (orange spots), field sample data, and field observations to show the extent of the WTC dust plume on September 16, 2001.

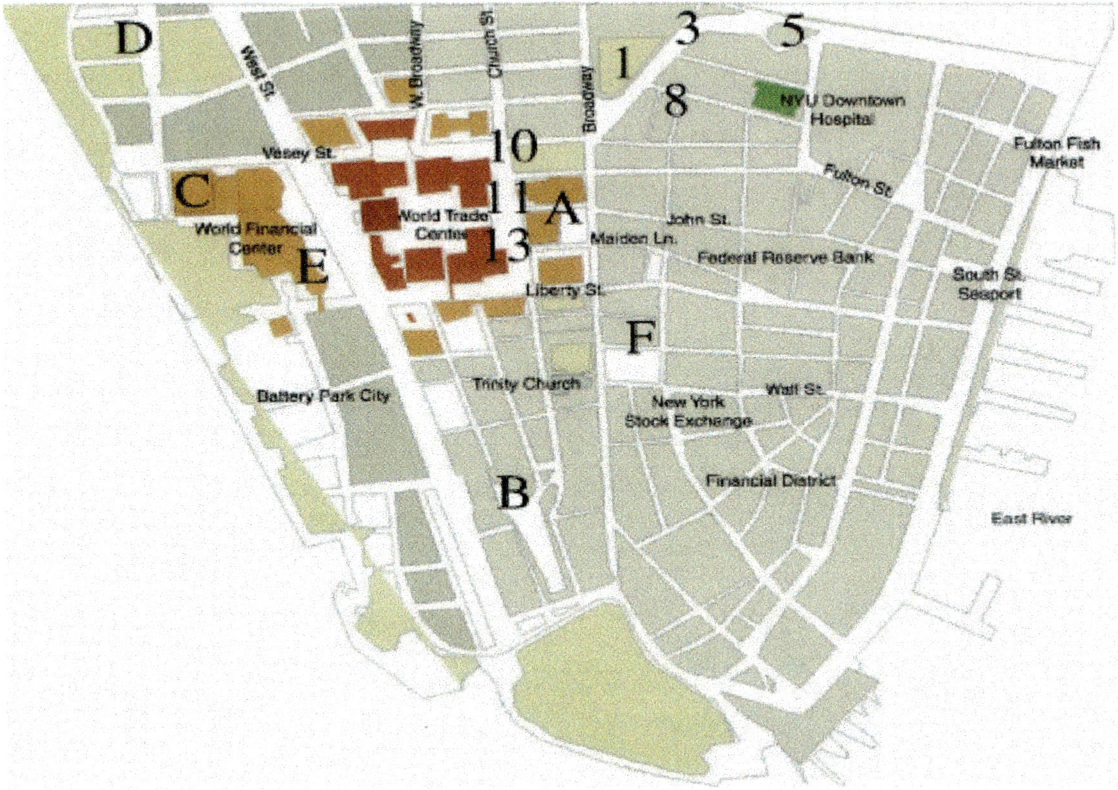

Plate 7.1. Sites where WTC dust samples were collected on September 12, 2001 (numbers) and September 13, 2001 (letters)

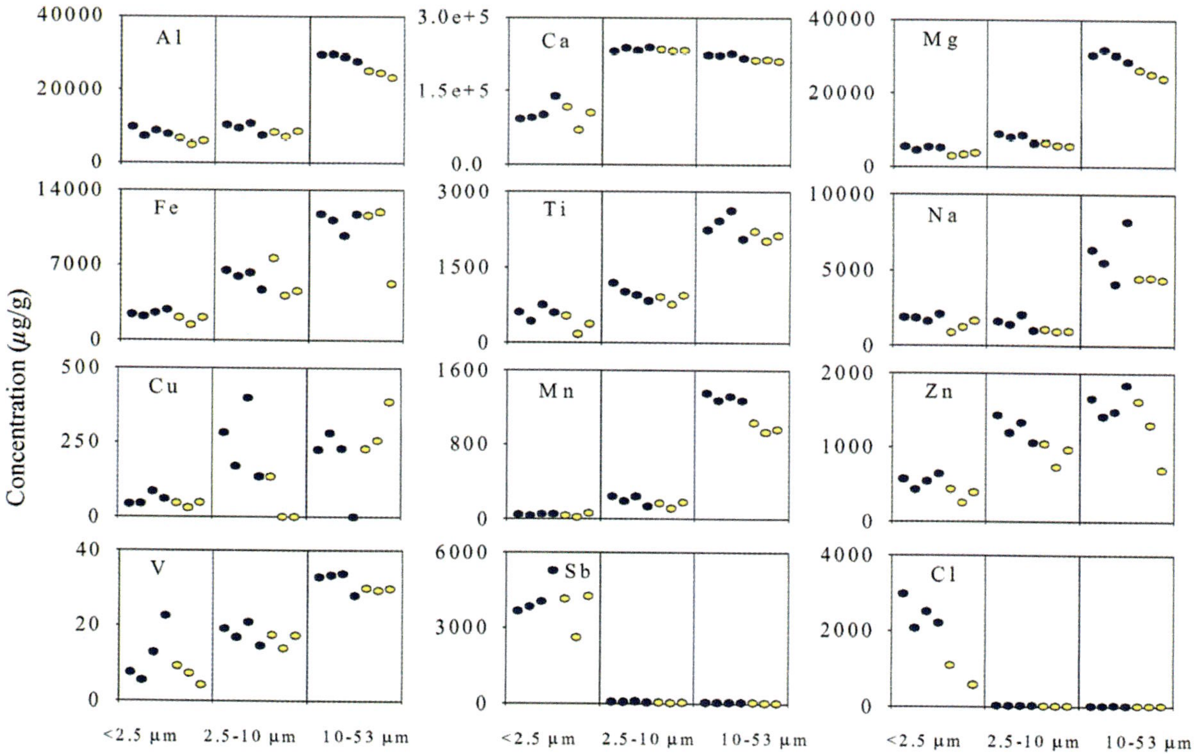

Plate 7.2. Elemental composition of selected outdoor (●) and indoor (○) dust samples by size classification.

CAMDEN, NJ (45 KM SW OF EML) 24

NEW BRUNSWICK, NJ (125 KM SW OF EML) 25

Plate 8.1. Monitoring site locations in the New York and New Jersey metropolitan areas. Data shown in Plate 1 are from (*): PS 154, Bronx, NY (5), PS 199, Queens, NY (9), Liberty State Park, NJ (10), Maspeth Library, Long Island, NY (12), Queens College, Queens, NY (18), Manhattanville Post Office, NY (19), Environmental Measurements Laboratory, NY (20), Jersey City, NJ (21), Fort Lee, NJ (22), Newark, NJ (23), Camden, NJ (24), and New Brunswick, NJ (25).

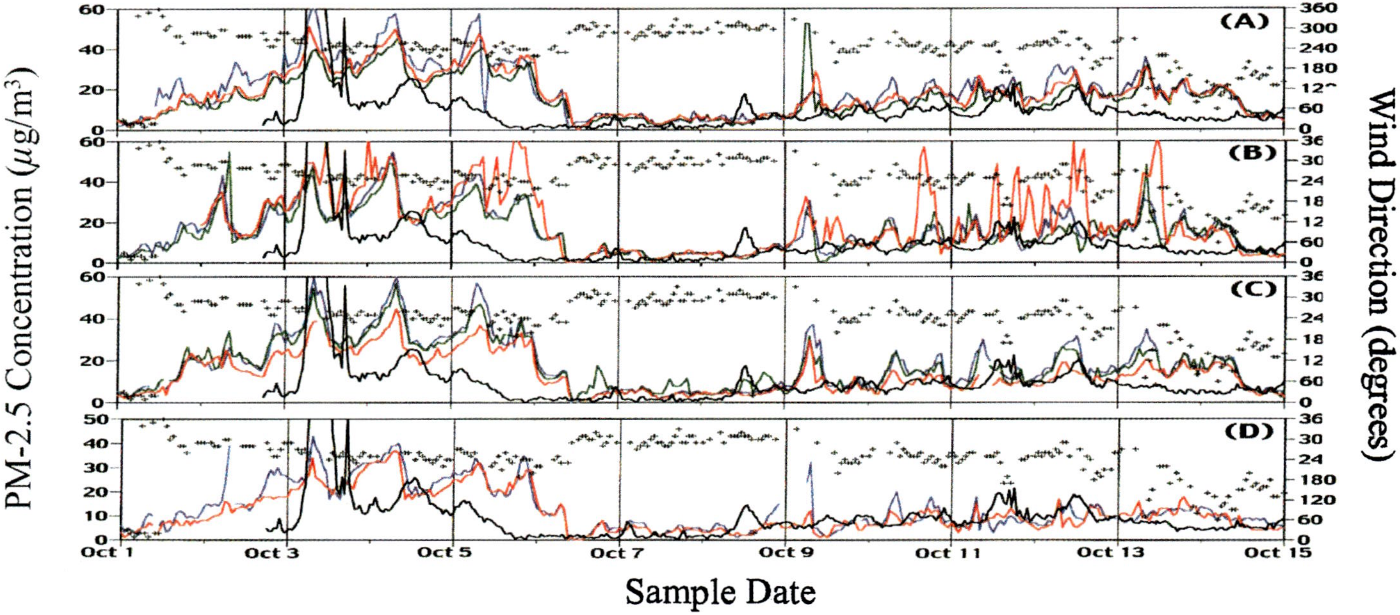

Plate 8.2. Comparison of measured PM-2.5 concentrations at EPA monitoring sites (colored lines) with the calculated PM-2.5 concentration at EML (black lines) along with wind directions measured in Central Park (black crosses). Sites north of EML; Fort Lee, NJ (blue), Manhattanville, NY (green), and PS 154 , Bronx, NY (red), are shown in (A). Sites west of EML; New Jersey City, NJ (blue), Newark, NJ (green), and Liberty State Park, NJ (red) are shown in (B). Sites East of EML; PS 199, Queens, NY (blue), Maspeth Library, NY (green) and Queens College, NY (red), are shown in (C). Sites south of EML; Camden, NJ (blue) and New Brunswick, NJ (red), are shown in (D).

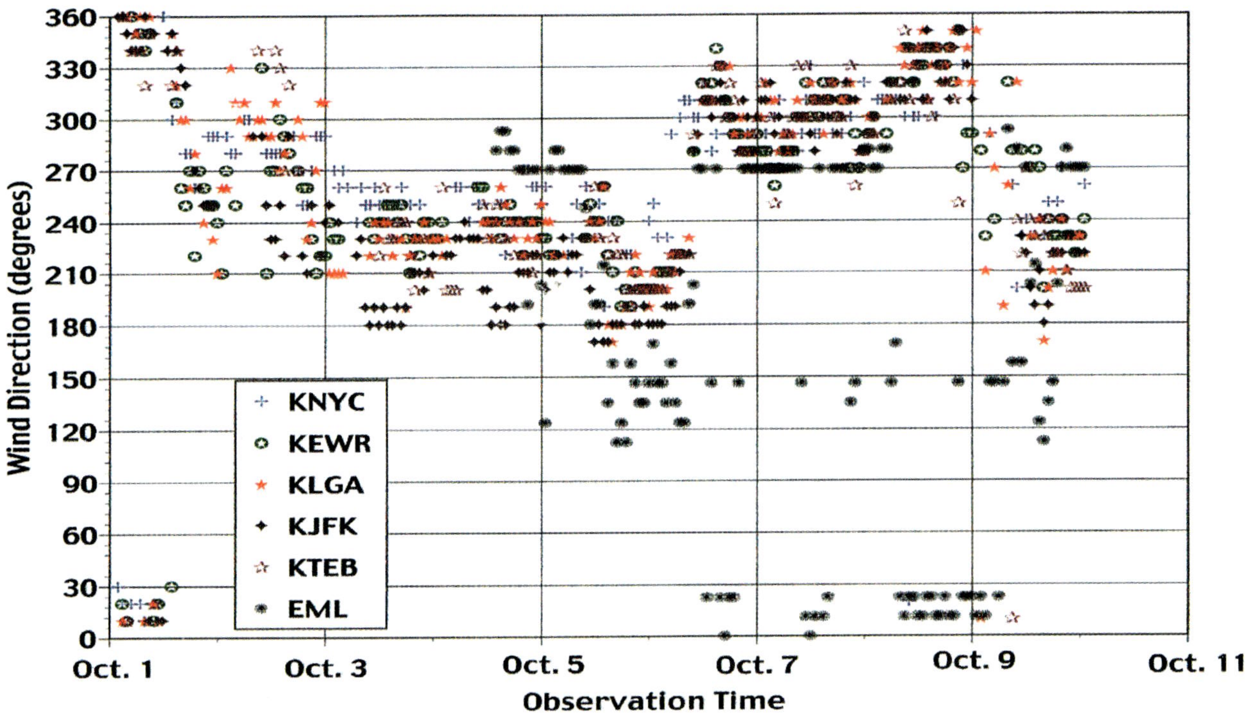

Plate 8.3. Comparison of wind directions from LaGuardia (KLGA), Newark (KEWR), JFK (KJFK) and Tetterborough (KTEB) airports, Central Park (KNYC) and the roof of EML (EML) for early October 2001.

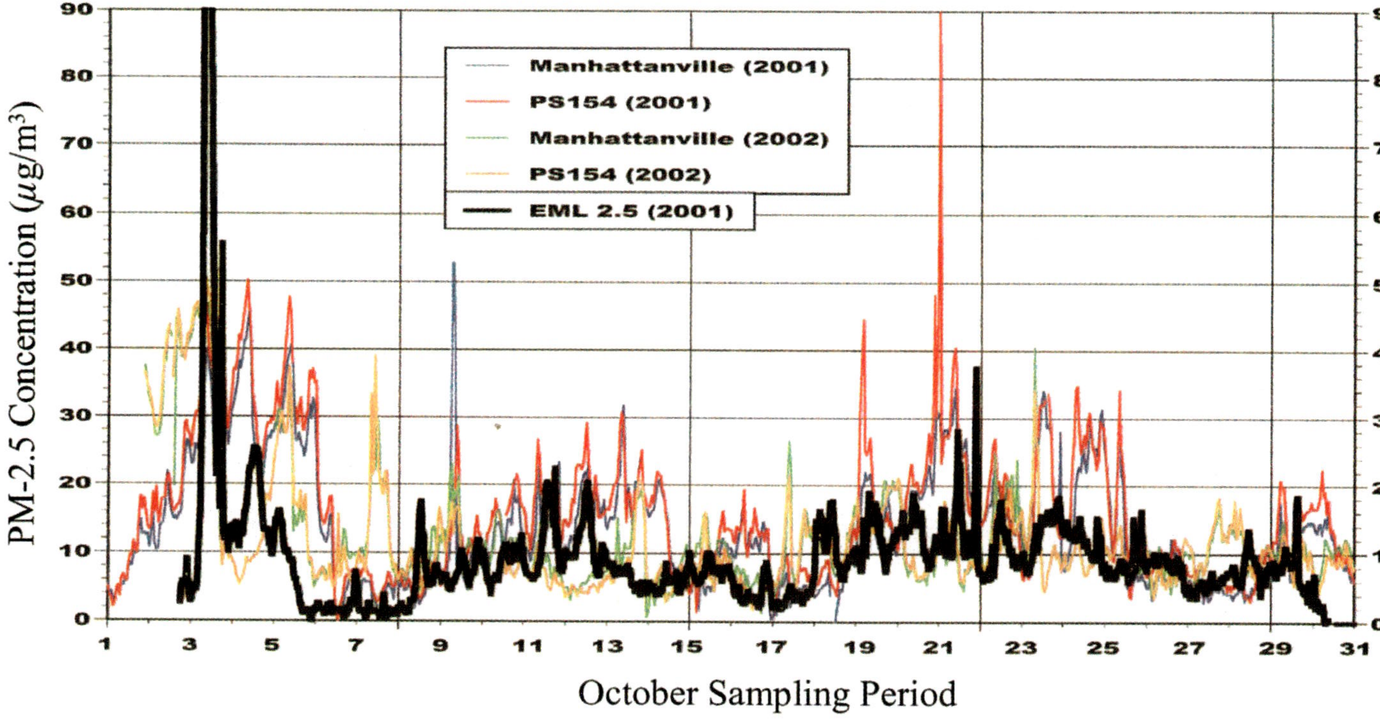

Plate 8.4. Comparison of October 2001 and 2002 PM-2.5 concentrations measured at the EPA monitoring sites at Manhattanville Post Office (Figure 2, site 19) and PS 154 (Figure 2, site 5) with the calculated PM-2.5 concentration at EML for October 2001.

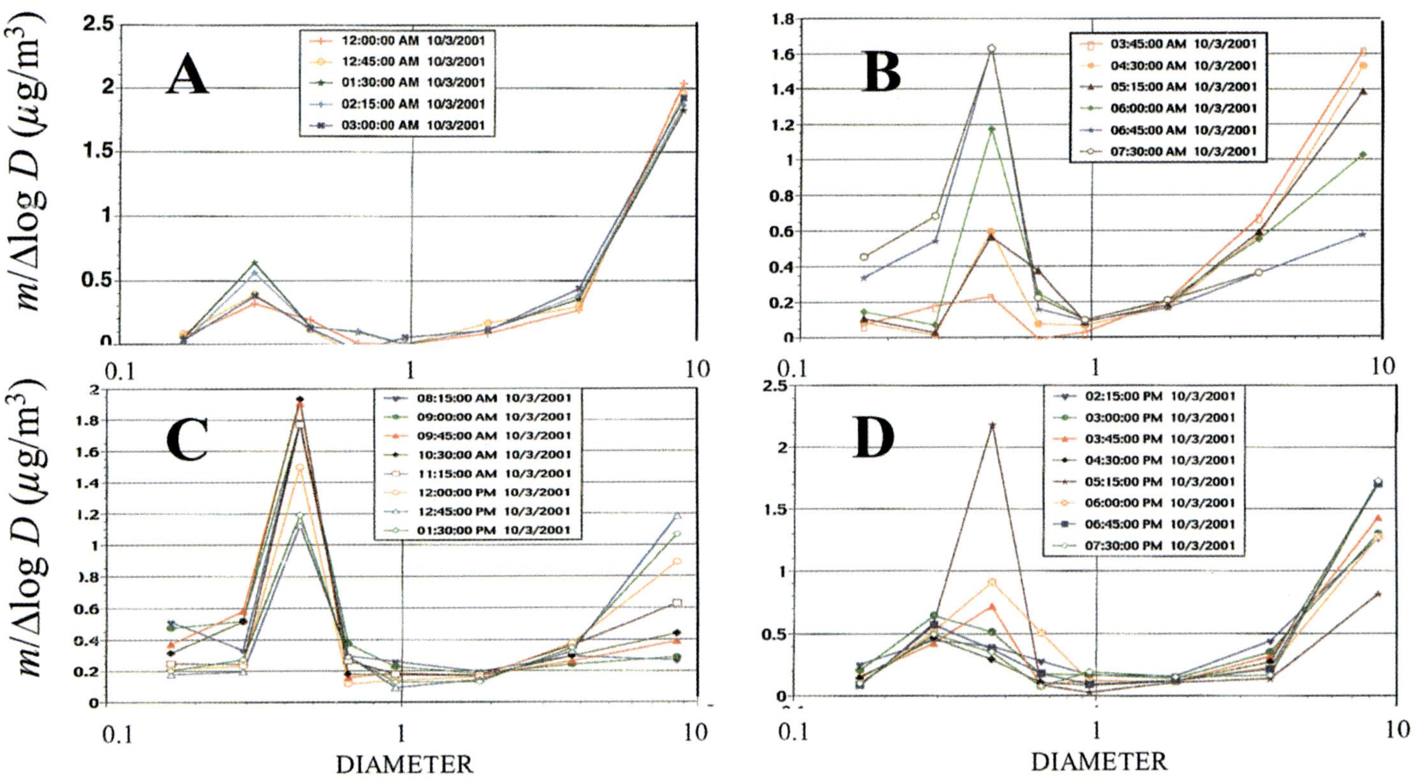

Plate 8.5. Normalized aerosol mass size distribution observed on October 3 when EML intercepted a plume from Ground Zero. (A) 12:00 am to 03:00 am. (B) 03:45 am to 07:30 am (C) 8:15 am to 1:30 pm (D) 2:15 pm to 7:30 pm.

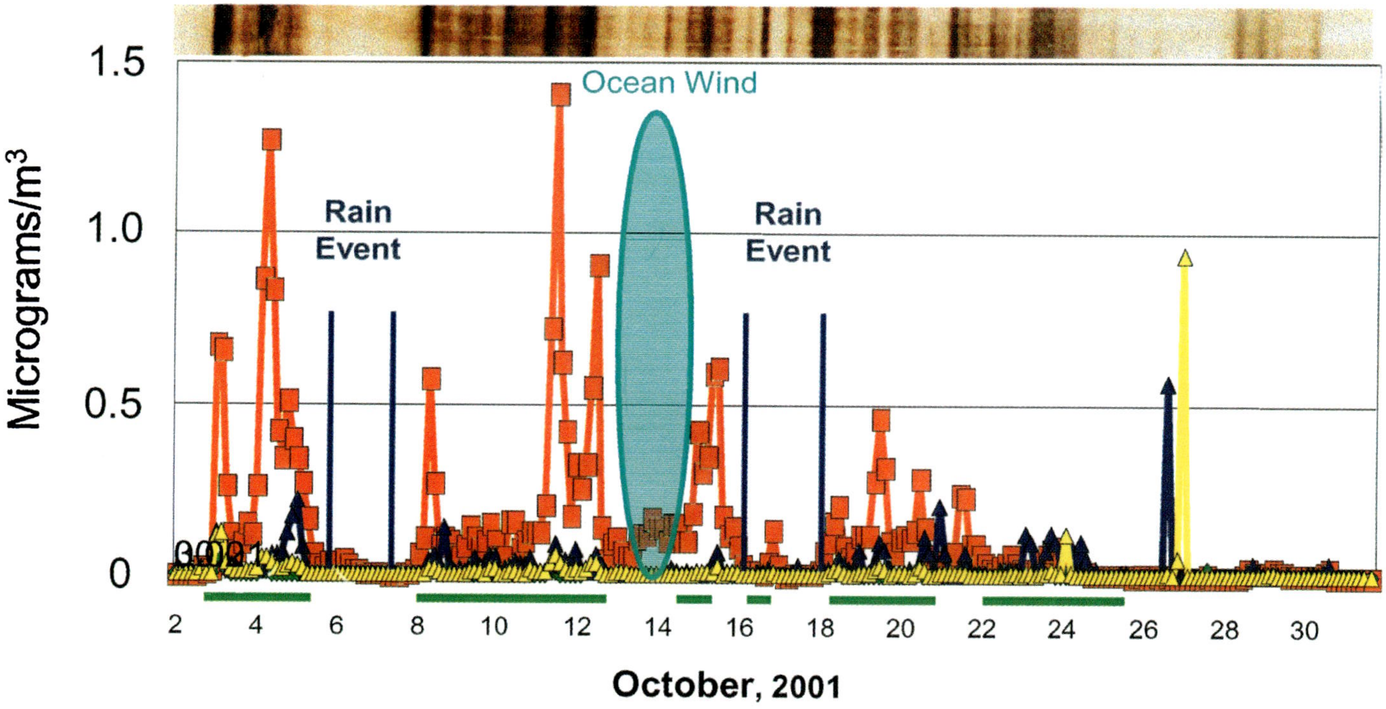

Plate 9.1. Typical crustal elements; silicon (——), calcium (——), iron (——), and aluminum (——) in the very fine particle mode (0.26 – 0.09 μm) of samples obtained at Varick Street during October, 2001. At top is shown the reflected light picture of this DRUM stage. Periods of favorable wind direction that would bring the WTC plume to Varick Street are indicated below the plot (——), Two rain events and an eastern ocean wind are indicated.

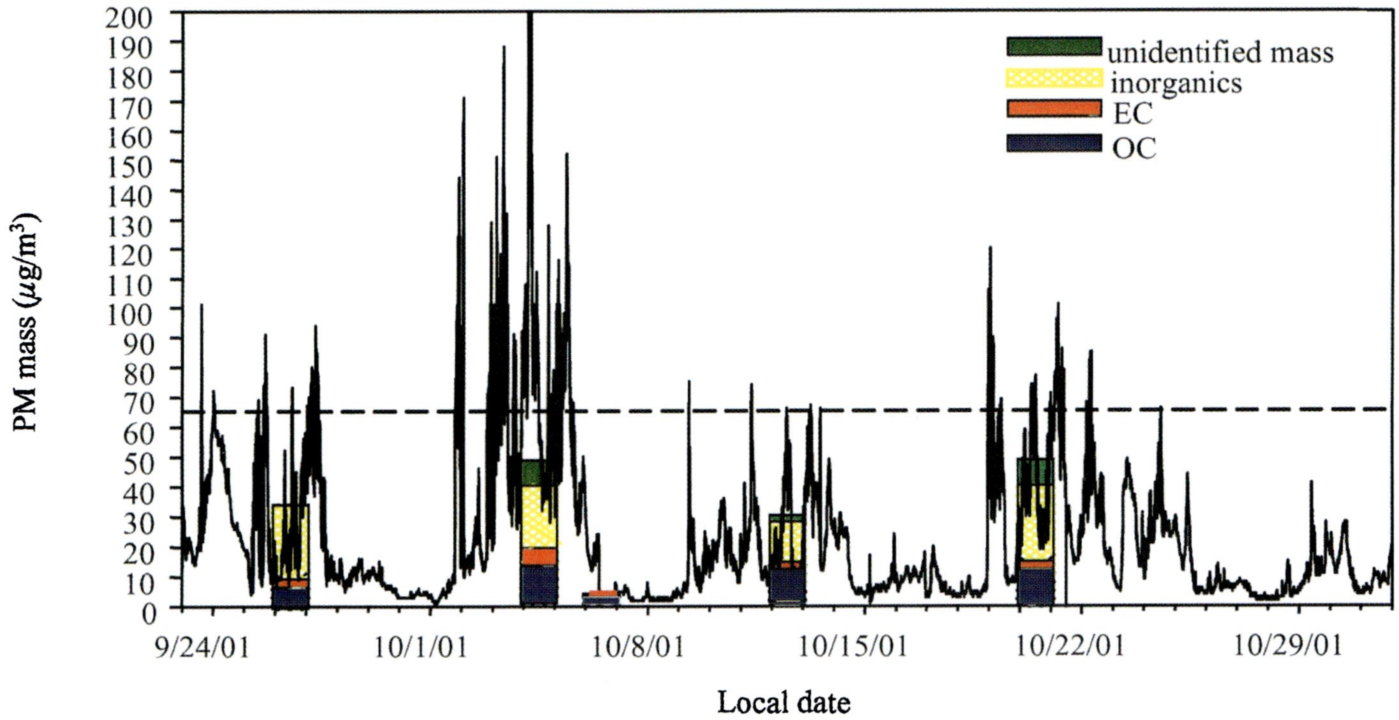

Plate 10.1. Chemical mass balance assignments for the sample sets obtained on 9/26, 10/4, 10/6, 10/12, and 10/20 superimposed on the real-time nephelometer data (solid line). The EPA revised primary 24-hr standard for PM-2.5 is shown by the dotted line at 65 g/μm^3.

Color insert page 18

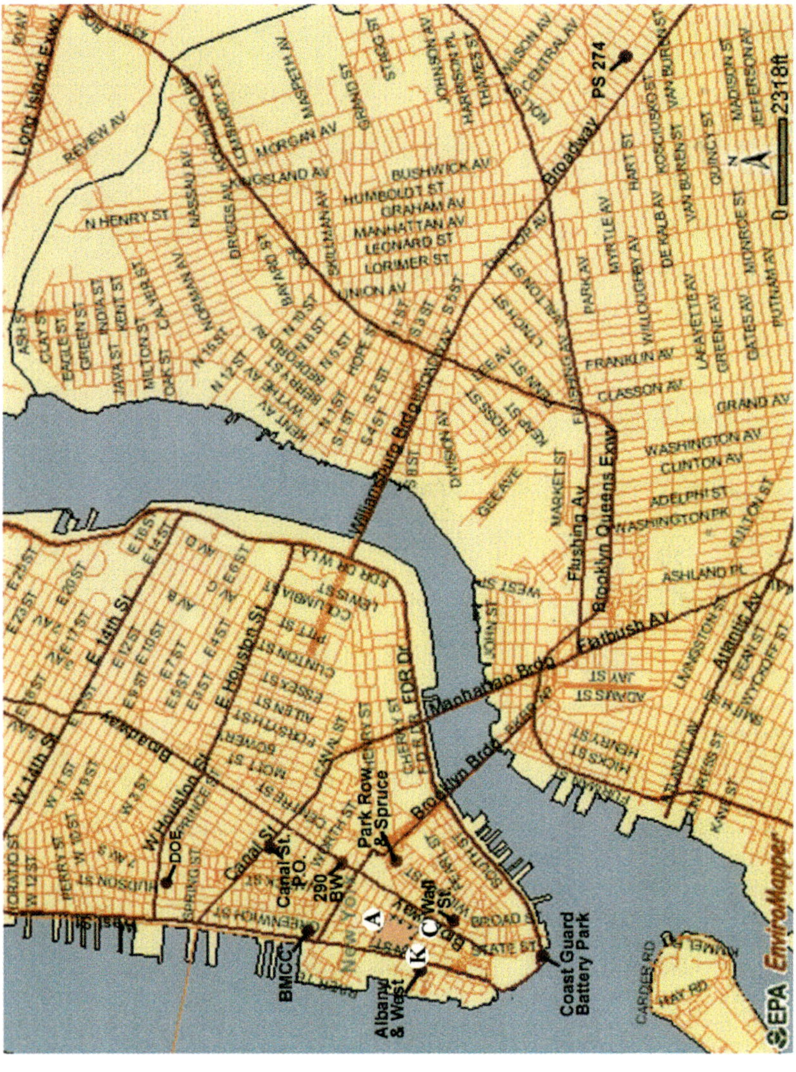

Plate 11.1. Particulate matter monitoring sites established in response to the WTC attack.

Plate 11.2. Spread of dense dust and smoke cloud over lower Manhattan, drifting to the east-southeast immediately after the September 11, 2001 collapse of the World Trade Center buildings. (NYPD file photo).

Plate 11.3. Spread of dense dust and smoke cloud over lower Manhattan, drifting to the east-southeast immediately after the September 11, 2001 collapse of World Trade Center buildings. (NYPD file photo).

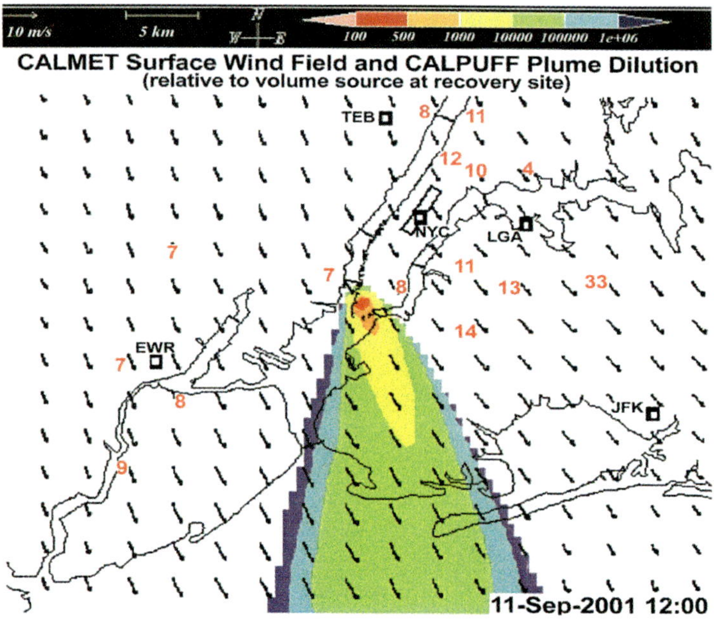

Plate 11.4. Modeled WTC plume dispersion for 12:00 PM September 11, 2001. The numbers are hourly PM-2.5 concentrations ($\mu g/m^3$).

Plate 11.5. Satellite photograph of the WTC plume lofting from Ground Zero at 11:43 AM on September 12, 2001. (Reproduced from Space Imaging) (11).

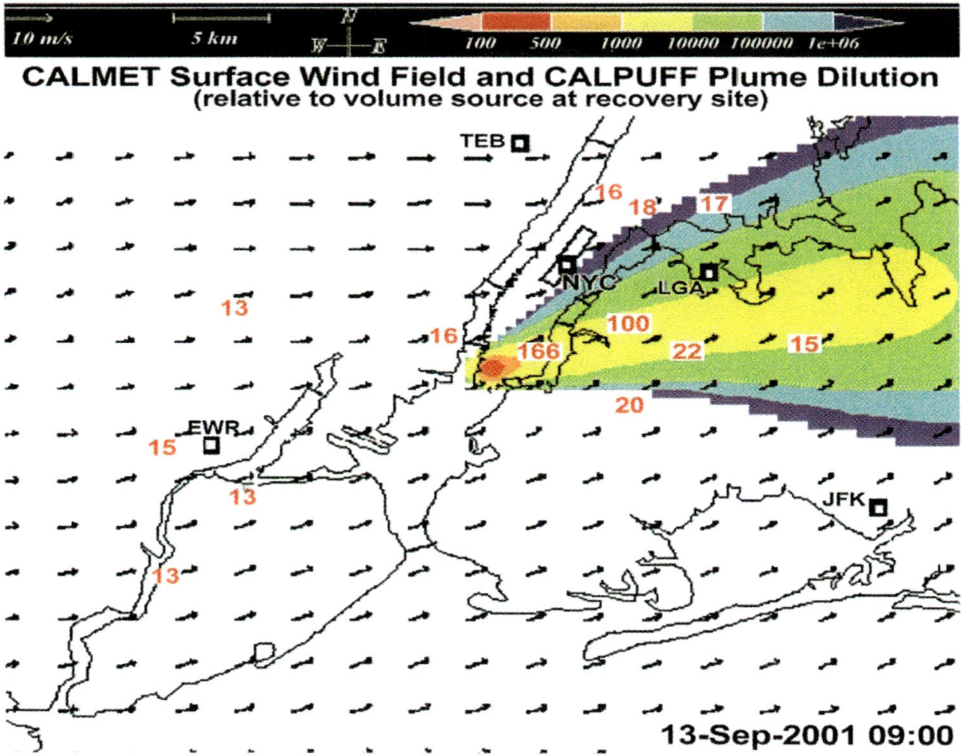

Plate11.6. Modeled WTC plume dispersion for 9:00 AM September 13, 2001. The numbers are hourly PM-2.5 concentrations (μg/m³).

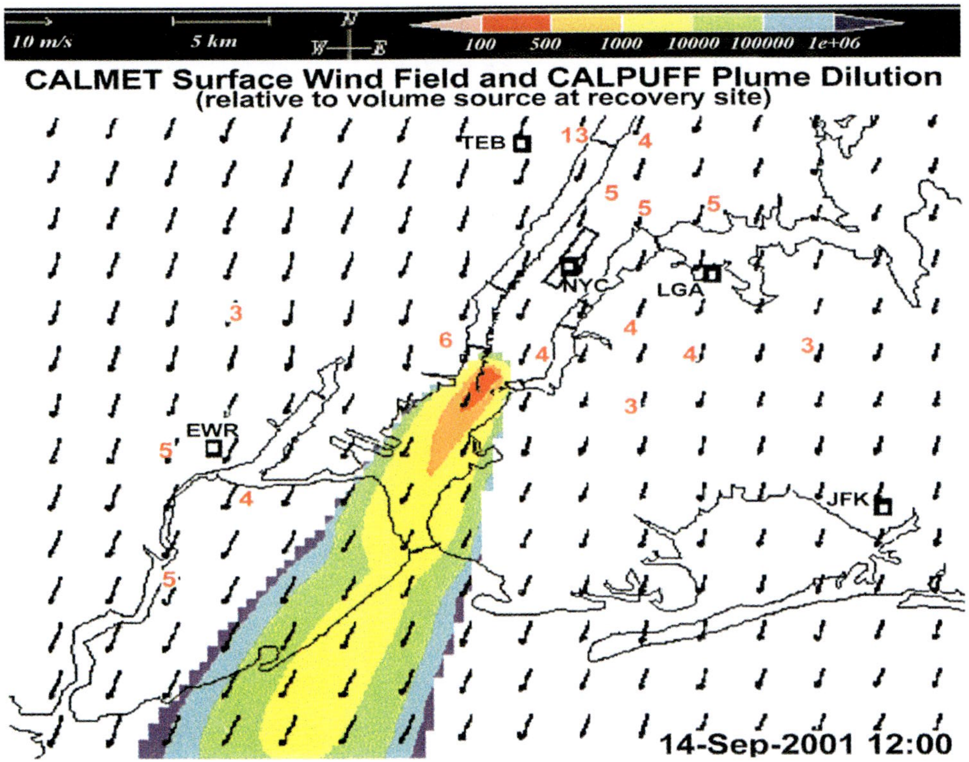

Plate 11.7. Modeled WTC plume dispersion for 12:00 noon September 14, 2001. The numbers are hourly PM-2.5 concentrations ($\mu g/m^3$).

Plate 11.8. Satellite photograph of WTC plume lofting from Ground Zero at 11:54 AM on September 15, 2001, and dispersing to the south-southwest out over the New York Harbor (Reproduced from Space Imaging) (11)

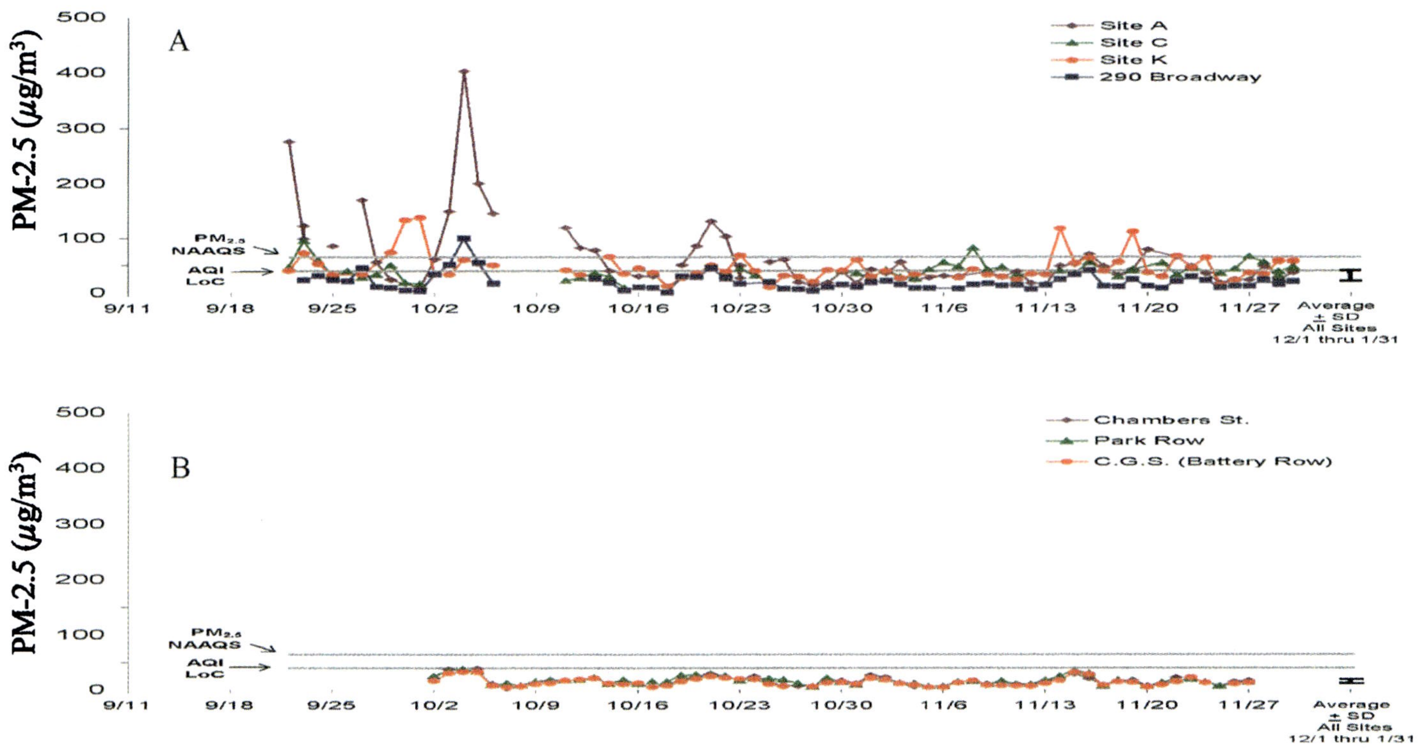

Plate 11.9. Daily PM-2.5 concentrations (A) at sites A, C, and K around Ground Zero perimeter and at 290 Broadway six blocks northeast of Ground Zero and (B) at several extended monitoring network sites in lower Manhattan within three to 10 blocks of WTC Ground Zero.

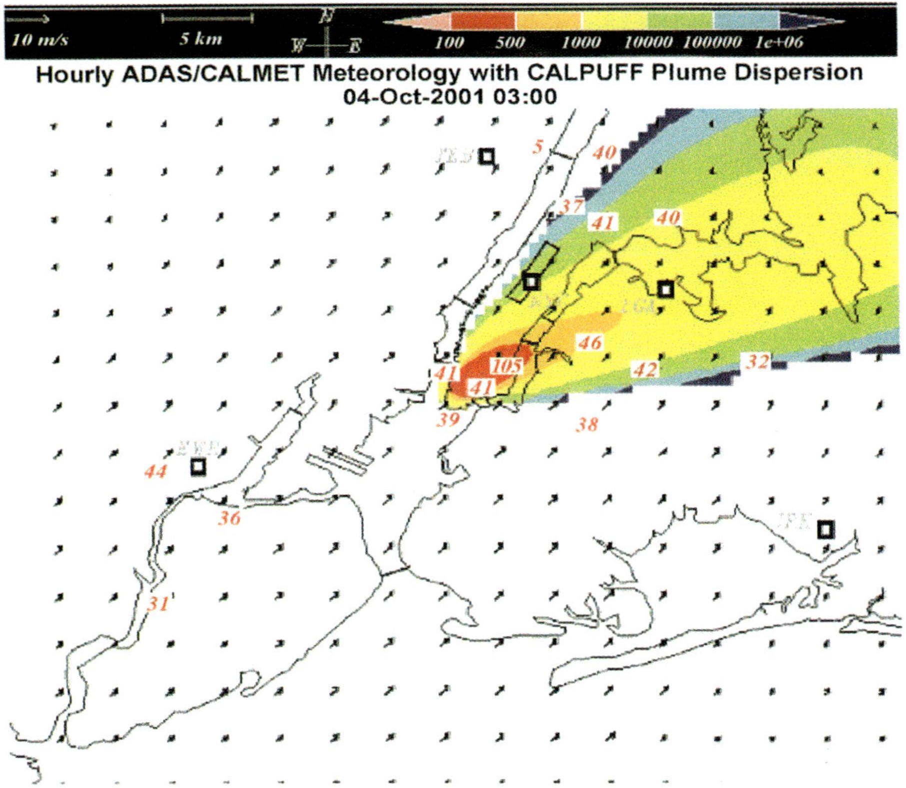

Plate 11.10. Modeled WTC plume dispersion for 3:00 to 4:00 AM October 4, 2001. The numbers are hourly PM-2.5 concentrations (μg/m³).

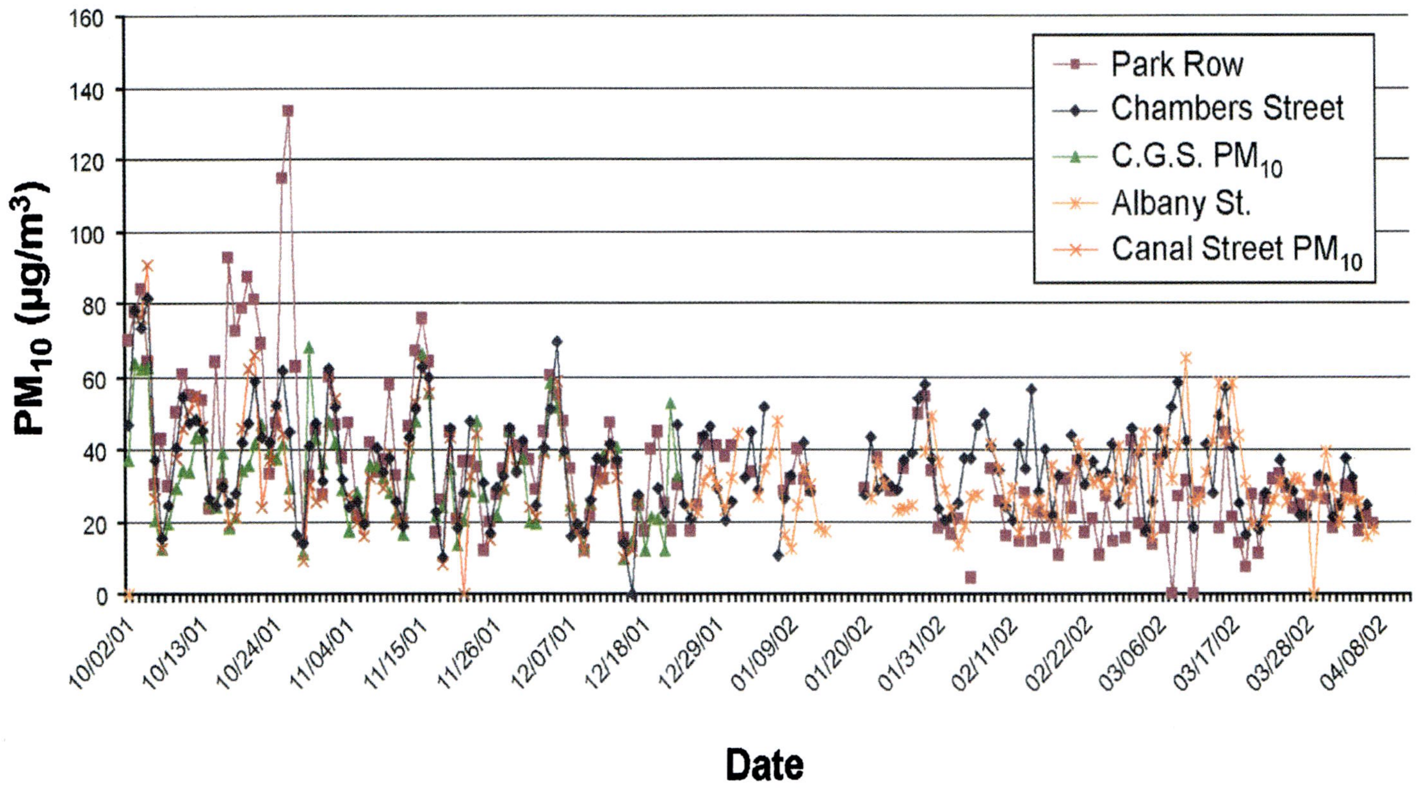

Plate 11.11. Daily average PM-10 concentrations measured at sites in lower Manhattan (as indicated in Figure 1).

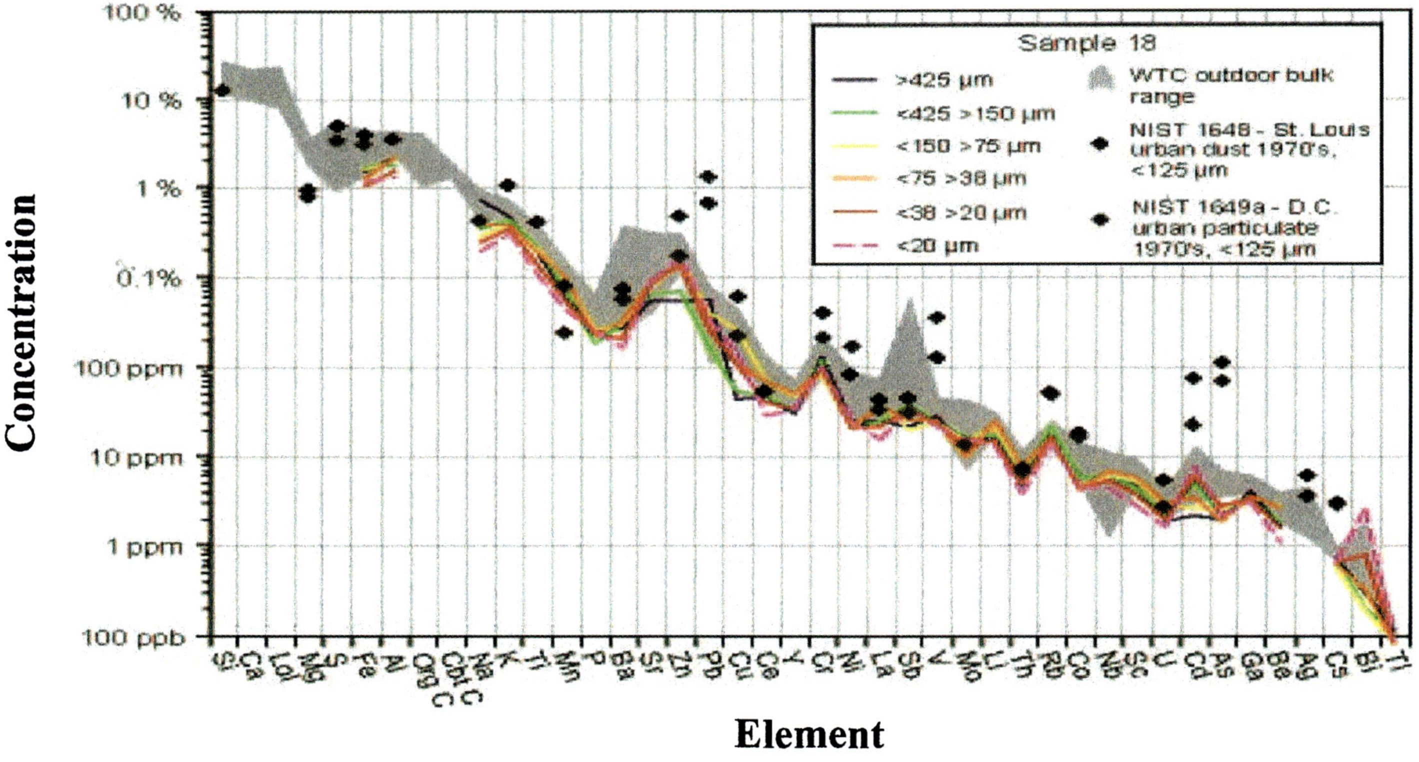

Plate 12.1. Concentrations of the elements in sieved size fractions of WTC dust sample 18 determined by ICP-MS compared to the WTC outdoor bulk dust composition ranges from Figure 1 (gray), and the concentrations of elements from NIST SRM-1648 and SRM-1649a sieved to < 125 µm(15, 16). LoI is loss on ignition; Org C is organic carbon; Cbt C is carbonate carbon.

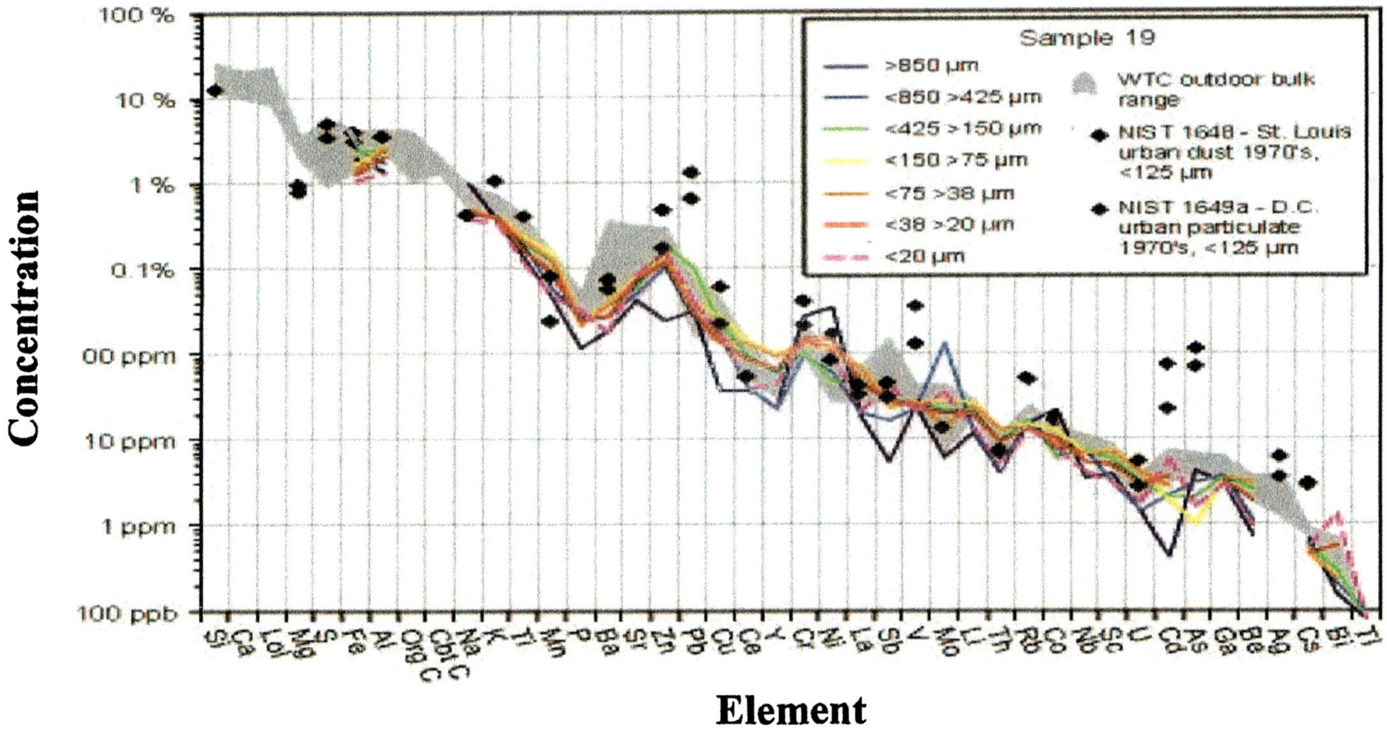

Plate 12.2. Concentrations of the elements in sieved size fractions of WTC dust sample 19 determined by ICP-MS compared to the WTC outdoor bulk dust composition ranges from Figure 1 (gray), and the concentrations of elements from NIST SRM-1648 and SRM-1649a sieved to < 125 µm(15, 16). LoI is loss on ignition; Org C is organic carbon; Cbt C is carbonate carbon.

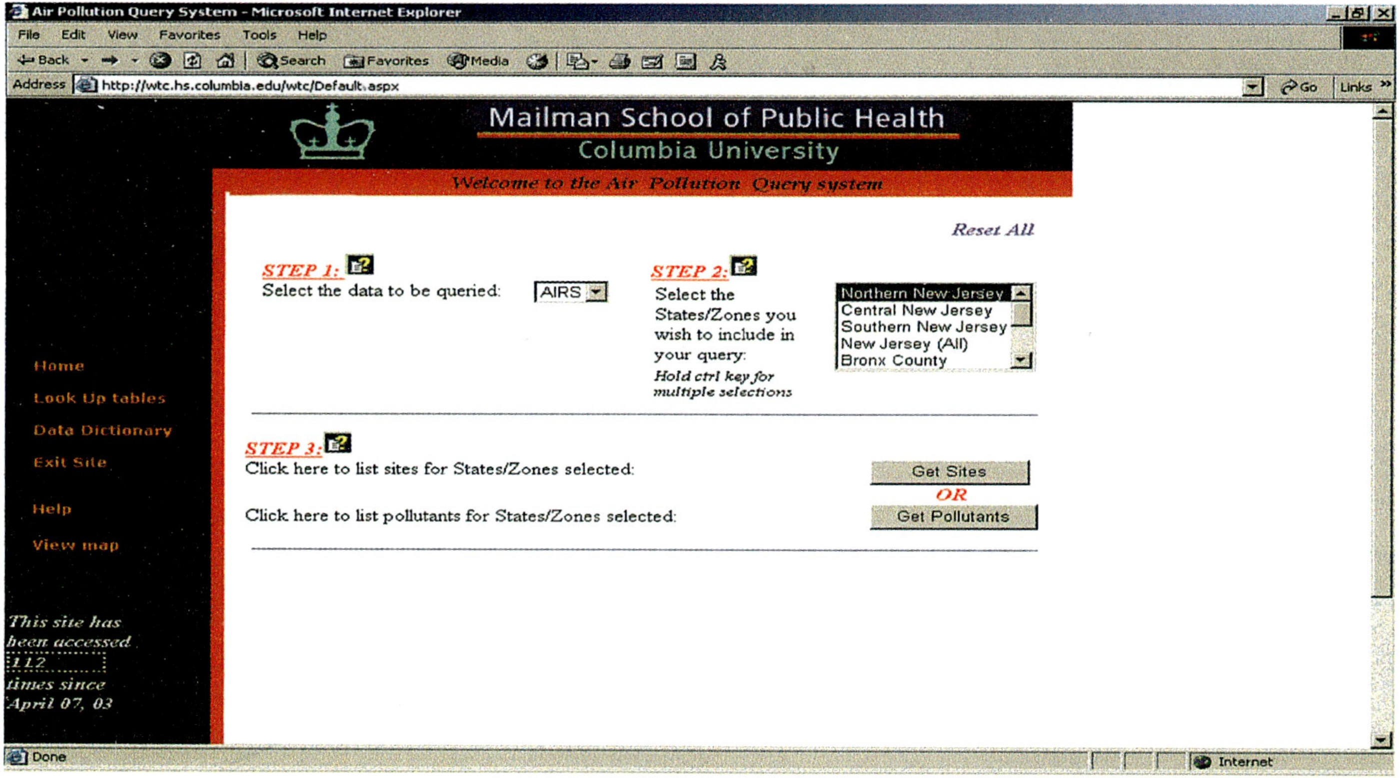

Plate 13.1. Example of the enumerated steps for searching the World Trade Center Environmental Contaminant Database.

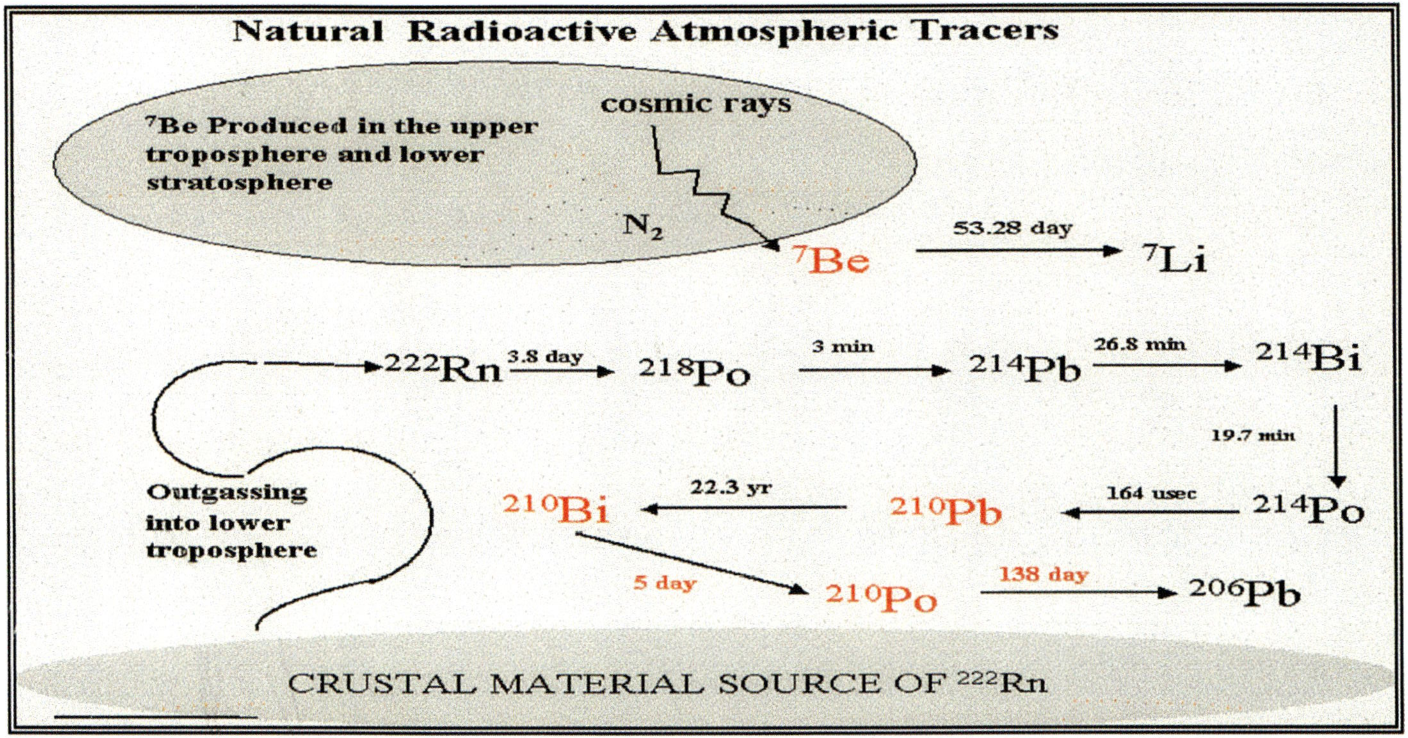

Plate 14.1. Sources of natural radioactivity in the atmosphere for ^{7}Be and ^{210}Pb and its daughters, ^{210}Bi and ^{210}Po (shown in red). Very short lived or unstable isotopic species are shown in black.

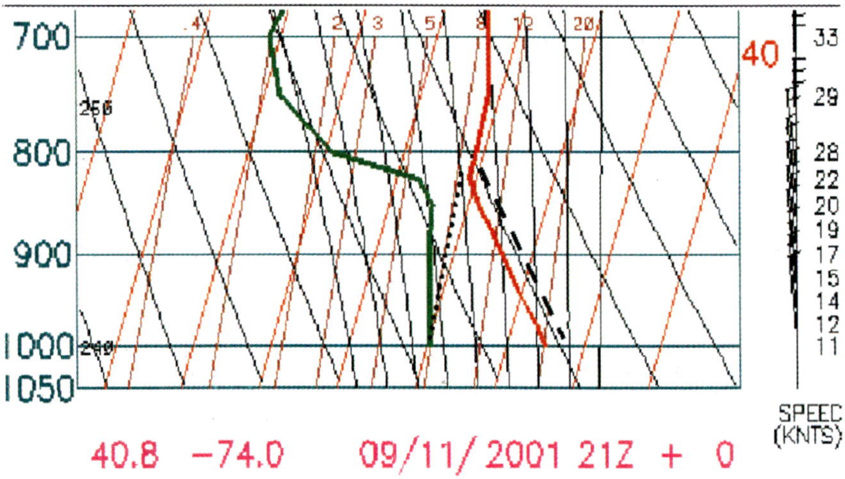

Plate 16.1. The environmental sounding for Central Park, NYC; September 11; 21Z (17EDT)

Indexes

Author Index

Subject Index

D

E

L

M

N

O

P

Q

R

S

T

U

V

W

X